Learning and Analytics in Intelligent Systems

Volume 9

Series Editors

George A. Tsihrintzis, University of Piraeus, Piraeus, Greece
Maria Virvou, University of Piraeus, Piraeus, Greece
Lakhmi C. Jain, Faculty of Engineering and Information Technology, Centre for
Artificial Intelligence, University of Technology Sydney, NSW, Australia;
University of Canberra, Canberra, ACT, Australia; KES International,
Shoreham-by-Sea, UK; Liverpool Hope University, Liverpool, UK

The main aim of the series is to make available a publication of books in hard copy form and soft copy form on all aspects of learning, analytics and advanced intelligent systems and related technologies. The mentioned disciplines are strongly related and complement one another significantly. Thus, the series encourages cross-fertilization highlighting research and knowledge of common interest. The series allows a unified/integrated approach to themes and topics in these scientific disciplines which will result in significant cross-fertilization and research dissemination. To maximize dissemination of research results and knowledge in these disciplines, the series publishes edited books, monographs, handbooks, textbooks and conference proceedings.

More information about this series at http://www.springer.com/series/16172

Lakhmi C. Jain · Sheng-Lung Peng ·
Basim Alhadidi · Souvik Pal
Editors

Intelligent Computing Paradigm and Cutting-edge Technologies

Proceedings of the First International
Conference on Innovative Computing
and Cutting-edge Technologies (ICICCT 2019),
Istanbul, Turkey, October 30–31, 2019

 Springer

Editors
Lakhmi C. Jain
University of Technology Sydney
Sydney, Australia

Basim Alhadidi
Al-Balqa' Applied University
Salt, Jordan

Sheng-Lung Peng
CSIE Department
National Dong Hwa University
New Taipei City, Taiwan

Souvik Pal
Department of Computer Science
Brainware University
Kolkata, West Bengal, India

ISSN 2662-3447   ISSN 2662-3455   (electronic)
Learning and Analytics in Intelligent Systems
ISBN 978-3-030-38500-2   ISBN 978-3-030-38501-9   (eBook)
https://doi.org/10.1007/978-3-030-38501-9

This Springer imprint is published by the registered company Springer Nature Switzerland AG
The registered company address is: Gewerbestrasse 11, 6330 Cham, Switzerland

Preface and Acknowledgement

With the proliferation of the challenging issues of Artificial Intelligence, machine learning, big data and data analytics, high-performance computing, network and device security, Internet of things (IoT), IoT-based digital ecosystem, and impact on society and communication have attracted a growing number of researchers. The main aim of this conference is to bring together leading academic scientists, researchers, and research scholars to exchange and share their experiences and research results on all aspects of intelligent ecosystems and data sciences. It also provides a premier interdisciplinary platform for researchers, practitioners and educators to present and discuss the most recent innovations, trends, and concerns as well as practical challenges encountered and solutions adopted in the fields of IoT and Analytics. This proceeding is about basics and high-level concepts regarding intelligent computing paradigm and communications in the context of distributed computing, big data, high-performance computing and Internet of things. International Conference on Innovative Computing and Cutting-edge Technologies (ICICCT 2019) is an opportunity to convey their experiences, to present excellent result analysis, future scopes, and challenges facing the field of computer science, information technology, and telecommunication. It brings together experts from industry, governments, universities, colleges, research institutes, and research scholars.

ICICCT 2019 is organized by Middle East Association of Computer Science and Engineering (MEACSE). ICICCT 2019 has been held during 30–31 October 2019 in Istanbul, Turkey. The conference brought together researchers from all regions around the world working on a variety of fields and provided a stimulating forum for them to exchange ideas and report on their researches. The proceeding of ICICCT 2019 consists of 45 selected papers which were submitted to the conferences and peer-reviewed by conference committee members and international reviewers. The Presenters came from different countries like Jordan, Algeria, Palestine, Turkey, Qatar, Egypt, Vietnam, India, Morocco, Tunisia, and Sudan. This conference became a platform to share the knowledge domain among different countries research culture.

We are sincerely thankful to Almighty to supporting and standing at all times with us, whether it is good or tough times and given ways to concede us. Starting from the call for papers till the finalization of chapters, all the team members have given their contributions amicably, which it is a positive sign of significant team works. On behalf of sponsors and conference committees, we would like to express our gratitude to all authors and the attendees for their contributions and participation for ICICCT 2019. Their high competence and expertise knowledge enable us to prepare this high-quality programme and make the conferences successful.

The editors and conference organizers are sincerely thankful to all the members of Springer especially Ms. Jayarani Premkumar for providing constructive inputs and allowing an opportunity to finalize this conference proceedings. We are also thankful to Thomas Ditzinger, Holger Schaepe, and Guido Zosimo-Landolfo for their support. We are equally thankful to all the reviewers who hail from different places in and around the globe and stand firm in maintaining high quality.

Finally, we would like to wish you have good success in your presentations and social networking. Your strong supports are critical to the success of this conference. We hope that the participants will not only enjoy the technical programme in the conference but also discover many beautiful places in Turkey. Wishing you a fruitful and enjoyable ICICCT 2019.

Lakhmi C. Jain
Sheng-Lung Peng
Basim Alhadidi
Souvik Pal

Organizing Committee

Conference General Chair

Thamer Al-Rousan   Al-Isra University, Jordan

Conference Honorary Chair

Lakhmi C. Jain   University of Technology Sydney, Australia
                        Liverpool Hope University, UK
                        KES International, UK

Programme Convenor

Basim Alhadidi   Al-Balqa' Applied University, Salt, Jordan

Programme Chair

Amer El-Khairy   Naif Arab University for Security Sciences,
                        Saudi Arabia

Programme Co-chair

Souvik Pal   Brainware University, India

International Advisory Board

Siddhartha Bhattacharyya   RCCIIT, India
Ahmed A. Elnger   Beni-Suef University, Egypt
Chinmay Chakraborty   BIT, Mesra, India
Anirban Das   University of Engineering and Management,
                                  India

Kusum Yadav	University of Hail, Kingdom of Saudi Arabia
Debashis De	Maulana Abul Kalam Azad University of Technology, India
Balamurugan Shanmugam	QUANTS IS and CS, India
Abdel-Badeeh M. Salem	Ain Shams University, Egypt
Prantosh Kumar Paul	Raiganj University, India
Paulo João	University of Lisbon, Portugal
Sudipta Das	IMPS College of Engineering and Technology, India
Ton Quang Cuong	University of Education, Vietnam National University, Hanoi

Programme Co-convenors

Osman Adiguzel	Firat University, Turkey
Bharat S. Rawal Kshatriya	Penn State University, USA
Sandor Szabo	University of Pécs, Hungary

Publications Chair

| Sheng-Lung Peng | National Dong Hwa University, Taiwan |

Technical Chairs

G. Suseedran	VELS University, India
Prasenjit Chatterjee	MCKV Institute of Engineering, India
Saravanan Krishnann	Anna University, India
Bikramjit Sarkar	JIS College of Engineering, India
Debabrata Samanta	Christ Deemed to be University, India
Saravanan Krishnann	Anna University, India
Abhishek Bhattacharya	Institute of Engineering and Management, India
S. Vijayarani	Bharathiar University, Coimbatore, India

Technical Programme Committee Members

Alti Adel	University UFAS of Setif, Algeria
Zoltan Gal	University of Debrecen, Hungary
Andrey Gavrilov	Novosibirsk State Technical University, Russia
Abhishek Bhattacharya	Institute of Engineering and Management, India
Prasenjit Chatterjee	MCKV Institute of Engineering, India
S. Vijayarani	Bharathiar University, Coimbatore, India
Kamel Hussein Rahouma	Minia University, Minia, Egypt
Jean M. Caldieron	Florida Atlantic University, USA
Arindam Chakrabarty	Rajiv Gandhi University, India
P. Vijayakumar	Anna University, India

Abraham G. van der Vyver	Monash University, South Africa
Ibikunle Frank	Covenant University, Nigeria
G. Suseedran	VELS University, India
Abdel-Badeeh M. Salem	Ain Shams University, Egypt
Saravanan Krishnann	Anna University, India
Maheshkumar H. Kolekar	University of Missouri, USA
Kalinka Regina Lucas Jaquie Castelo Branco	Universidade de Sao Paulo–USP, Brazil
Ken Revett	Loughborough University, England
Santanu Singha	JIS College of Engineering, India
Sourav Samanta	University of Technology, India
Paulo João	University of Lisbon, Portugal
Jaiyeola T. Gbolahan	Obafemi Awolowo University, Nigeria
Sattar B. Sakhan	Babylon University, Iraq
Shaikh Enayet Ullah	University of Rajshahi, Bangladesh
Suresh Sankaranarayanan	University of West Indies, Jamaica
Tarig Osman Khider	University of Bahri, Sudan
Virendra Gawande	College of Applied Sciences, Oman
Zhao Shuo	Northwestern Polytechnic University, China

About the Book

The conference proceeding book is a depository of knowledge enriched with recent research findings. The main focus of this volume is to bring all the computing and communication related technologies in a single platform, so that undergraduate and postgraduate students, researchers, academicians, and industry people can easily understand the intelligent and innovative computing systems, big data and data analytics, IoT-based ecosystems, high-performance computing, and communication systems. This book is a podium to convey researchers' experiences, to present excellent result analysis, future scopes, and challenges facing the field of computer science, information technology, and telecommunication. The book also provides a premier interdisciplinary platform for researchers, practitioners, and educators to present and discuss the most recent innovations, trends, and concerns as well as practical challenges encountered and solutions adopted in the fields of Computer Science and Information Technology. The book will provide the authors, research scholars, and listeners with opportunities for national and international collaboration and networking among universities and institutions for promoting research and developing the technologies globally. The readers will have the chance to get together some of the world's leading researchers, to learn about their most recent research outcome, analysis and developments, and to catch up with current trends in industry academia. This book aims to provide the concepts of related technologies regarding intelligent and innovative computing systems, big data and data analytics, IoT-based ecosystems, high-performance computing, and communication systems and novel findings of the researchers through its Chapter Organization. The primary audience for the book incorporates specialists, researchers, graduate understudies, designers, experts, and engineers who are occupied with research- and computer science-related issues. The edited book will be organized in independent chapters to provide readers with great readability, adaptability and flexibility.

Contents

About the Editors

Lakhmi C. Jain, BE (Hons), ME, PhD, Fellow (Engineers Australia) is with the University of Technology Sydney, Australia, Liverpool Hope University, UK, and KES International, UK.

Professor Jain founded the KES International for providing a professional community with the opportunities for publications, knowledge exchange, cooperation, and teaming. Involving around 5,000 researchers drawn from universities and companies worldwide, KES facilitates international cooperation and generate synergy in teaching and research. KES regularly provides networking opportunities for the professional community through one of the largest conferences of its kind in the area of KES. http://www.kesinternational.org/organisation.php.

Sheng-Lung Peng is a full professor of the Department of Computer Science and Information Engineering at National Dong Hwa University, Taiwan. He received the BS degree in Mathematics from National Tsing Hua University and the MS and PhD degrees in Computer Science and Information Engineering from the National Chung Cheng University and National Tsing Hua University, Taiwan, respectively. His research interests are in designing and analysing algorithms for combinatorics, bioinformatics, and networks.

Dr. Peng has edited several special issues at journals, such as Soft Computing, Journal of Internet Technology, Journal of Computers, and MDPI Algorithms. He is also a reviewer for more than 10 journals such as IEEE Transactions on Emerging Topics in Computing, Theoretical Computer Science, Journal of Computer and System Sciences, Journal of Combinatorial Optimization, Journal of Modelling in Management, Soft Computing, Information Processing Letters, Discrete Mathematics, Discrete Applied Mathematics, Discussions Mathematical Graph Theory, and so on. He has about 100 international conferences and journal papers.

Dr. Peng is now the director of the Library and Information Center of NDHU and an honorary professor of Beijing Information Science and Technology University of China. He is a secretary general of the Institute of Information and Computing Machinery (IICM) in Taiwan. He is also a director of the ACM-ICPC

Contest Council for Taiwan. Recently, he is elected as a supervisor of Chinese Information Literacy Association and of Association of Algorithms and Computation Theory (AACT). He has been serving as a secretary general of Taiwan Association of Cloud Computing (TACC) from 2011–2015 and of AACT from 2013–2016. He was also a convener of the East Region of Service Science Society of Taiwan from 2014–2016.

Basim Alhadidi is presently a full professor at the Computer Information Systems Department at Al-Balqa' Applied University, Jordan. He earned his PhD in 2000, in Engineering Science (Computers, Systems and Networks). Dr. Alhadidi received his MSc in 1996 in Engineering Science (Computer and Intellectual Systems and Networks). He published many research papers on many topics such as: computer networks, image processing, and artificial intelligence. He is a reviewer for several journals and conferences. He was appointed in many conferences as chair, keynote speaker, reviewer, track chair, and track co-chair.

Souvik Pal, Ph.D, MCSI; MCSTA/ACM, USA; MIAENG, Hong Kong; MIRED, USA; MACEEE, New Delhi; MIACSIT, Singapore; MAASCIT, USA, is an associate professor at the Department of Computer Science and Engineering, Brainware University, Kolkata, India. Dr. Pal received his B. Tech degree in Computer Science and Engineering from West Bengal University of Technology, Kolkata. He has received his M. Tech and PhD. degree in Computer Engineering from KIIT University, Bhubaneswar, India. He has worked as an assistant professor in Nalanda Institute of Technology, Bhubaneswar, and JIS College of Engineering, Kolkata (NAAC "A" Accredited College). He has also worked as a head of the Computer Science Department in Elitte College of Engineering, Kolkata.

Dr. Pal has published several research papers in Scopus-indexed International journals and conferences. He is editor of **07** Elsevier/Springer/CRC Press/Apple Academic Press Books and author of **1** Cloud Computing book. He is the organizing Chair and Plenary Speaker of RICE Conference in Vietnam and organizing co-convener of ICICIT, Tunisia. He has been invited as a Keynote Speaker in ICICCT, Turkey. He has served in many conferences as chair and keynote speaker, and he also chaired international conference sessions and presented session talks internationally. Dr. Pal also serves as a reviewer and editorial board member for many journals and IEEE/Springer/ACM conferences. His research area includes cloud computing, big data, wireless sensor network (WSN), Internet of things, and data analytics.

Brexit Twitter Sentiment Analysis: Changing Opinions About Brexit and UK Politicians

Muntazar Mahdi Chandio$^{(\boxtimes)}$ and Melike Sah

Computer Engineering Department, Near East University,
Nicosia, North Cyprus, Turkey
{muntazar, melike.sah}@neu.edu.tr

Abstract. In this paper, we analyze weather social sentiments can be utilized for the prediction of election results. In particular, we analyze Twitter sentiments about Brexit and United Kingdom (UK) politicians. Twitter is the essential social network for sentiments analyzing and it provides useful information for mining data. Through periods, we collected Twitter data about Brexit and UK politicians using Twitter Application Program interface (API). First, we cleaned and pre-processed Tweet data for sentiment analysis. Then, we create a Twitter search and sentiment visualization interface using python. Python provides useful libraries for sentiment analysis and graphical presentations. Finally, we analyze the changing opinions about Brexit and UK politicians using sentiments. In particular, in advance, we were able to correctly predict the UK parliament voting results in January 2019. In this paper, we discuss Twitter data collection, Twitter sentiment search/visualization interface and detailed sentiment analysis results about Brexit and UK politicians.

Keywords: Twitter · Sentiment analysis · Brexit · Graph visualization · Python · Social media

1 Introduction

This is a new era of technology which connects people to each other no matter how far they are. This credit goes to social media. Social media is the platform of sharing and receiving information, data, as well as communication system of people. They share their psychology, thinking, ideas, behaviors and sentiments. It is very powerful weapon of increasing literature and business. People use social media to gain education and power for a better life and health. There are many useful social media platform but twitter is the most reliable platform for sentiment analysis because there are more the 336 million worldwide active users (statista.com), more than 100 million daily active users (twitter-statistics last update 6-24-18) and 500 million posts every day (last update 6-24-18). People show their opinions and they are participated on different topics through the twitter posts (tweets) which is useful knowledge base for sentiment analysis. The Twitter data can be received from Twitter in a secure and easy way. We can receive the bulk amount of data through twitter API (Application Programing Interface).

© Springer Nature Switzerland AG 2020
L. C. Jain et al. (Eds.): ICICCT 2019, LAIS 9, pp. 1–11, 2020.
https://doi.org/10.1007/978-3-030-38501-9_1

In this research, we analyze opinions, thoughts and perspectives of people about the general topics, politics and political parties. We use worldwide twitter data for general opinion analysis and political perspective and specific data for analysis of political parties such as UK politics. We created an interface to search for specific keywords in a particular Twitter datasets. Then, according to the matching tweets to the Twitter search, we present the sentiments analysis of those matching tweets (such as sentiments results are presented as pie chart and bar chat). In this research, we analyze twitter post (tweets) and show the result in three ways; positive, negative and neutral. The analysis results are compared for different political parties and visualized as pie chart and bar chart graphs. Result show the total number of tweets, number of positive, number of negative and number of neutral tweet posts for each searched keyword in the political domain. In addition, we also the results of the sentiment analysis as bar charts; the percentage of tweets positive in green, negative in red and neutral in blue color. In this way, the analyzed tweets show the fairness of the elections based on the post-twitter data.

We apply the sentiment analysis on UK election tweets data. We focused on British exit (Brexit) from the Europe Union and see reaction and support of people. Now a days Brexit is the hard issue and people want to see which policy England will apply for separation. We know that the parliament already reject the bill of Prime minister of UK and showed disagreement. Since it is a hot topic, we choose it for researching the reaction of people. In this research, we only analyzed two political party of UK that has the most seats in the last election, it means current prime minister of UK Theresa May and opposition leader Jeremy Corbyn. In certain intervals, we collected Twitter data about Brexit and politicians to be utilized in this research. For sentiment analysis, we use the open source python code with different module in single class (single program); "Textblob" module for sentiments analysis such as calculating polarity of tweets; "Matplotlib" module for calculating the percentage and drawing pie chart graphs in three different colors. We also show PNN (positive, negative and neutral) tweets and present percentage of each part on the pie chart.

Analysis of sentiment results for Brexit show that there is a change in opinions of public about Brexit and EU as well as leading politicians. We observe that positivity about Brexit is decreasing as well as peoples' positive thoughts about EU is also decreasing. Positive opinions of people about both British politicians Theresa May and Jeremy Corby are also losing power. However, it seems that Jeremy Corby is gaining more positive comments in Tweeter compared to Theresa May. It means Theresa May lost her parliament election voting about Brexit bill of separation from EU and still in the case of pending. It is a very difficult processes of separation of UK. Our analysis of Tweet sentiments show that UK public stand with Jeremy Corbyn as compare to Theresa May.

2   Related Work

There are many research paper and articles about the social media and election prediction through the microblogging sites and twitter but no one can give an easy results and visualization to understand for normal people. In this study, we present the positivity and negativity of political leaders and opinions about them. This kind of results helps us to predict the election results and popularity of the politicians. There are many papers about election prediction such as (Jaidka et al. 2018) the election prediction of three different countries India, Pakistan and Malaysia. The accuracy of results is awesome. They only show volumetric performance, supervised and unsupervised model, as well as present the results on histogram graph chart and expression. But these methods do not give an open result that an average people can understand. However, in our research, we visualize the clear number of results and clear number of tweets. Some research paper compare two or more than two parties such as USA (Bovet et al. 2016) Trump versus Clinton on a very large scale twitter data of 0.73 million. They presented good results and prediction but result was reversed, Clinton being more popular than Trump. They did not show the number of tweets for each candidate and also presented a line graph which does not show number of tweets. The paper of (Livne et al. 2011) also which discusses USA elections, however they give few details about their approach and did not compare candidates in histogram charts. United Kingdom predictions (Boutet et al. 2012) have the same problem like (Livne et al. 2011) and even not gave an enough information about the prediction of results. In Ireland analysis (Birmingham and Smeaton 2011), researchers gave very low number of datasets, unclear approach and low fragmentation on line graph and histogram chart graph. As compare to these papers, we give clear visualization and number of tweets on pie chart and analyze tweet by tweet that mean the average people can easily read and understand the approach.

3   System Architecture

There are many tools for sentiment analysis. The most popular tools are MATLAB, Python, and Java and C#. Due to large number of libraries available in python as well as easy coding capabilities, we choose to use python (Mohammed Innat). The sentiments analysis algorithm consists of 4 modules. The procedure in each model starts with importing data with pandas, since the power of pandas is data preprocessing. Then we use NLTK and Textblob modules for analyzing the text of CSV files and calculating the polarity of each text separately, which outputs a numeric value (-1 to $+1$). Finally, Matplotlib is utilized for plotting the results on a pie chat and bar chat with different colors and different formats positive, negative and neutral (greater than zero, less than zero and equal to zero) (Fig. 1).

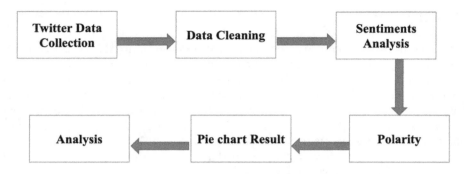

Fig. 1. System architecture

3.1 Sentiment Analysis

Sentiments analysis is the invented science of psychology and sociology and both are the scientific study of people emotions, relationships, opinions, and behaviors (wiki). Psychologist apply sentiments process through the hypothesis but data scientist apply through the data. In other words, it is the computational process which identifies and categories the opinions, thoughts and ideas through the text data. The sentiments analysis process also refer the NLP (Natural language processing). It is internal action process between human and computer. It also analyzes the treasure of natural language data. Sentiments analysis are expressed in two different categories: polarity and subjectivity. The polarity measure the text data is positive (>0) or negative (<0) or neutral (0). Classifying a sentence as subjective or objective, known as subjectivity classification. Subjectivity measures from (0.0 to 1.0). Where 0.0 is very objective and 1.0 is very subjective. In this research, we only calculate the sentiments polarity from twitter data (tweets data is in CSV format). Polarity showed three different colors positive for green color, negative in red color and neutral in blue color. Polarity calculated through the python code using library of Textblob and python module Natural Language Tool Kit (NLTK) which explained later.

3.2 Twitter Data Collection

There are many ways to collecting the data from twitter but in our suggestion python is the easiest and simple way to collection the data. Using the python code library tweepy we access the data through twitter API. API provide the keys for accessing the data of twitter. There are four keys which used for authentication and accessing the twitter account consumer key, consumer secret key, token key, token secret key. There are three different steps to collect the cleaned data from Twitter. First streaming the data from Twitter and saved in CSV file. Second collect the tweets text from one CSV file to save in other CSV file. Third removed duplication from tweets data. These steps are also showed in the Figs. 2, 3 and 4.

Fig. 2. Data streaming from twitter using Python idle

Fig. 3. Collected CVS file

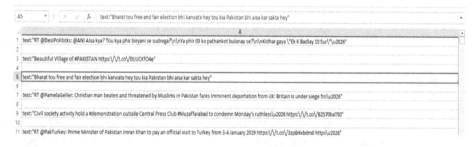

Fig. 4. Cleaned tweet data

4 Brexit Tweet Sentiment Analysis

In our work, we analyze sentiment analysis of two political leaders of Britain one Theresa May and second Jeremy Corbyn. We also analyze changing opinions of people about the Europe Union and Brexit. There are four different kind of datasets, two datasets downloaded from (dataworld.com) and two datasets are collected from the twitter API. In particular, we queried the Tweeter API with a combination of keywords such as Brexit, Theresa May, Jeremy Corby, European Union (EU) and collected daily tweet data in January 2019 and February 2019. In this research, we collected data from twitter before and after the parliament election of UK (13 Jan to 20 Jan 2019) and in the first week of February. In Table 1, we show the number of collected tweets on the analyzed dates after pre-processing and cleaning. It can be seen that on the day of Parliament voting (17th of January), the tweet activities were increased considerably.

In Figs. 5, 6, 7 and 8, we demonstrate visual analysis of sentiments about Brexit, European Union (EU) and UK politicians. Each keyword search is applied to four time intervals that we collected tweet data. In particular, we present four pie charts as oppose to tweet data in 30 May 2017, in 31 May 2017, in January 2019 (January tweets are combined together) and in February 2019 (February tweets are combined together).

Table 1. Brexit tweet datasets.

Date collected	Number of Tweets
30.5.2017	793,352
31.5.2017	1,048,576
13.1.2019	15,937
14.1.2019	4,961
15.1.2019	1,491
16.1.2019	48,578
17.1.2019	197,175
18.1.2019	31,453
20.1.2019	33,893
January total Tweets	333,510
2.1.2019	13231
2.2.2019	10286
February total tweets	23,517

We observe that people were more positive about Brexit in 2017, whereas in January 2019 and especially after parliament voting in February 2019, their positivity was dropped around 5%. Similarly, even after the Brexit referendum, positivity about EU was high around 38% in 2017. However, before British parliament voting in January 2019, the positivity was dropped around 3%, and after the parliament voting in January 2019, the positivity about EU was also dropped and kept around 30%. When we observe changes in UK politicians, we observe that Therasa May tweet sentiments were dropped considerably. In 2017, people were more positive about Theresa May around 30%. Before and after the parliament voting in January 2019, the positivity about Theresa May dropped to 28% and 23% respectively. The opposition party leader Jermy Corby also has more positive sentiments in 2017 with around 40%. In January positivity about Jeremy Corby was kept stable with around 40%. But after the parliament voting in February 2019, the positivity about Jeremy Corby was dropped to 29%.

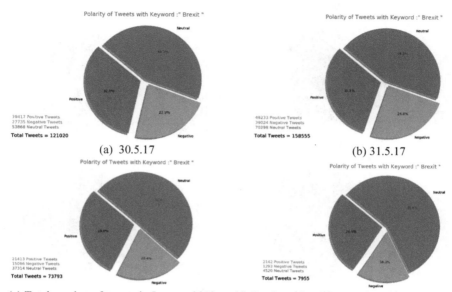

(a) 30.5.17

(b) 31.5.17

(c) Total number of tweets in January 2019   (d) Total number of tweets in February 2019

Fig. 5. Sentiment analysis for "Brexit" search term

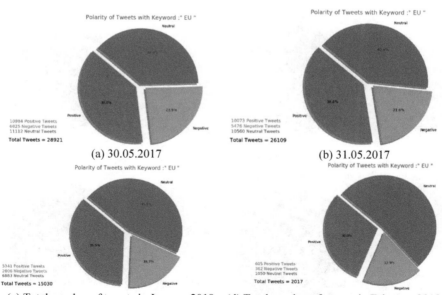

(a) 30.05.2017

(b) 31.05.2017

(c) Total number of tweets in January 2019   (d) Total number of tweets in February 2019

Fig. 6. Sentiment analysis for "EU" search term

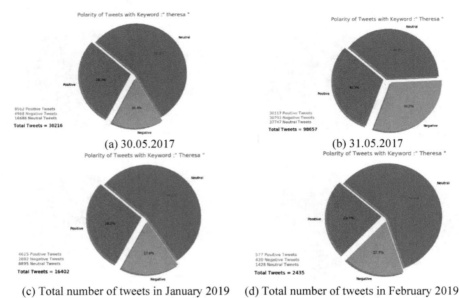

(a) 30.05.2017   (b) 31.05.2017

(c) Total number of tweets in January 2019   (d) Total number of tweets in February 2019

Fig. 7. Sentiment analysis for "Theresa May" search term

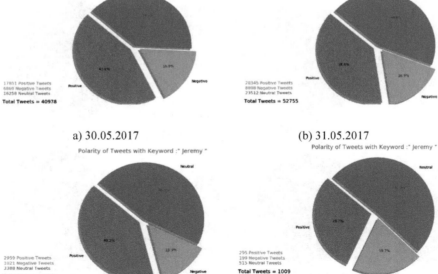

a) 30.05.2017   (b) 31.05.2017

c) Total number of tweets in January 2019   (d) Total number of tweets in February 2019

Fig. 8. Sentiment analysis for "Jeremy Corby" search term

In Table 2, we also analyze sentiment data quantitatively. 'P' represents positive, 'N' represented negative, 'NT' represents neutral, 'T' for total, 'PA' for positive average, 'MXP' for maximum positivity, 'MNP' for minimum positivity, 'NA' negative average, and 'MXN' maximum negativity and 'MNN' minimum negativity.

$$PA = \frac{\max positive\, tweets}{total\, tweets} * 100 \tag{1}$$

$$NA = \frac{\max negative\, tweets}{total\, tweets} * 100 \tag{2}$$

With the help of Eqs. (1) and (2), we can calculate the positive average and negative average of the sentiments analysis results which shown in given Table 2. Analysis of the results on these datasets shows that Theresa May received a maximum of positive average 29.7% and maximum negative average 26.45% tweets compared to maximum positive average 40.6% and maximum negative average 16.65% of Jeremy.

Table 2. Quantitative tweet sentiment analysis.

Dates	Brexit	EU (Europe Union)	Theresa May	Jeremy Corbyn
30.5.2017	P = 32.6 = 39417 N = 22.9 = 27735 NT = 44.5 = 53868 T = 100 = 121020	P = 38 = 10984 N = 23.6 = 6825 NT = 38.4 = 11112 T = 100 = 28921	P = 28.3 = 8562 N = 16.4 = 4968 NT = 55.2 = 16686 T = 100 = 30216	P = 43.6 = 17851 N = 16.8 = 6869 NT = 39.7 = 16258 T = 100 = 40978
31.5.2017	P = 31.1 = 49233(1.5) N = 24.6 = 39024 NT = 44.3 = 70298 T = 100 = 158555	P = 38.6 = 10073(0.6) N = 21 = 5476 NT = 40.4 = 10560 T = 100 = 26109	P = 30.5 = 30117(2.2) N = 31.2 = 30793 NT = 38.3 = 37747 T = 100 = 98657	P = 38.6 = 20345(5) N = 16.9 = 8898 NT = 44.6 = 23512 T = 100 = 52755
01.2019	P = 29 = 21413(3.6) N = 20.4 = 15066 NT = 50.6 = 37314 T = 100 = 73793	P = 35.5 = 5341(2.5) N = 18.7 = 2806 NT = 45.8 = 6883 T = 100 = 15030	P = 28.2 = 4625(0.1) N = 17.6 = 2882 NT = 54.2 = 8895 T = 100 = 16402	P = 40.2 = 2959(3.4) N = 13.9 = 1021 NT = 46 = 3388 T = 100 = 7268
02.2019	P = 26.9 = 2142(5.7) N = 16.3 = 1293 NT = 56.8 = 4520 T = 100 = 7955	P = 30 = 605(8) N = 17.9 = 362 NT = 52.1 = 1050 T = 100 = 2017	P = 23.7 = 577(4.5) N = 17.7 = 430 NT = 58.6 = 1428 T = 100 = 2435	P = 29.2 = 295(13.8) N = 17.9 = 199 NT = 51 = 515 T = 100 = 1009
Overall	MXP = 32.6 MNP = 26.9 PA = 31.05 MXN = 24.6 MNN = 16.3 NA = 23	MXP = 38.6 MNP = 30 PA = 37.46 MXN = 24.6 MNN = 16.3 NA = 21.4	MXP = 30.5 MNP = 23.7 PA = 29.7 MXN = 31.2 MNN = 16.4 NA = 26.45	MXP = 43.6 MNP = 29.2 PA = 40.6 MXN = 17.9 MNN = 13.9 NA = 16.65

We observed that positive and negative average of Jeremy is better than Theresa. In other case EU receives 37.46% maximum positive and 21.4 maximum negative average as compared to Brexit 31% maximum positive and 23% maximum negative average which is not good as compared to EU.

5 Challenges and Conclusions

In this research faced different kind of challenges and most difficult challenge is collecting the data (tweets) from twitter because when the same search keyword is queried every day, the twitter disconnect your connection and stop the data collecting process. After data collection, another challenging task is data cleaning. Duplication of tweets is also major issue, since the same tweets appear in search results and affect healthy result analysis. Multi language or non-English tweets this one is typical issue during the sentiments analysis because non English tweet count as neutral. As a conclusion, in this research we gave a way to predict useful results about politics because every person directly or indirectly connected with politics through votes. The microblogging are mostly using weapon in technology and twitter is useful resource for prediction. In parliament voting in January 2019, we were expecting a reject due to positive thoughts about EU as well as changing opinions about Brexit. Positive average of Brexit was 31.05% and negative average was 23 as well as positive average of EU was 37.46 and negative average was 21.4. It means that positive average of EU was around 6.4% higher than the Brexit. On the other hand, Theresa May's positive and negative average was 29.7 and 26.45 respectively compared to Jeremy Corbyn's 40.6 positive average and 16.65 negative average. It means that people are more supportive of Jeremy Corbyn as compared to Theresa May according to the Tweet datasets we collected in January and February 2019. Our findings are correlating with the rejection decision that was made in February 2019.

References

The Statistical portal. https://www.statista.com/statistics/274564/monthly-active-twitter-users-in-the-united-states

Twitter via SMS FAQ. Archived 6 April 2012, at the Wayback Machine. Accessed 13 Apr 2012

Kuhlman, D.: A Python Book: Beginning Python, Advanced Python, and Python Exercises. Archived from the original on 23 June 2012

Textblob. https://media.readthedocs.org/pdf/textblob/dev/textblob.pdf

The Matplotlib development team. https://matplotlib.org/tutorials/index.html#introductory

Dhar, V.: Data science and prediction. Commun. ACM **56**(12), 64 (2013). https://doi.org/10.1145/2500499

Tansley, S., Tolle, K.M.: The Fourth Paradigm: Data-Intensive Scientific Discovery. Microsoft Research (2009). ISBN 978-0-9825442-0-4

Mak, W.: https://data.world/wwymak/uk-election-tweets-2017-may-3/workspace/file?filename=2017-5-31.csv

Obar, J.A., Wildman, S.: Social media definition and the governance challenge: an introduction to the special issue. Telecommun. Policy **39**(9), 745–750 (2015). https://doi.org/10.1016/j.telpol.2015.07.014.SSRN2647377

The definitive history of social media. The Daily Dot, 11 September 2016. Accessed 5 Feb 2018

Gilbertson, S.: Twitter Vulnerability: Spoof Caller ID to Take over Any Account. Wired, 11 June 2007. Accessed 5 Feb 2011

Twitter/OpenSource. Twitter.com. Archived from the original on 15 April 2013. Accessed 18 Apr 2013

Xi, H., Scott, D.: Dependent types in practical programming. In: Proceedings of ACM SIGPLAN Symposium on Principles of Programming Languages, pp. 214–227. ACM Press (1998). CiteSeerX 10.1.1.41.548

Moujahid, A., 21 April 2014. http://adilmoujahid.com/posts/2014/07/twitter-analytics/

Bird, S., Klein, E., Loper, E.: Natural Language Processing with Python. O'Reilly Media Inc., Sebastopol (2009)

Jaidka, K., Ahmed, S., Skoric, M., Hilbert, M.: Predicting elections from social media: a three-country, three-method comparative study. Asian J. Commun. **29**(3), 252–273 (2018). https://doi.org/10.1080/01292986.2018.1453849

Bermingham, A., Smeaton, A.F.: On using Twitter to monitor political sentiment and predict election results. In: Sentiment Analysis Where AI Meets Psychology (SAAIP), pp. 2–10. ACL, Chiang Mai (2011)

Livne, A., Simmons, M.P., Adar, E., Adamic, L.A.: The party is over here: structure and content in the 2010 election. In: Proceedings of the International AAAI Conference on Weblogs and Social Media, pp. 17–21. Association for the Advancement of Artificial Intelligence, Barcelona (2011)

Boutet, A., Kim, H., Yoneki, E.: What's in your tweets? I know who you supported in the UK 2010 general election. In: Proceedings of the International AAAI Conference on We Blogs and Social Media, pp. 411–414. Association for the Advancement of Artificial Intelligence, Dublin (2012)

Bovet, A., Morone, F., Makse, H.A.: Predicting election trends with Twitter: Hillary Clinton versus Donald Trump (2016). arXiv preprint arXiv:1610.01587

A Clustering Algorithm for Multi-density Datasets

Ahmed Fahim[1,2(✉)]

[1] Faculty of Sciences and Humanitarian Study,
Prince Sattam Bin Abdulaziz University, Al-Aflaj, Saudi Arabia
a.abualeala@psau.edu.sa, ahmmedfahim@yahoo.com
[2] Faculty of Computers and Information, Suez University, Suez, Egypt

Abstract. DBSCAN algorithm discovers clusters of various shapes and sizes. But it fails to discover clusters of different density. This is due to its dependency on global value for Eps. This paper introduces an idea to deal with this problem. The offered method estimates local density for a point as the sum of distances to its k-nearest items, arranges items in ascending order according to their local density. The clustering process is started from the highest density point by adding un-clustered points that have similar density as first point in cluster. Also, the point is assigned to current cluster if the sum of distances to its Minpts-nearest neighbors is less than or equal to the density of first point (core point condition in DBSCAN). Experimental results display the efficiency of the proposed method in discovering varied density clusters from data.

Keywords: Clustering methods · Data analysis · Data mining · Knowledge discovery · Un-supervised learning

1 Introduction

Clustering methods are very charming for the mission of class identification in datasets. Clustering is the way toward gathering the items in a dataset into significant subclasses dependent on certain criteria. Many clustering methods have been proposed.

In partitioning methods like k-means [1], PAM "Partitioning Around Medoids" [2], CLARA "Clustering LARge Applications" [2]; each cluster is represented by single data point called mean or median. These methods take number of clusters as input parameter, and handle convex shaped clusters of similar size only. K-means starts by selecting randomly or heuristically k initial starting points. Each point represents a cluster center. Then, it assigns each of n-k points to the closest cluster. After the distribution of points, it computes new centers as the mean of points in each cluster and redistributes the points again over the new centers, this process is iterated until no point change its cluster or maximum number of iterations is reached or the difference between the last two values of the objective function is less than some threshold. K-means is very efficient handling large datasets since it require linear time complexity, but it can't handle clusters of varied shapes or sizes. It is efficient handling convex shaped cluster of similar sizes.

L. C. Jain et al. (Eds.): ICICCT 2019, LAIS 9, pp. 12–27, 2020.
https://doi.org/10.1007/978-3-030-38501-9_2

PAM algorithm starts by selecting k centroids from the input dataset - randomly or heuristically - and attach each data point to the nearest centroid. It tries to minimize the objective function by swapping centroid by one of the n-k non-centroid point. This process is repeated as the clustering result improves, i.e. the objective function is minimized. This algorithm is strong to noise and outliers and good for small datasets because of its time complexity.

Since PAM is not suitable for handling large dataset, an improvement has been done by proposing CLARA algorithm. CLARA operates on samples. This algorithm can handle larger dataset than that of PAM, but it restricts the result to fixed sample. CLARANS "Clustering Large Applications based on RANdomized Search" [3] is based on graph, where each node is a clustering result. It does not check all the neighbors of a node to find best clusters, it selects number of neighbors to check improving result, so it is more efficient than CLARA, it does not restrict the search to fixed sample of data point as CLARA do.

In hierarchical methods like single link [4], average link [5], complete link [6] built a dendrogram; like tree structure. This dendrogram may be built from top down (called divisive algorithm) or from bottom up (agglomerative algorithm). These methods but suffer from the chain effect and take $o(n^2)$ complexity.

The single link method starts with n singleton clusters, and every time it selects the two most similar clusters to merge (have the shortest distance between them). The merge process stops when the level of dissimilarity is reached. This method suffers from the chain effect.

Average link method is agglomerative like single link. It merges the two clusters that have minimum average distance among the points in them. This method solves the problem of chain effect in single link method. Complete link method is an agglomerative method that selects the two clusters with minimum distance among the far pair of points to merge them. The time complexity for these three agglomerative methods is $o(n^2)$.

There are other popular methods such as CHAMELEON [7], BIRCH "Balanced iterative reducing and clustering using hierarchies" [8] and CURE "Clustering using representatives" [9]. These methods handle varied shaped clusters well but are not appropriate to varied density clusters.

CHAMELON method uses dynamic model to select the clusters to be merged. This method operates on sparse graph and is applicable only on similarity space. It is a two phases clustering algorithm, that firstly, partitions the sparse graph to get large number of relatively small clusters. This is done to enable it to estimate internal closeness and connectivity for each small cluster.

BIRCH method uses a compact summary of cluster called cluster feature. And arranges cluster feature in cluster feature tree (CF). This algorithm prefers spherical shaped cluster like partitioning methods, however this algorithm handles large datasets.

CURE algorithm operates on random sample taken from the large input dataset. Then it partitions the sample and partially clusters each partition. This algorithm is a middle ground between all point and single point approaches. The time complexity of CURE is $o(n^2)$.

In density based methods like DBSCAN "Density-based spatial clustering of applications with noise" [10], and OPTICS "Ordering Points To Identify the Clustering Structure" [11] don't need number of clusters, and introduced the idea of density; this idea is based on counting points in a neighborhood radius of a central point, each item is classified into core item if it has more than or equal to Minpts in its neighborhood radius "Eps", or border item if it resides in neighborhood radius of core item, and it doesn't have Minpts in its neighborhood radius. Otherwise item is classified as noise.

DENCLUE [12] method uses influence functions. This algorithm fails to discover clusters of varied density since it uses two input parameters σ and ζ which are like Eps and Minpts in DBSCAN respectively.

Recently an improved version of DENCLUE has been proposed [13]. DENCLUE-SA uses simulated annealing algorithm instead of hill climbing algorithm in finding density attractors. This version improves the execution time of DENCLUE. Another version is called DENCLUE-GA; uses genetic algorithm instead of hill climbing algorithm in finding the density attractors. This version improves the clusters compactness. These two versions cannot handle clusters with varied density.

DBSCAN is the leader algorithm in this category and has received a great attention from the researchers all over the world. It discovers clusters of diverse shapes and sizes of similar density and handles noise well. The main weakness of DBSCAN is that it cannot discover clusters of various density due to its universal parameter Eps.

This paper introduces an idea taken from the DBSCAN algorithm to discover clusters of various density without the use of Eps but use Minpts as in DBSCAN. This idea determines the highest and lowest density acceptable within each cluster. The proposed method arranges data points in dataset based on k-density and starts forming clusters from the highest dense point (called cluster initiator point Ci) toward the lowest density acceptable within cluster. The results of the proposed method prove the ability of it to determine clusters of various shapes, sizes and density even so contiguous clusters.

This research is arranged as follows; Sect. 2 represents preceding works linked to the topic of research. Section 3 introduces the main definitions used in the proposed method that are like DBSCAN's definitions. Section 4 presents the proposed method. The experimental results are clarified in Sect. 5. Lastly, Sect. 6 accomplishes the research.

2 Related Works

DBSCAN is a well-known thickness based grouping technique [10] and has numerous attractive highlights including discover clusters of varied shapes, sizes and strength to noise and outliers. However, because of its dependency on universal parameter Eps, where Eps represents neighborhood radius for each object in dataset, DBSCAN cannot identify clusters with multi-level of density. To solve this trouble, many varied papers have been suggested, these papers may be classified into two categories; the first allows varied values for Eps or Minpts and the second uses alternative density definition without Eps.

Many algorithms have been proposed about the idea of allowing Eps to be varied according to the local density of each region. Reader may be referred to [14], where the author sorts the data points in dataset in ascending order based on the distance to the k-nearest neighbors. Initially the smallest distance will be assigned to the Eps, and DBSCAN is applied on the data using the current Eps, the algorithm discovers the most dense clusters, then it moves to next level of density by using another value for Eps that will be greater than the previous value, and the DBSCAN will be applied again on the data but using the new value for Eps, the process will be continue until all data are clustered. The knn represents the maximum number of points in Eps of any core point, and Minpts represents the minimum number of points in Eps of any core points. Our proposed method is different where it uses knn to find local density of point as the sum of distances between a point and its knn (density function based on distance).

GMDBSCAN [15] uses grid technique and is based on spatial index (sp-tree), for each grid cell it computes local Minpts according to the density of grid cell, and Eps value is fixed for all cells and applies DBSCAN on data in each grid cell. This method uses many parameters that affect the resulting clusters. Using different local Minpts doesn't guarantee discovering varied density clustering since Eps is fixed for all cells. The other Problem of GMDBSCAN is a time consuming to perform well on large datasets, and sometimes it gives the output after a long time [16].

GMDBSCAN-UR [16] is adaptation for GMDBSCAN, it is a grid technique. It selects number of well scattered points that capture the shape and extent of the dataset as representative points from each grid cell. This algorithm solves the problem of time consuming of GMDBSCAN. This method allows one of the two parameters of DBSCAN (Minpts or Eps) to be varied from cell to cell and the other is fixed so its performance is better than the performance of GMDBSCAN. But it doesn't produce accurate results with varied density clusters.

VDBSCAN [17] partitions the k-dist plot based on seeing sharp changes at the value of k-dist, and for each partition it selects a suitable value for Eps and applies DBSCAN on each partition. This method output accurate results only when dataset contains clusters of different uniform density. But when gradient in density exist the result of this method is not accurate. It may split dense cluster or merge sparser clusters unless they are well separated.

DBSCAN-DLP [18] partitions input dataset into diverse density level sets based on some statistical characteristic of density difference, then computes Eps value for each thickness level set and applies DBSCAN clustering on each thickness level set with corresponding Eps. This method is suitable for clusters of uniform density; the variance in density of points in the same cluster should be very small and less than threshold.

In [19] the authors introduced a mathematical idea to select several values for Eps form the k-dist plot and apply the DBSCAN algorithm on the data for each value with ignoring the previously clustered point. They use spline cubic interpolation to find inflection points on the curve where the curve changes its concavity. This method leads to split some clusters.

DSets-DBSCAN [20] runs DSets grouping first, where the parameters of DBSCAN are resolved from the first grouping extracted by DSets. All the previous algorithms estimate the value of Eps based on some local density criteria and apply the DBSCAN to discover clusters from dataset with multi-levels of density.

The second types of algorithms that redefine the density are like K-DBSCAN [21] and Multi Density DBSCAN [22]. K-DBSCAN [21] estimates the l-density for each data point, and arranges l-density values in ascending order, then uses k-means algorithm to divide the data set into k subsets; each subset represent a level of density. Finally, it applies a modified version of DBSCAN (uses difference between density levels of points beside distance between points while clustering) on each subset to get clusters of different density. K-DBSCAN needs more input parameters (l and k used in k-means algorithm) than the DBSCAN algorithm and tuning parameters is not a simple problem.

Multi Density DBSCAN [22] does not use Eps at all but use k-nearest neighbor to estimate DSTp as the average distance to its kth neighbor and AVGDST as the average distance to kth neighbor for points in the cluster. So, this method allows small difference in thickness within the cluster. This method uses sensitive threshold parameter to govern the variance in thickness accepted within the cluster that affects the result.

KDDBSCAN [23] uses mutual k-nearest neighborhood to define k-deviation density of point and uses threshold called density factor (instead of Eps) to find direct density reachable neighbors for core points to expand cluster.

CBLDP (Clustering Based on Local Density of Points) [24] algorithm discovers clusters with varied density, it depends on k-nearest neighbors and density rank for each point, it classifies points into attractors, attracted and noise, a cluster composed of attractors and attracted (border) items. This algorithm requires four input parameters, and this large number. Tuning these parameters is not always easy task.

3   Multi-density Based Notion of Clusters

K-nearest neighbor represents the density of region very well. In dense region the distance to the kth neighbor of a central item tends to be small, while in sparse region this distance tends to be large. From here, we present the dataset as a directed graph, where each item represents a vertex, and there is an edge from vertex p_i to vertex p_j if p_j belongs to the k-nearest neighbors of p_i; i is not equal to j, j is ranging from 1 to k, and i is ranging from 1 to N; where N is the size of dataset. For each item p, we estimate its density to all its k-nearest neighbors separately. i.e. the algorithm keeps density(p, q_1), density(p, q_2), ..., density(p, q_k). Where density(p, q_k) = kden(p). This idea can be represented mathematically as follows:

$$\text{density}(p_i, q_1) = \text{dist}(p_i, q_1). \tag{1}$$

$$\text{density}(p_i, q_j) = \text{density}(p_i, q_{j-1}) + \text{dist}(p_i, q_j). \tag{2}$$

$$\text{kden}(p_i) = \sum_{j=1}^{k} \text{dist}(p_i, q_j). \tag{3}$$

$$\text{dist}(p_i, q_j) = \sqrt{\sum_{m=1}^{d} (p_{i,m} - q_{j,m})^2}. \tag{4}$$

$$q_j \in knn(p_i), j = 1, \ldots, k, i = 1, \ldots, N. \tag{5}$$

As in Eq. (3), since kden(p_i) is the sum of distances among p_i and all its k neighbors, so as the value of kden(p_i) decrease the local density of p_i increase. There is a reverse proportional between kden(p_i) and local density of p_i. For example, see the following Fig. 1 the local density of p_1 is larger than that of p_2, while value of kden(p_1) is smaller than value of kden(p_2). In DBSCAN there is no difference in local density between p_1 and p_2 since it only counts the points in Eps neighborhood. Note that in DBSCAN, Eps is fixed for all data point as in Fig. 1.

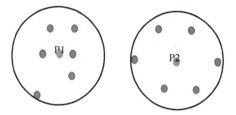

Fig. 1. Density based on total of k distances from a central point.

The proposed method uses the following definitions:-

Definition 1. k nearest neighbors of a point p is a set of size k that contains the k-nearest points from p; $knn(p) = \{q_i \in \text{dataset} \mid \text{dist}(p, q_i) \leq \text{dist}(p, q_k),$ $i = 1, 2, \ldots, k\}$.

Definition 2. k density of point p is represented as in Eq. (3).

Definition 3. Cluster initiator (Ci) is the first point from which the cluster is grown and has the highest local density. It has the smallest value for kden(Ci) among all other point that will be assigned to the cluster.

Definition 4. Directly density-reachable, item p is directly density reachable from item q wrt. kden(Ci), MinPts if:-

1. $p \in \{a_j \mid \text{density}(q, a_j) \leq \text{kden}(Ci), 1 \leq j \leq k\}$
2. $\text{density}(q, a_{Minpts}) \leq \text{kden}(Ci)$.

Definition 5. Noise, a very small clusters are treated as noise.

These definitions are similar to that of DBSCAN. Reader may be referred to [10]. The proposed method uses the definition of density reachable, density connected and cluster but replace Eps by kden(Ci) as in [10].

4   CMDD: Clustering Multi-density Datasets

This section presents the proposed algorithm CMDD (clustering Multi-density Dataset). This algorithm solves the problem of DBSCAN; it handles varied density clusters well. This algorithm doesn't use the Eps parameter of DBSCAN. But it use an alternative parameter called kden(Ci) that is computed easily from the k-nearest neighbor graph. It uses the Minpts parameter of DBSCAN in the same way as used in DBSCAN. The difference between DBSCAN and the suggested algorithm is that the suggested algorithm introduces alternative definition for density, it doesn't count the point in Eps neighborhood, instead; it computes the density to k-nearest neighbor for each point as in (3). And the kden(Ci) represents the highest thickness acceptable within the cluster and also control the lowest thickness acceptable within the cluster as in Definition 4; core point condition.

To find the clusters, the CMDD estimates the kden(pi) for all points in the input dataset, then it arranges them in an ascending order based on kden(pi). After that it starts from the first one in ordered dataset considering the kden(first point) as the maximum local density allowed in the cluster and expands cluster such that all point that are density reachable from the first point are assigned to the cluster. When there is no more points can be attached to the current cluster it moves toward the following unclassified point that has kden(new point) larger than that of the previous classified point.

The following lines depict the basic steps in the proposed algorithm.

```
CMDD(Dataset, Minpts, k)
//All points in dataset are unclassified
For i = 1 to Dataset.size
    Find the k-nearest neighbors for pi
    For j = 1 to k
        Compute density(pi,qj) as in Eq. (2)
    Next j
    Arrange Dataset points in ascending order based on
    kden(pi) as in Eq. (3)
Next i
CClusterId=NextClusId(Noise);
For i= 1 to OrderedDataset.volume
    CPoint = OrderedDataset.getpoint(i);
    IF CPoint.ClusId = UNCLUSTERED Then
        IF ExpandCluster(OrderedDataset, CPoint,
            CClusterId, kden(CPoint), MinPts) Then
            CClusterId = nextId(CClusterId);
        ENDIF
    ENDIF
Next i
End // CMDD
```

Dataset is the input dataset to be clustered. Minpts and k are global input parameters used to control the thickness acceptable within the cluster. Experimentally Minpts is equal to 3 or 4. The method OrderedDataset.getpoint(i) returns the ith data point from the ordered dataset. The basic method used by CMDD is ExpandCluster which is described as follows:

```
ExpandCluster(OrderedDataset, CPoint, CClusterId,
                    kden(CPoint), Minpts) : Boolean;
Ci=CPoint;
Seeds1:= OrderedDataset.regionQuery1(CPoint,k);
  IF density(CPoint, seedsMinpts)>kden(Ci) Then
       // is not core item
       OrderedDataset.changeClusId(CPoint,NOISE);
       RETURN UNTRUE;
  OTHERWISE
   //items in seeds1 are reachable from CPoint
   OrderedDataset.changeClusIds(seeds1,ClusId);
   Seeds1.delete(CPoint);
   While seeds1 != Empty do
      currentP1 := seeds1.firstitem();
      result1 = OrderedDataset.regionQuery1(currentP,
                                             k);
      IF density(currentP,resultMinpts)<= kden(Ci)Then
        For i = 1 TO result1.volume
            If density(currentP, resulti)>kden(Ci)Then
              Exit for
            End if
             resultP1 := result1.getitem(i);
            IF resultP1.ClusId IN {UNCLUSTERED, NOISE}
               Then
               IF resultP1.ClusId = UNCLUSTERED Then
                    Seeds1.appenditem(resultP1);
               Endif;
               OrderedDataset.changeClusId(resultP1,ClusId);
            End if; // UNCLUSTERED or NOISE
        Next i
      Endif//density(currentP1,resultMinpts)<= kden(Ci)
      Seeds1.delete(currentP1);
   ENDWhile; // seeds1 is not Empty
  RETURN not false;
  ENDIF
END; // function ExpandCluster
```

The cluster is expanded from cluster initiator (Ci), which is the most dense core point. So all its k nearest neighbors is form the seed list. Any point p in seed list satisfies the relation density(p, q_{Minpts}) $\leq$ kden(Ci) is core point, and the algorithm assigns its neighbors satisfying the relation density(p, q_j) $\leq$ kden(Ci) to the current cluster, where $j = 1, 2, ..., k$. The core points from these neighbors are appended to the seed list. A call of OrderedDataset.regionQuery1(Point, k) returns the k-nearest neighborhood of Point in OrderedDataset as a list of points and they have been founded before. District inquiries can be upheld proficiently by spatial access strategies, for example, R*-trees. The border points are not added to the seed list.

5 Experimental Results

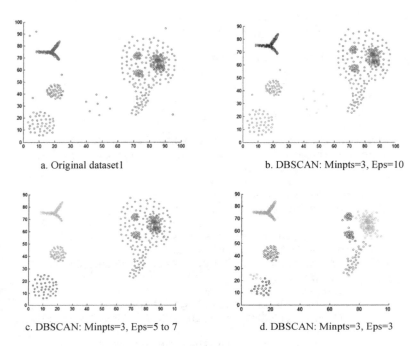

a. Original dataset1

b. DBSCAN: Minpts=3, Eps=10

c. DBSCAN: Minpts=3, Eps=5 to 7

d. DBSCAN: Minpts=3, Eps=3

Fig. 2. DBSCAN results from dataset1.

This section validates the effectiveness of the offered CMDD method with data clustering experiments. We have utilized diverse datasets containing groups of fluctuated densities, shapes, and sizes. The proposed strategy prevailing with regards to finding clusters despite the fact that the high closeness of one another.

Figure 2a shows the first dataset, that contains varied density clusters and 6 noise points. DBSCAN can't find the correct clusters from the data as shown in Fig. 2b it assigns two noise points to the nearest cluster, discards only four points as noise and merges four clusters in one. When we decreased Eps to 7, it discards the sparse cluster as a noise as shown in Fig. 2c. As Eps decreased it removes more points as outlier. When we set Eps = 3 it discovers 9 clusters and discards 89 points as noise and outliers as shown in Fig. 2d there is no suitable global value for Eps to be used to discover the actual clusters in this dataset. That is the problem the proposed algorithm solves.

The proposed algorithm discovers the actual clusters from the data since it controls the variance in density allowed within each cluster. Figure 3 shows the results from CMDD using different values for its input parameters k and Minpts, as variance in density allowed within the cluster increase it tends to produce bigger clusters (number of points in each cluster increase) as shown in Fig. 3a when the variance in density allowed within the cluster decrease it splits some clusters that contain points with different density as shown in Fig. 3b.

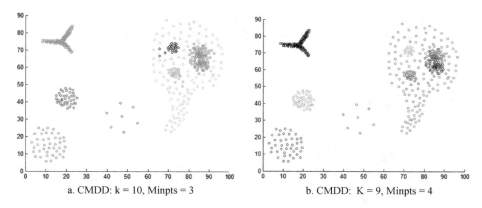

a. CMDD: k = 10, Minpts = 3   b. CMDD: K = 9, Minpts = 4

Fig. 3. CMDD results from dataset1.

Table 1 shows the kden(Ci) value that represent the maximum density allowed within each cluster in dataset1. Note that when the variance in density allowed decreased the algorithm split cluster2 to produce 9 clusters as in Fig. 3b.

The second dataset used in experiments is shown in Fig. 4a when we allow large variance in density the CMDD discovers 6 clusters as in Fig. 4b, it merges the two upper left clusters. When we decreased the allowed variance in density it split the upper left cluster into two clusters and removed 8 points as outliers; since they are below the density allowed within the nearest cluster as shown in Fig. 4c.

Table 1. The maximum density allowed within each cluster.

	K = 10, Minpts = 3	K = 9, Minpts = 4
Cluster1	6.419	5.522
Cluster2	6.900	5.631
Cluster3	15.269	6.885
Cluster4	15.458	12.899
Cluster5	20.246	13.391
Cluster6	26.904	17.010
Cluster7	34.633	22.597
Cluster8	131.811	29.630
Cluster9		115.000

When we applied DBSCAN on the dataset2 it failed to discover the actual clusters in data. Figure 4d shows that it merges the four left clusters into one and the two right clusters into one to discover two clusters only. When we decreased the Eps it removed 38 points from the sparse cluster in right side (removed almost the cluster) as shown in Fig. 4e. when we decreased the Eps to 1 it discovered only one correct cluster on right side of Fig. 4f and removed the surrounded cluster as outliers, also it discovered portion of two upper left clusters, but the two lower left clusters are still merged. Again, there is no global suitable value for Eps can be used in the current dataset2.

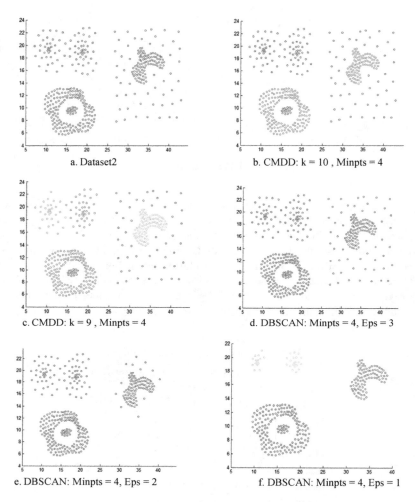

Fig. 4. CMDD and DBSCAN results from dataset2.

Table 2. The kden(Ci) in each cluster.

	K = 10, Minpts = 4	K = 9, Minpts = 4
Cluster1	5.940	5.039
Cluster2	6.459	5.508
Cluster3	6.490	5.546
Cluster4	6.858	5.695
Cluster5	18.628	5.867
Cluster6		16.373

The proposed method succeeded in the matter. Table 2 shows the kden(Ci) value for each discovered cluster in dataset2 that shown in Fig. 4b and c

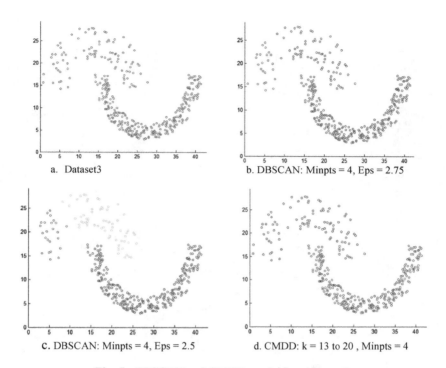

Fig. 5. DBSCAN and CMDD results from dataset3.

The third dataset is shown in Fig. 5a that contains only two clusters. Applying DBSCAN on it with Eps = 2.75 produce result as shown in Fig. 5b that merging portion of sparse cluster with the dense one and remove single point as noise. When we decreased the Eps to 2.5 we get the result as shown in Fig. 5c where it discovered the dense cluster correctly but split the sparse one into two and removed 5 points as outliers from the two sparse clusters. DBSCAN can't discover the sparse cluster.

The proposed algorithm discovered the clusters in the data correctly as shown in Fig. 5d when we set k = 13 the calculated kden(Ci) was 9.114 for the dense cluster and 16.366 for the sparse one.

Figure 6a shows the fourth dataset used to test the proposed method. DBSCAN gets 7 clusters from the data and allows large variance in density within the five clusters on right of Fig. 6b, while the proposed method discovers 9 clusters as shown in Fig. 6c where it gets a very dense cluster inside the upper left cluster, and a sparse cluster among the left clusters and the lower right clusters. These two clusters are impossible to be getting from DBSCAN. In addition to the quality of right side clusters is better than that result from DBSCAN. When we increased the value of k to 30 to allow larger variance in density within the clusters we get the result as shown in Fig. 6d where it merges the two upper left clusters into one, and the sparse cluster becomes bigger. The final number of clusters becomes 8.

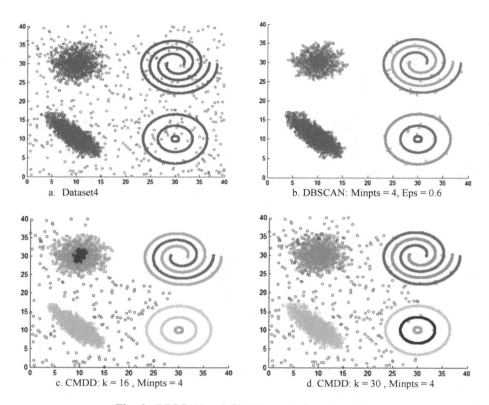

a. Dataset4

b. DBSCAN: Minpts = 4, Eps = 0.6

c. CMDD: k = 16 , Minpts = 4

d. CMDD: k = 30 , Minpts = 4

Fig. 6. DBSCAN and CMDD results from dataset4.

Figure 7a depicts the fifth dataset in our experiments that contains four clusters of varied density. The proposed methods discovered the four clusters efficiently as shown in Fig. 7b. While DBSCAN failed to discover them, when we set Eps = 0.5 it removed the sparse cluster on the right as outliers, see Fig. 7c, and merged the left three clusters into one. When we set Eps = 0.4 it removed the middle clusters on left and returned the other two clusters correctly as shown in Fig. 7d when we set Eps = 0.25 it only discovered the dense cluster on the top left side of Fig. 7a and removed the other clusters as outliers.

The kden(Ci) value evaluated for each cluster in fifth dataset is kden(C1) = 1, kden (C2) = 1.449, kden(C3) = 2.539, kden(C4) = 6.105.

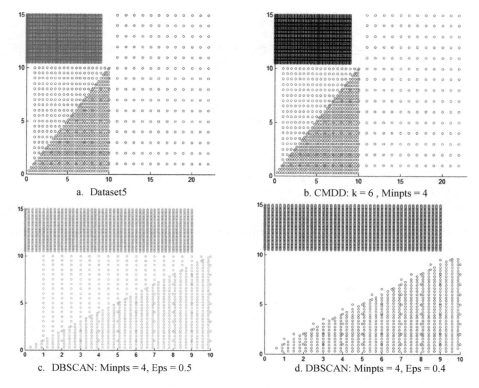

a. Dataset5

b. CMDD: k = 6 , Minpts = 4

c. DBSCAN: Minpts = 4, Eps = 0.5

d. DBSCAN: Minpts = 4, Eps = 0.4

Fig. 7. CMDD and DBSCAN results from dataset2.

6 Conclusion

In this paper we have proposed a clustering method that is able to handle varied density clusters well. The method performed simple modification on the DBSCAN method. The idea is to compute the maximum density allowed within each cluster and use Minpts to control the minimum density allowed within each cluster. The method based on the k-nearest neighbor to compute the maximum density in each cluster and uses the density to the ith neighbor to decide whether the point can be assigned to the current cluster or not, and ith neighbor must be greater than or equal to Minpts.

The experimental results presented here is evidence that the proposed algorithm is more efficient than the DBSCAN algorithm in handling datasets with varied density clusters. The proposed algorithm has the same time complexity as that of DBSCAN algorithm which is o(n log n).

Acknowledgements. This project was supported by the Deanship of Scientific Research at Prince Sattam Bin Abdulaziz University under the research project no. 2017/01/7120.

References

1. Fahim, A.M., Salem, A.M., Torkey, F.A., Ramadan, M.: An efficient enhanced k-means clustering algorithm. J. Zhejiang Univ. Sci. A **7**(10), 1626–1633 (2006)
2. Kaufman, L., Rousseeuw, P.J.: Finding groups in data: an introduction to cluster analysis. In: Partitioning Around Medoids (Program PAM). Wiley (1990)
3. Ng, R.T., Han, J.: Efficient and effective clustering methods for spatial data mining. In: Proceedings of the 20th International Conference on Very Large Databases, Santiago, Chile, pp. 145–155 (1994)
4. Sibson, R.: SLINK: an optimally efficient algorithm for the single-link cluster method. Comput. J. **16**(1), 30–34 (1973)
5. Seifoddini, H.K.: Single linkage versus average linkage clustering in machine cells formation applications. Comput. Ind. Eng. **16**(3), 419–426 (1989)
6. Defays, D.: An efficient algorithm for a complete link method. Comput. J. **20**(4), 364–366 (1977)
7. Karypis, G., Han, E.H., Kumar, V.: CHAMELEON: a hierarchical clustering algorithm using dynamic modeling. Computer **32**(8), 68–75 (1999)
8. Zhang, T., Ramakrishnan, R., Livny, M.: BIRCH: an efficient data clustering method for very large databases. In: Proceedings of the 1996 ACM SIGMOD International Conference on Management of Data, SIGMOD 1996, pp. 103–114. ACM, New York (1996)
9. Guha, S., Rastogi, R., Shim, K.: Cure: an efficient clustering algorithm for large databases. In: Haas, L.M., Tiwary, A. (eds.) Proceedings ACM SIGMOD International Conference on Management of Data, Seattle, Washington, USA, 2–4 June 1998, pp. 73–84. ACM Press (1998)
10. Ester, M., Krigel, H.P., Sander, J., Xu, X.: A density-based algorithm for discovering clusters in large spatial databases with noise. In: Proceedings of the International Conference on Knowledge Discovery and Data Mining, pp. 226–231 (1996)
11. Ankerst, M., Breunig, M.M., Kriegel, H.P.: OPTICS: ordering points to identify the clustering structure. In: Proceedings of ACM SIGMOD, pp. 49–60 (1999)
12. Hinneburg, A., Keim, D.A.: An efficient approach to clustering in large multimedia databases with noise. In: Proceedings of the 4th International Conference on Knowledge Discovery and Data Mining, New York, September 1998, pp. 58–65 (1998)
13. Idrissi, A., Rehioui, H., Laghrissi, A., Retal, S.: An improvement of DENCLUE algorithm for the data clustering. In: 2015 5th International Conference on Information and Communication Technology and Accessibility (ICTA), Marrakech, pp. 1–6 (2015)
14. Fahim, A.: Homogeneous densities clustering algorithm. Int. J. Inf. Technol. Comput. Sci. (IJITCS) **10**(10), 1–10 (2018)
15. Chen, X., Min, Y., Zhao, Y., Wang, P.: GMDBSCAN: multi-density DBSCAN cluster based on grid. In: Proceedings of the IEEE International Conference on e-Business Engineering, ICEBE 2008, China, October 2008, pp. 780–783 (2008)
16. Alhanjouri, M.A., Ahmed, R.D.: New density-based clustering technique: GMDBSCAN-UR. Int. J. Adv. Res. Comput. Sci. **3**(1), 1–9 (2012)
17. Liu, P., Zhou, D., Wu, N.: VDBSCAN: varied density based spatial clustering of applications with noise. In: Proceedings of the ICSSSM 2007: 2007 International Conference on Service Systems and Service Management, China, June 2007
18. Xiong, Z., Chen, R., Zhang, Y., Zhang, X.: Multi-density DBSCAN algorithm based on density levels partitioning. J. Inf. Comput. Sci. **9**(10), 2739–2749 (2012)

19. Louhichi, S., Gzara, M., Abdallah, H.: A density based algorithm for discovering clusters with varied density. In: 2014 World Congress on Computer Applications and Information Systems (WCCAIS), Hammamet, pp. 1–6 (2014)
20. Hou, J., Gao, H., Li, X.: DSets-DBSCAN: a parameter-free clustering algorithm. IEEE Trans. Image Process. **25**(7), 3182–3193 (2016)
21. Debnath, M., Tripathi, P.K., Elmasri, R.: K-DBSCAN: identifying spatial clusters with differing density levels. In: Proceedings of the 2015 International Workshop on Data Mining with Industrial Applications, DMIA 2015, Paraguay, September 2015, pp. 51–60 (2015)
22. Ashour, W., Sunoallah, S.: Multi density DBSCAN. In: Lecture Notes in Computer Science (including subseries Lecture Notes in Artificial Intelligence and Lecture Notes in Bioinformatics): Preface, vol. 6936, pp. 446–453 (2011)
23. Jungan, C., Jinyin, C., Dongyong, Y., Jun, L.: A k-deviation density based clustering algorithm. Math. Probl. Eng. **2018**, 1–16 (2018)
24. Fahim, A.: A clustering algorithm based on local density of points. Int. J. Mod. Educ. Comput. Sci. (IJMECS) **9**(12), 9–16 (2017)

Generate a New Types of Fuzzy Ψ_i-Operator

Raghad Almohammed$^{(\boxtimes)}$ and Luay A. AL-Swidi

Department of Mathematics, College of Education for Pure Sciences,
University of Babylon, Hillah, Iraq
ralkarblae@yahoo.com, drluayhall@yahoo.com

Abstract. In recent years, fuzzy set technology has become a serious element in the field of computer vision and artificial intelligence. As a consequence of intensive research activity, it now competes with classical methods in terms of quality and performance. This mathematical branch has been shown to be very useful in modeling several problems which arise in branches of applied sciences such as Economics, Artificial Intelligence, and Computer Science. An example in that last field is its role in image processing, where a fuzzy operator is used to solve image processing problems such as image smoothing, extracting edges, sharpening images. The aim of this research is to introduce different types of fuzzy operators that help programmers to solve the problems of image processing, encryption or communication. We provided four types of the fuzzy Ψ-operators based on the definition of fuzzy local functions and study of the advantages and differences between them. We also, introduced different types of fuzzy topology that are easy to handle in processing programs to help specify the properties of digital images.

Keywords: Fuzzy topological space · Fuzzy ideal · Fuzzy local function · Fuzzy Ψ_i-operator

1 Introduction

In 1965, Zadeh introduced the fundamental concept of a fuzzy set, which provided a logical explanation for some of the fuzzy phenomena in our real world [1]. At that time his definition received criticisms from the research community. However, the remarkable advancement in the field of computer and artificial intelligence has motivated scientists to review the fuzzy set concept which currently plays a significant role in branches of computer science such as digital images, network topology, encryption and communication [2, 3]. Fuzzy functions are important functions in computer science because of the flexibility of movement points. We know that the characteristic function takes the value of 1 or 0 and this is the basic building block of computers of all kinds and, therefore, the cornerstone of electronic science and communications. The importance of the fuzzy function lies in the fact that the range of the function is the closed interval between 0 and 1. This means that infinity points exist between the zero and one, and the value of the function will be based on the infinity of the points close to 1 and the infinity of the points close to 0. Consequently, both the credibility and lack of credibility of the solution are infinite and here lies the fuzziness we need to address directly in computer science, especially in fields of cryptography and communication.

© Springer Nature Switzerland AG 2020
L. C. Jain et al. (Eds.): ICICCT 2019, LAIS 9, pp. 28–39, 2020.
https://doi.org/10.1007/978-3-030-38501-9_3

The solution to an encrypted message which is difficult to solve is possible through the characteristics of the use of fuzzy functions directly or indirectly in deciphering operations. It is also used as the basis for building a complex mathematical method in the algorithms of encryption and decoding, as well as building keys to those systems. In this research we focused on introducing new types of fuzzy operators and studying their properties, some of their characteristics and the relationships between them to solve the problems of encryption or communication and other branches of computer science that depend directly or indirectly on the fuzzy functions. In Sect. 2, we briefly review the fuzzy set theory and provide some basic definitions. In Sect. 3, we introduce different types of the Ψ_i–operator by using types of a fuzzy local function and studying its advantages. We discuss types of Ψ_i–operators to identify new fuzzy topology. Finally, we present our conclusions and suggestions for future work in this field.

2 Preliminaries

Throughout this paper $(1_x, \tau, I)$ means a fuzzy ideal topological space, denoted by FITS where $(1_x, \tau)$ a fuzzy topology space in Change [4], and I is collection fuzzy ideal define in Sarkar [5]. Any fuzzy set denoted by $\hat{A}$. 1_x Is the universal fuzzy set. The collection of all fuzzy set of 1_x denoted by Γ^X. A fuzzy point denoted by P_x^δ, $(0 < \delta \le 1)$. Also $P_x^\delta \in \hat{A}$ iff $\delta \le f_{\hat{A}}(x)$ where $f_{\hat{A}}(x)$ the memberships define by

$$f_{\hat{A}}(x) = \begin{cases} f(x) & \text{for} \quad x \in \hat{A} \\ 0 & \text{for} \quad x \notin \hat{A} \end{cases} \forall x \in X \ [6].$$

$\hat{A}$ is fuzzy neighborhood of P_x^δ iff $\exists$ fuzzy open set $\hat{V}$ s.t $P_x^\delta \in \hat{V} \subseteq \hat{A}$. all neighborhood of P_x^δ denoted by $\mathcal{N}\left(P_x^\delta\right)$. Also the collection of all open set of the point P_x^δ denoted by $\tau(P_x^\delta)$. Any fuzzy set is said to be quasi - coincident with other fuzzy set symbolizes them $\hat{A} q \hat{B}$ iff there exist $y \in X \ni f_{\hat{A}}(y) + g_{\hat{B}}(y) > 1$. Therefore, $f_{\hat{A}}(x) + g_{\hat{B}}(x) \le 1 \ \forall x \in X$ implies hereafter $\hat{A}$ is not quasi- coincident with $\hat{B}$ symbolizes them $\hat{A} \bar{q} \hat{B}$. $\hat{A}$ is called q - neighborhood (short, q_nbd) of a fuzzy point P_x^δ iff there exist the fuzzy open set $\hat{U}$ s.t $P_x^\delta q \hat{U}$ and $\hat{U} \le \hat{A}$. we will denote the set of all q_nbd of P_x^δ by q-$\mathcal{N}(P_x^\delta)$. cl $\hat{A}$, int$\left(\hat{A}\right)$ respectively, denote the fuzzy closure, fuzzy interior [7].

Definition 2.1: Let $\hat{A}, \hat{B}$ any fuzzy set in Γ^X. The stander of union, intersection, complement, difference are from,

1. $\hat{A} \cup \hat{B} = \{(x, \max\{f_{\hat{A}}(x), g_{\hat{B}}(x)\}), \forall x \in X\}$.
2. $\hat{A} \cap \hat{B} = \{(x, \min\{f_{\hat{A}}(x), g_{\hat{B}}(x)\}), \forall x \in X\}$.
3. $1_x - \hat{A} = \{(x, 1 - f_{\hat{A}}(x)), \forall x \in X\}$.

4. The stander difference has two defines [8]. To standardize terminology we will write the definition as follows

i. $\left(\widehat{A} - \widehat{B}\right)_{max} = \left\{\left(x, \max\left\{f_{\widehat{A}}(x) - g_{\widehat{B}}(x), 0\right\}\right), \forall x \in X\right\}.$

ii. $\left(\widehat{A} - \widehat{B}\right)_{min} = \left\{\left(x, \min\left\{f_{\widehat{A}}(x), g_{\left(1_x - \widehat{B}\right)}(x)\right\}\right), \forall x \in X\right\}.$

Noted that for $\widehat{A} = 1_X$ we have $\left(1_X - \widehat{B}\right)_{max} = \left(1_X - \widehat{B}\right)_{min}.$

In this paper we will use four different types of the fuzzy local function.

Definition 2.2: $\widehat{A}^{*1}(\mathcal{T}, I) = \vee\{P_x^\delta; \forall \widehat{\mathcal{U}}\epsilon q - \mathcal{N}\left(P_x^\delta\right), \exists y \in X \text{ s.t } f_{\widehat{\mathcal{U}}}(y) + g_{\widehat{A}}(y) - 1 > h_I(y) \text{ for every } I \in I\}$ [5].

Definition 2.3: $\widehat{A}^{*2}(\mathcal{T}, I) = \vee\{P_x^\delta; \forall \widehat{\mathcal{U}}\epsilon q - \mathcal{N}\left(P_x^\delta\right), \exists y \in X \text{ s.t } \min\left(f_{\widehat{\mathcal{U}}}(y), g_{\widehat{A}}(y)\right) > h_I(y) \text{ for every } I \in I\}$ [9].

Definition 2.4: $\widehat{A}^{*3}(\mathcal{T}, I) = \vee\{P_{\dot{x}}^\delta; \forall \mathcal{U}\epsilon q - \mathcal{N}(P_{\dot{x}}^\delta), \dot{x} \in X \text{ s.t } f_{\widehat{\mathcal{U}}}(\dot{x}) + g_{\widehat{A}}(\dot{x}) - 1 > h_I(\dot{x}) \text{ for every } I \in I\}$ [9].

Definition 2.5: $\widehat{A}^{*4}(\mathcal{T}, I) = \vee\{P_{\dot{x}}^\delta; \forall \widehat{\mathcal{U}}\epsilon q - \mathcal{N}\left(P_{\dot{x}}^\delta\right), \dot{x} \in X \text{ s.t } \min\left(f_{\widehat{\mathcal{U}}}(\dot{x}), g_{\widehat{A}}(\dot{x})\right) > h_I(\dot{x}) \text{ for every } I \in I\}$ [9].

The closure operator define as $cl^{*i}\left(\widehat{A}\right) = \widehat{A} \cup \widehat{A}^{*i}.$

3 Fuzzy Ψ_i-Operator

In this section, we will introduce different types of Ψ_i-operator and examine the relationships between them. Also, we exploit these types in the creation of fuzzy topology.

Definition 3.1: Let $(1_x, \tau, I)$ FITS. Fuzzy Ψ_i-operator is defined as $\Psi_i(\widehat{A}) = 1_X - \left(1_X - \widehat{A}\right)^{*i}$ for any $\widehat{A} \epsilon \Gamma^X$. Where $(*i)$ is the fuzzy local function $i = 1, 2, 3, 4.$

Example 3.1: Let $(1_x, \tau, I)$ FITS. Let $X = \{1, 2, 3\}$, $A = \{1, 3\}$, $B = \{1, 2\}$, $C = \{1, 3\}$ and $I = \{1\}$. The membership of $\widehat{A}$, $\widehat{B}$, $\widehat{C}$ and I are:

$$f_{\widehat{A}}(x) = \begin{cases} \frac{1}{10} & \text{if } x = 1 \\ \frac{4}{10} & \text{if } x = 3 \end{cases} \forall x \in A, g_{\widehat{B}}(x) = \frac{x}{10} \ \forall x \in B, K_{\widehat{C}}(x) = \frac{3}{10} \ \forall x \in C,$$
$$h_I(x) = \frac{9}{10} \quad \forall x \in I.$$

$\widehat{A} = \{(1,0.1),(2,0),(3,0.4)\}.$

$\widehat{B} = \{(1,0.1),(2,0.2),(3,0)\}, \widehat{C} = \{(1,0.3),(2,0),(3,0.3)\}.$

$I = \{0_X, I\} \cup \{\zeta; \zeta \leq I\}$ Where $I = \{(1,0.9),(2,0),(3,0)\}.$

Put $\tau = \{0_X, 1_X, \widehat{B}, \widehat{C}, \widehat{B} \cap \widehat{C}, \widehat{B} \cup \widehat{C}\}$

$\Psi_1\left(\widehat{A}\right) = \{(1,0.3),(2,0),(3,0.3)\}.$

$\Psi_2(\widehat{A}) = \{(1,0.1),(2,0),(3,0)\}.$

$\Psi_3\left(\widehat{A}\right) = \{(1,1),(2,0),(3,0.3)\}.$

$\Psi_4\left(\widehat{A}\right) = \{(1,1),(2,0),(3,0)\}.$

Clear $\Psi_i(\widehat{A})$ where i = 1, 2 is open, since $\left(1_X - \widehat{A}\right)^{*i}$ is closed this mean Ψ_i-opetator: $\Gamma^X \longrightarrow \tau$ for i = 1, 2.

But $\Psi_i\left(\widehat{A}\right)$ Where i = 3, 4 is not open. So we will expand the domain of the operator to include all the co domain, where it is Ψ_i-opetator: $\Gamma^X \longrightarrow \Gamma^X$ for i = 3, 4.

The following theorem contains basic results of Fuzzy Ψ_i-operator which are known in general topology, and also shows the difference between two structures [10].

Theorem 3.1: Let$(1_x, \tau, I)$ FITS. $\widehat{A}, \widehat{B}$ any two fuzzy set in Γ^X then,

1. $\widehat{A} \subseteq \widehat{B} \Rightarrow \Psi_i(\widehat{A}) \subseteq \Psi_i(\widehat{B}).$
2. $\widehat{A} \cap \Psi_i(\widehat{A}) = \text{int}^{*i}\left(\widehat{A}\right).$
3. $\Psi_i(\widehat{A}) \subseteq \Psi_i(\Psi_i(\widehat{A}))$
4. $\Psi_i(\widehat{A}) = \Psi_i(\Psi_i(\widehat{A}))$ Iff $(\widehat{A}^{*i})^{*i} = \widehat{A}^{*i}.$
5. $\Psi_i(\widehat{0}) = 1_X - 1_X^{*i},$ and $\Psi_i(1_X) = 1_X.$
6. $\Psi_i(\widehat{0}) \subseteq \Psi_i(\widehat{A}) \forall \widehat{A} \in I.$
7. $(1_X - \widehat{A}) \in I$ then $\Psi_i\left(\widehat{A}\right) = 1_X.$
8. $\Psi_i(\widehat{A}) \cap \Psi_i(\widehat{B}) = \Psi_i(\widehat{A} \cap \widehat{B}).$
9. $\Psi_i(\widehat{A}) \cup \Psi_i(\widehat{B}) \subseteq \Psi_i(\widehat{A} \cup \widehat{B}).$

Proof

(1) $\widehat{A} \subseteq \widehat{B}$ This leads to $1_X - \widehat{B} \subseteq 1_X - \widehat{A}.$ By proportion of $*i$ we get $\left(1_X - \widehat{B}\right)^{*i} \subseteq \left(1_X - \widehat{A}\right)^{*i},$ thus $\Psi_i(\widehat{A}) \subseteq \Psi_i(\widehat{B}).$

(2) $\widehat{A} \cap \Psi_i(\widehat{A}) = \widehat{A} \cap \left(1_X - \left(1_X - \widehat{A}\right)^{*i}\right) = 1_X - \left(\left(1_X - \widehat{A}\right) \cup \left(1_X - \widehat{A}\right)^{*i}\right) =$
$1_X - \text{cl}^{*i}\left(1_X - \widehat{A}\right) = \text{int}^{*i}\left(\widehat{A}\right).$

(3) $\Psi_i(\widehat{A}) = 1_X - \left(1_X - \widehat{A}\right)^{*i} \subseteq 1_X - \left(\left(1_X - \widehat{A}\right)^{*i}\right)^{*i} = 1_X - (1_X - (1_X - \widehat{A})^{*i}))^{*i} = \Psi_i(\Psi_i(\widehat{A}))$

(4) $\Psi_i(\Psi_i(\widehat{A})) = 1_X - \left(1_X - (1_X - \left(1_X - \widehat{A}\right)^{*i})\right)^{*i} = 1_X - \left(\left(1_X - \widehat{A}\right)^{*i}\right)^{*i} =$

$1_X - \left(1_X - \widehat{A}\right)^{*i}$ since $(\widehat{A}^{*i})^{*i} = \widehat{A}^{*i}$ thus $\Psi_i\left(\Psi_i\left(\widehat{A}\right)\right) = \Psi_i(\widehat{A})$.

Again, let $\Psi_i(\widehat{A}) = \Psi_i(\Psi_i(\widehat{A}))$. This implies $1_X - \left(1_X - \widehat{A}\right)^{*i} = 1_X - (1_X -$

$(1_X - \left(1_X - \widehat{A}\right)^{*i}))^{*i} = 1_X - \left(\left(1_X - \widehat{A}\right)^{*i}\right)^{*i}$ thus $\left(1_X - \widehat{A}\right)^{*i} = \left(\left(1_X - \widehat{A}\right)^{*i}\right)^{*i}$

for all $\widehat{A} \epsilon \Gamma^X$. Therefore $\left(\widehat{A}\right)^{*i} = \left(\left(\widehat{A}\right)^{*i}\right)^{*i}$.

(5) Obviously, $\Psi_i(\widehat{0}_X) = 1_X - (1_X - 0_X)^{*i} = 1_X - (1_X)^{*i}$..   So, $\Psi_i(1_X) = 1_X - (1_X - 1_X)^{*i} = 1_X - (0_X)^{*i} = 1_X - 0_X = 1_X$.

(6) $\Psi_i(\widehat{0}) = 1_X - (1_X)^{*i}$. We know $\left(1_X - \widehat{A}\right) \subseteq 1_X$ this implies $\left(1_X - \widehat{A}\right)^{*i} \subseteq 1_X^{*i}$ thus $1_X - (1_X)^{*i} \subseteq 1_X - \left(1_X - \widehat{A}\right)^{*i} = \Psi_i(\widehat{A})$. This also true of every $\widehat{A} \in \Gamma^X$. The converse is not always true by the following example 3.1 evident, $\Psi_1\left(\widehat{A}\right) = \{(1, 0.3), (2, 0), (3, 0.3)\}$ but $\Psi_1(0_X) = \{(1, 0.1), (2, 0), (3, 0)\}$.

(7) Let $(1_X - \widehat{A}) \in I$ this implies $\left(1_X - \widehat{A}\right)^{*i} = 0_X$, take the complement we get

$(1_X - \left(1_X - \widehat{A}\right)^{*i}) = 1_X$.

(8) $\Psi_i(\widehat{A}) \cap \Psi_i(\widehat{B}) = (1_X - \left(1_X - \widehat{A}\right)^{*i}) \cap (1_X - \left(1_X - \widehat{B}\right)^{*i}) = 1_X - \left(\left(1_X - \widehat{A}\right)^{*i} \cup \right.$

$\left(1_X - \widehat{B}\right)^{*i}) = 1_X - \left(\left(1_X - \widehat{A}\right) \cup \left(1_X - \widehat{B}\right)\right)^{*i} = 1_X - \left(1_X - \left(\widehat{A} \cap \widehat{B}\right)\right)^{*i} =$
$\Psi_i(\widehat{A} \cap \widehat{B})$.

(9) $\Psi_i(\widehat{A}) \cup \Psi_i(\widehat{B}) = (1_X - \left(1_X - \widehat{A}\right)^{*i}) \cup (1_X - \left(1_X - \widehat{B}\right)^{*i}) = 1_X - \left(\left(1_X - \widehat{A}\right)^{*i} \cap \right.$

$\left(1_X - \widehat{B}\right)^{*i}) \subseteq 1_X - \left(\left(1_X - \widehat{A}\right) \cap \left(1_X - \widehat{B}\right)\right)^{*i} = 1_X - \left(1_X - \left(\widehat{A} \cup \widehat{B}\right)\right)^{*i}$ by

proportion of $*i$, thus $\Psi_i(\widehat{A}) \cup \Psi_i(\widehat{B}) \subseteq \Psi_i(\widehat{A} \cup \widehat{B})$. The following example shows that the converse is not true.

Example 3.2: Let $(1_x, \tau, I)$ FITS. Let $X = \{1, 2, 3\}$, $A = \{1, 3\}$, $B = \{1, 2\}$, $D = \{1, 2\}$, $H = \{2, 3\}$ and $I = \{1,2,3\}$. The membership of $\widehat{A}, \widehat{B}, \widehat{D}, \widehat{H}$ and I are:

$f_{\widehat{A}}(x) = \frac{x}{5}$ $\forall x \in A$, $g_{\widehat{B}}(x) = \frac{4+X}{10}$ $\forall x \in B$, $K_{\widehat{D}}(x) = \frac{X^2+2X}{10}$ $\forall x \in D$, $L_{\widehat{H}}(x) = \frac{11-X}{10} \forall x \in H$, $h_I(x) = \frac{2X-1}{10}$ $\forall x \in I$.
$\widehat{A} = \{(1, 0.2), (2, 0), (3, 0.6)\}$, $\widehat{B} = \{(1, 0.5), (2, 0.6), (3, 0)\}$,

$\widehat{D} = \{(1,0.3),(2,0.8),(3,0)\}, \widehat{H} = (1,0),(2,0.9),(3,0.8)\},$
$I = \{0_X, I\} \cup \{\zeta; \zeta \le I\}$ Where $I = \{(1,0.1),(2,0.3),(3,0.5)\}.$
Put $\tau = \{0_X, 1_X, \widehat{D}, \widehat{H}, \widehat{D} \cap \widehat{H}, \widehat{D} \cup \widehat{H}\}$
$\Psi_1(\widehat{A}) = 0_X$
$\Psi_1(\widehat{B}) = \{(1,0.3),(2,0.8),(3,0)\}.$
$\Psi_1(\widehat{A}) \cup \Psi_1(\widehat{B}) = \{(1,0.3),(2,0.8),(3,0)\}.$
$\Psi_1(\widehat{A} \cup \widehat{B}) = \{(1,0.3),(2,0.9),(3,0.8)\}.$ This means $\Psi_1(\widehat{A}) \cup \Psi_1(\widehat{B}) \subseteq \Psi_1(\widehat{A} \cup \widehat{B}).$

Remark 3.1: If $I = \Gamma^X$ then $\Psi_i\left(\widehat{A}\right) = 1_X.$

Proposition 3.1: Let $(1_x, \tau, I)$ FITS.Then, for any fuzzy set $\widehat{A}$ of Γ^X,

1- $\Psi_1\left(\widehat{A}\right) = \text{int}(\Psi_3(\widehat{A}))$

2- $\Psi_2\left(\widehat{A}\right) = \text{int}(\Psi_4(\widehat{A}))$

3- $\Psi_2\left(\widehat{A}\right) \subseteq \Psi_1(\widehat{A}) \subseteq \Psi_3(\widehat{A})$

4- $\Psi_2\left(\widehat{A}\right) \subseteq \Psi_4(\widehat{A}) \subseteq \Psi_3(\widehat{A})$

Proof

(1) By Theorem [3.24, 9] $\widehat{A}^{*1} = \text{cl}\widehat{A}^{*3}$ this also true for $(1_x - \widehat{A})^{*1} = \text{cl}(1_x - \widehat{A})^{*3}$ take the complement we get $(1_x - (1_x - \widehat{A})^{*1}) = (1_x - \text{cl}(1_x - \widehat{A})^{*3})$ this leads $\Psi_1\left(\widehat{A}\right) = \text{int}(\Psi_3(\widehat{A})).$

(3) By proposition [3.2, 9] $\widehat{A}^{*1} \subseteq \widehat{A}^{*2}$ we get $(1_x - \widehat{A})^{*1} \subseteq (1_x - \widehat{A})^{*2}$, so $(1_x - (1_x - \widehat{A})^{*2}) \subseteq (1_x - (1_x - \widehat{A})^{*1})$ thus $\Psi_2\left(\widehat{A}\right) \subseteq \Psi_1(\widehat{A}).$ Easy to proof $\Psi_1(\widehat{A}) \subseteq \Psi_3(\widehat{A})$ by using lemma [3.23, 9] $\widehat{A}^{*3} \subseteq \widehat{A}^{*1}.$

Theorem 3.2: Let $(1_x, \tau, I)$ FITS. If $\widehat{U} \in \tau$ Then $\widehat{U} \subseteq \Psi_i(\widehat{U})$ where i = 1, 3.

Proof. Let $\widehat{U} \in \tau$, we know that $\Psi_1(\widehat{U}) = 1_x - \left(1_x - \widehat{U}\right)^{*1}$

So, $\left(1_X - \widehat{U}\right)^{*1} \subseteq \text{cl}\left(1_X - \widehat{U}\right) = \left(1_X - \widehat{U}\right)$, since $\left(1_X - \widehat{U}\right)$ is closed, Thus $1_X - \left(1_X - \widehat{U}\right) \subseteq 1_x - \left(1_x - \widehat{U}\right)^{*1}$ this means $\widehat{U} \subseteq \Psi_1(\widehat{U})$. Also, by proposition 3.1 we get $\widehat{U} \subseteq \Psi_1(\widehat{U}) = \text{int}(\Psi_3(\widehat{U})) \subseteq \Psi_3(\widehat{U})$, thus $\widehat{U} \subseteq \Psi_3(\widehat{U}).$

But this is not true if i = 2, 4 by using example 3.1, $\widehat{C} = \{(1,0.3), (2,0),(3,0.3)\} \in \tau, \Psi_2(\widehat{C}) = \{(1,0.1),(2,0),(3,0)\}.$ This mean $\Psi_2(\widehat{C}) \subseteq \widehat{C}.$

Remark 3.2: P_x^δ is quasi-coincident with $\widehat{A}$, if and only if $\delta > f_{(1_x - \widehat{A})}(x)$ [14].This means $P_x^\delta q\, \widehat{A}$ iff $P_x^\delta \notin (1_x - \widehat{A})$.

Lemma 3.1: Let $(1_x, \tau, I)$ FITS. For all $\delta \in (0, 1]$, $P_x^\delta \in (1_x - \widehat{A})$ if and only if $P_x^{1-\delta} \notin$ / $\in \widehat{A}$ for all except that $\delta = f_{1_x - \widehat{A}}(x)$. Since $\delta = 1 - f_{1_x - \widehat{A}}(x) = f_{\widehat{A}}(x)$.

The following theory, which represents the definition of Ψ-operator in topology general, can be used as an equivalent definition of fuzzy Ψ-operator in fuzzy topological space.

Theorem 3.3: Let $(1_x, \tau, I)$ FITS, for all $\widehat{A} \in \Gamma^X$ the following properties hold for all $(0 < \lambda \leq 1)$ except $\lambda \neq$ membership of $\Psi_i(\widehat{A})$. Where i = 1,2,3,4

1. $\Psi_1(\widehat{A}) = \{P_x^\delta; \exists \widehat{\mathcal{U}} \in \tau(P_x^\delta) \forall x \in X, \left(f_{\widehat{\mathcal{U}}}(x) - g_{\widehat{A}}(x)\right)_{max} \leq h_j(x), \text{for some } j \in I\}.$

2. $\Psi_2(\widehat{A}) = \{P_x^\delta; \exists \widehat{\mathcal{U}} \in \tau(P_x^\delta) \ni \forall x \in X, \left(f_{\widehat{\mathcal{U}}}(x) - g_{\widehat{A}}(x)\right)_{min} \leq h_j(x) \text{ for some } j \in I\}.$

3. $\Psi_3(\widehat{A}) = \{P_{\dot{x}}^\delta; \exists \widehat{\mathcal{U}} \in \mathcal{N}(P_{\dot{x}}^\delta) \ni \dot{x} \in X, \left(f_{\widehat{\mathcal{U}}}(\dot{x}) - g_{\widehat{A}}(\dot{x})\right)_{max} \leq h_j(\dot{x}) \text{ for some } j \in I\}.$

4. $\Psi_4(\widehat{A}) = \{P_{\dot{x}}^\delta; \exists \widehat{\mathcal{U}} \in \mathcal{N}(P_{\dot{x}}^\delta) \ni \dot{x} \in X, \left(f_{\widehat{\mathcal{U}}}(\dot{x}) - g_{\widehat{A}}(\dot{x})\right)_{min} \leq h_j(\dot{x}) \text{ for some } j \in I\}.$

Proof

(1) $P_x^\delta \in \Psi_1(\widehat{A})$ iff $P_x^\delta \in (1_x - \left(1_x - \widehat{A}\right)^{*1})$ iff by lemma 3.1 $P_x^{1-\delta} \notin \left(1_x - \widehat{A}\right)^{*1}$ for all except that $\delta = f_{(1_x - \left(1_x - \widehat{A}\right)^{*1})}(x)$ iff there exists $\widehat{\mathcal{V}} \in q\_\mathcal{N}(P_x^{1-\delta})$ such that $\forall x \in X, f_{\widehat{\mathcal{V}}}(x) + g_{(1_x - \widehat{A})}(x) - 1 \leq h_j(x)$ for some $j \in I$, iff $\forall x \in X$, $\left(f_{\widehat{\mathcal{V}}}(x) - g_{\widehat{A}}(x)\right)_{max} \leq h_j(x)$ for some $j \in I$. iff there exists open set $\widehat{\mathcal{U}} \subseteq \widehat{\mathcal{V}}$ such that $P_x^{1-\delta} q\, \widehat{\mathcal{U}}$ iff $P_x^{1-\delta} \notin (1_x - \widehat{\mathcal{U}})$ by lemma3.1 $P_x^\delta \in \widehat{\mathcal{U}}$ thus $\widehat{\mathcal{U}} \in \tau(P_x^\delta)$, $\forall x \in X, f_{\widehat{\mathcal{U}}}(x) + g_{1_x - \widehat{A}}(x) - 1 \leq h_j(x)$ for some $j \in I$ iff $\forall x \in X$, $\left(f_{\widehat{\mathcal{U}}}(x) - g_{\widehat{A}}(x)\right)_{max} \leq h_j(x)$ for some $j \in I$.

Proof 2, 3, 4 similarly.

Corollary 3.1:

1. $\Psi_1(\widehat{A}) = \vee\{\widehat{\mathcal{U}}; \widehat{\mathcal{U}} \in \tau(P_x^\delta) \ni \forall x \in X, \left(f_{\widehat{\mathcal{U}}}(x) - g_{\widehat{A}}(x)\right)_{max} \leq h_j(x) \text{ for some } j \in I\}.$

2. $\Psi_2(\widehat{A}) = \vee\{\widehat{\mathcal{U}}; \widehat{\mathcal{U}} \in \tau(P_x^\delta) \ni \forall x \in X, \left(f_{\widehat{\mathcal{U}}}(x) - g_{\widehat{A}}(x)\right)_{min} \leq h_j(x) \text{ for some } j \in I\}.$

3. $\Psi_3(\widehat{A}) = \vee\{\widehat{\mathcal{U}}; \widehat{\mathcal{U}} \in \mathcal{N}(P_{\dot{x}}^\delta) \ni \dot{x} \in X, \left(f_{\widehat{\mathcal{U}}}(\dot{x}) - g_{\widehat{A}}(\dot{x})\right)_{max} \leq h_j(\dot{x}) \text{ for some } j \in I\}.$

4. $\Psi_4(\widehat{A}) = \vee\{\widehat{\mathcal{U}}; \widehat{\mathcal{U}} \in \mathcal{N}(P_{\dot{x}}^\delta) \ni \dot{x} \in X, \left(f_{\widehat{\mathcal{U}}}(\dot{x}) - g_{\widehat{A}}(\dot{x})\right)_{min} \leq h_j(\dot{x}) \text{ for some } j \in I\}.$

Proof. Clear from Theorem 3.3 since $P_x^\delta \in \Psi_i(\widehat{A})$ iff $\exists \widehat{\mathcal{U}} \in \tau(P_x^\delta)$ or $\mathcal{N}(P_x^\delta)$ such that $\forall x \in X$, $\left(f_{\widehat{\mathcal{U}}}(x) - g_{\widehat{A}}(x)\right)_{\max \text{ or } \min} \leq h_j(x)$, for some $j \in I$ respectively iff $P_x^\delta \in$ $\vee\{\widehat{\mathcal{U}}; \widehat{\mathcal{U}} \in \tau(P_x^\delta)$ or $\mathcal{N}(P_x^\delta) \ni \forall x \in$ X, $\left(f_{\widehat{\mathcal{U}}}(x) - g_{\widehat{A}}(x)\right)_{\max \text{ or } \min} \leq h_j(x)$, for some $j \in I\}$ respectively.

Theorem 3.4: Let $(1_x, \tau)$ FTS. I_1, I_2 Are fuzzy ideals. The following statement are hold:

1. If $I_1 \subseteq I_2$, then $\Psi_i(\widehat{A})(I_1) \subseteq \Psi_i\left(\widehat{A}\right)(I_2)$.
2. $\Psi_i(\widehat{A})(I_1 \cap I_2) = \Psi_i(\widehat{A})(I_1) \cap \Psi_i\left(\widehat{A}\right)(I_2)$.
3. $\Psi_i(\widehat{A})(I_1) \cup \Psi_i\left(\widehat{A}\right)(I_2) \subseteq \Psi_i(\widehat{A})(I_1 \cup I_2)$.
4. $\Psi_i\left(\left(\widehat{A} - I_1\right)_{\min}\right) = \Psi_i\left(\widehat{A}\right) \subseteq \left(\widehat{A} \cup I_1\right)$.

Proof

(1) $I_1 \subseteq I_2$ by proposition of fuzzy local function, we get $\left(\widehat{A}\right)^{*i}(I_2) \subseteq \left(\widehat{A}\right)^{*i}(I_1)$. This also true of $\left(1_X - \widehat{A}\right)^{*i}(I_2) \subseteq \left(1_X - \widehat{A}\right)^{*i}(I_1)$ by take the complement we have $\Psi_i(\widehat{A})(I_1) \subseteq \Psi_i\left(\widehat{A}\right)(I_2)$.

(2) $\Psi_i\left(\widehat{A}\right)(I_1) \cap \Psi_i\left(\widehat{A}\right)(I_2) = [(1_X - \left(1_X - \widehat{A}\right)^{*i})(I_1)] \cap [(1_X - \left(1_X - \widehat{A}\right)^{*i})$ $(I_2)] = 1_X - [\left(\left(1_X - \widehat{A}\right)^{*i}\right)(I_1)) \cup (\left(1_X - \widehat{A}\right)^{*i})(I_2))] = 1_X - \left[\left(1_X - \widehat{A}\right)^{*i}\right.$ $(I_1 \cap I_2)] = \Psi_i(\widehat{A})(I_1 \cap I_2)$.

(3) $I_1 \subseteq (I_1 \cup I_2)$ and $I_2 \subseteq (I_1 \cup I_2)$ thus $\Psi_i(\widehat{A})(I_1) \subseteq \Psi_i(\widehat{A})(I_1 \cup I_2)$ and $\Psi_i(\widehat{A})(I_2) \subseteq$ $\Psi_i(\widehat{A})(I_1 \cup I_2)$ this implies $\Psi_i(\widehat{A})(I_1) \cup \Psi_i(\widehat{A})(I_2) \subseteq \Psi_i(\widehat{A})(I_1 \cup I_2)$. The converse is not always true. By example 3.1 let $\widehat{A} = \{(1, 0.9), (2, 1), (3, 0.4)\}$, $I_1 = \{(1, 0.9), (2, 0), (3, 0)\}$, $I_2 = \{(1, 0), (2, 0.8), (3, 0.5)\}$. evident, $\Psi_2(\widehat{A})(I_1 \cup I_2)$ $= 1_X$, but $\Psi_2(\widehat{A})(I_1) \cup \Psi_2\left(\widehat{A}\right)(I_2) = \{(1, 0.1), (2, 0.2), (3, 0)\}$.

(4) $\Psi_i\left(\left(\widehat{A} - I_1\right)_{\min}\right) = 1_X - \left(1_X - \left(\widehat{A} \cap (1_X - I_1)\right)\right)^{*i} = 1_X - \left((1_X - \widehat{A}) \cup I_1\right)^{*i} =$ $1_X - \left((1_X - \widehat{A}) \cup I_1\right)^{*i} = 1_X - ((1_X - \widehat{A})^{*i} \cup (I_1)^{*i}) = 1_X - \left(1_X - \widehat{A}\right)^{*i} = \Psi_i(\widehat{A})$.

Evident, $\widehat{A} \subseteq \left(\widehat{A} \cup I_1\right)$ this implies $\Psi_i\left(\widehat{A}\right) \subseteq \Psi_i\left(\widehat{A} \cup I_1\right)$. The converse is not true. Example 3.2 shows that after the change only fuzzy ideal, $I = \{0_X, I\} \cup \{\zeta; \zeta \leq I\}$ Where $I = \{(1, 0), (2, 0.7), (3, 0.5)\}$. Let $I_1 \in I$ such that $I_1 = \{(1, 0), (2, 0.7), (3, 0.3)\}$.

$\Psi_2(\widehat{A}) = 0_X$ but $\Psi_2(\widehat{A} \cup I_1) = \{(1,0),(2,0.9),(3,0.8)\}$. In the case of stander difference is max become a relationship partial only, thus. $\Psi_i\left(\left(\widehat{A}-I_1\right)_{max}\right) \subseteq \Psi_i\left(\widehat{A}\right) \subseteq \left(\widehat{A} \cup I_1\right)$. the following example explain that

Example 3.3: Let $(1_x, \tau, I)$ FITS, and $X = \{1, 2, 3\}$ the membership of $\widehat{B} = \{1,3\}, \widehat{C} = \{1,2\}, I = \{1,3\}$ are:

$$g_{\widehat{B}}(x) = \begin{cases} 1 & \text{if } x=1 \\ \frac{1}{10} & \text{if } x=3 \end{cases}, \quad f_{\widehat{C}}(x) = \begin{cases} \frac{9}{10} & \text{if } x=1 \\ \frac{8}{10} & \text{otherwise} \end{cases}, h_I(x) = \begin{cases} \frac{9}{10} & \text{if } x=1 \\ 1 & \text{otherwise} \end{cases}.$$
$$\forall x \in X$$

$\widehat{B} = \{(1,1),(2,0),(3,0.1)\}, \qquad \widehat{C} = \{(1,0.9),(2,0.8),(3,0)\}, \qquad I = \{0_X, I\} \cup \{\zeta; \zeta \leq I\}$ Where $I = \{(1,0.9),(2,0),(3,1)\}$. Put $\tau = \left\{0_X, 1_X, \widehat{B}, \widehat{C}, \widehat{B} \cup \widehat{B}, \widehat{B} \cap \widehat{C}\right\}$. Let $\widehat{A} = \{(1,0.1),(2,0.1),(3,1)\}, I_1 = \{(1,0.8),(2,0),(3,1)\}$. Evident, $\Psi_2\left(\widehat{A}_1\right) = \{(1,1),(2,0),(3,0.1)\}$ but $\Psi_2\left(\widehat{A}_1 - I_1\right)_{max} = \{(1,0.9),(2,0),(3,0)\}$.

Since Ψ_2 - operator is smaller operator, this means $\Psi_i\left(\left(\widehat{A}_1 - I_1\right)_{max}\right) \subseteq \Psi_i\left(\widehat{A}_1\right)$ for all i.

It is well- known that not all of the existences in the general topology can be achieved in the fuzzy topology. The following lemma [If $\left(\widehat{A}-\widehat{B}\right) \cup \left(\widehat{B}-\widehat{A}\right) \in I$ then $\Psi_i(\widehat{A}) = \Psi_i(\widehat{B})$] is not an achieved of the fuzzy topology space, the following example shows this,

Example 3.4: Let $(1_x, \tau, I)$ FITS, and $X = \{1, 2\}$ the membership of $\widehat{H}, \widehat{C}, I$ are:
$$g_{\widehat{H}}(x) = \begin{cases} 0 & \text{if } x \text{ is odd} \\ \frac{7}{10} & \text{if } x \text{ is even} \end{cases}, f_{\widehat{C}}(x) = \begin{cases} \frac{8}{10} & \text{if } x=1 \\ 0 & \text{otherwise} \end{cases}, h_I(x) = \begin{cases} \frac{1}{2} & \text{if } x \text{ is odd} \\ \frac{3}{10} & \text{if } x \text{ is even} \end{cases}. \quad \forall x \in X.$$
$\widehat{H} = \{(1,0),(2,0.7)\}, \widehat{C} = \{(1,0.8),(2,0)\}$
$I = \{0_X, I\} \cup \{\zeta; \zeta \leq I\}$ Where $I = \{(1,0.5),(2,0.3)\}$.

Put $\tau = \left\{0_X, 1_X, \widehat{H}, \widehat{C}, \widehat{H} \cup \widehat{C}\right\}$ and let $\widehat{A} = \{1,0.3),(2,0.1)\}, \widehat{B} = \{(1,0),(2,0.3)\}$.

Note $\left(\widehat{A}-\widehat{B}\right) \cup \left(\widehat{B}-\widehat{A}\right) \in I$, but $\Psi_1(\widehat{A}) \neq \Psi_1(\widehat{B})$ Since $\Psi_1(\widehat{A}) = \{(1,0.8),(2,0)\}, \Psi_1(\widehat{B}) = 0_X$. We have referred of this lemma of its importance in general topology in the definition of Baire concept. But it can be achieved in fuzzy topology in case $\left(\widehat{A} \cup I\right) = \Psi_i\left(\widehat{A}\right)$.

The following theory shows the relationship between the Fuzzy Ψ_i-operator and interior,

Theorem 3.5: Let $(1_x, \tau, I)$ FITS, for all $\widehat{A} \in \Gamma^X$ the following are hold:

1. $\text{int}\left(\widehat{A}\right) \subseteq \Psi_1(\widehat{A}) = \text{int}\,\Psi_1(\widehat{A})$.
2. $\text{int}\,\Psi_2\left(\widehat{A}\right) = \Psi_2(\widehat{A})$.
3. $\text{int}\left(\widehat{A}\right) \subseteq \text{int}\,\Psi_3(\widehat{A}) \subseteq \Psi_3(\widehat{A})$.
4. $\text{int}\,\Psi_4\left(\widehat{A}\right) \subseteq \Psi_4(\widehat{A})$.

Proof. By Theorem 3.3 and define fuzzy interior [7].

Remark 3.2: If $I = \{0_X\}$ then,

1. $\text{int}\left(\widehat{A}\right) = \Psi_1(\widehat{A})$ [11].
2. $\Psi_2(\widehat{A}) \subseteq \text{int}\left(\widehat{A}\right)$.
3. $\text{int}\left(\widehat{A}\right) \subseteq \Psi_3(\widehat{A})$.

Through the theory and remark above we note that there is no relationship between $\Psi_4(\widehat{A})$ and $\text{int}\left(\widehat{A}\right)$. The following example explain that,

Example 3.5: Let $(1_x, \tau, I)$ FITS, and $X = \{1, 2, 3\}$ the membership of $\widehat{B} = \{2, 3\}, \widehat{C} = \{1, 3\}, I = \{3\}$ are:

$$g_{\widehat{B}}(x) = \begin{cases} \frac{3}{10} & \text{if } x \text{ is even} \\ \frac{2}{10} & \text{if } x \text{ is odd} \end{cases} \quad , \quad f_{\widehat{C}}(x) = \frac{2x-1}{10} \quad \forall x \in \widehat{C} \quad , \quad h_I(x) = \frac{7}{10} \quad , \quad \forall x \in I.$$

$\widehat{B} = \{(1,0), (2,0.3), (3,0.2)\}, \widehat{C} = \{(1,0.1), (2,0), (3,0.5)\}$

$I = \{0_X, I\} \cup \{\zeta; \zeta \leq I\}$ Where $I = \{(1,0), (2,0), (3,0.7)\}$. Put $\tau = \left\{0_X, 1_X, \widehat{B}, \widehat{C}, \widehat{B} \cup \widehat{C}, \widehat{B} \cap \widehat{C}\right\}$. And let $\widehat{A} = \{(1,0.8), (2,0.1), (3,0.6)\}$, evident, $\Psi_1\left(\widehat{A}\right) = \{(1,0.1), (2,0), (3,0.5)\}, \Psi_2\left(\widehat{A}\right) = \{(1,0), (2,0), (3,0.2)\}$,

$\Psi_3\left(\widehat{A}\right) = \{(1,0.1), (2,0), (3,1)\}, \Psi_4\left(\widehat{A}\right) = \{(1,0), (2,0), (3,1)\}$. so, $\text{int}\left(\widehat{A}\right) = \{(1,0.1), (2,0), (3,0.5)\}$. that means $\Psi_2\left(\widehat{A}\right) = \text{int}\,\Psi_2\left(\widehat{A}\right) \subseteq \text{int}\,(\widehat{A})$ but not relationship between $\Psi_4(\widehat{A})$ and $\text{int}\left(\widehat{A}\right)$.

Remark 3.3: Let τ_1, τ_2 be two fuzzy topologies such that τ_2 is finer than τ_1. Then, $\Psi_i\left(\widehat{A}\right)(\tau_1,\ I)\subseteq\Psi_i(\widehat{A})(\tau_2,\ I)$.

Proof. Let $(1_X - \widehat{A}) \in \Gamma^X$ then, $\left(1_X - \widehat{A}\right)^{*i}(\tau_2)\subseteq\left(1_X - \widehat{A}\right)^{*i}(\tau_1)$ [Theorem 3.9, 5]. $(1_X - \left(1_X - \widehat{A}\right)^{*i})(\tau_1)\subseteq(1_X - \left(1_X - \widehat{A}\right)^{*i})(\tau_2)$, thus $\Psi_i\left(\widehat{A}\right)(\tau_1,\ I)\subseteq\Psi_i(\widehat{A})(\tau_2,\ I)$.

The following theorem shows that using the four different types of fuzzy Ψ-operators can generate coarse and fine fuzzy topology spaces. These can be used in the field of digital topology and image processing. It is important also in all applications of artificial intelligence dealing with spatial structures.

Theorem 3.6: $\tau^{*i}(I) = \{\widehat{A} \in \Gamma^X; \widehat{A}\subseteq\Psi_i(\widehat{A})\}$ is a fuzzy topological space generated by Ψ_i-operator.

Proof. Since $0_X\subseteq\Psi_i(0_X), 1_X = \Psi_i(1_X)$, thus $0_X, 1_X \in \tau^{*i}(I)$. Let $\widehat{A}, \widehat{B} \in \tau^{*i}(I)$ we get $\widehat{A} \cap \widehat{B}\subseteq\Psi_i(\widehat{A}) \cap \Psi_i(\widehat{B}) = \Psi_i(\widehat{A} \cap \widehat{B})$. Let $\widehat{A}_\lambda \in \tau^{*i}(I)\ni\lambda \in \Lambda$. This implies $\cup \widehat{A}_\lambda\subseteq\cup \Psi_i(\widehat{A}_\lambda)\subseteq\Psi_i(\cup \widehat{A}_\lambda)$, thus $\cup \widehat{A}_\lambda \in \tau^{*i}(I)$.

4   Discussion and Conclusion

Through our research, we noticed that, Ψ-operator in the general topology can be defined through the open set where $\Psi(A) = \{x;\ \exists\mathcal{U} \in \tau(x)$ such that $(\mathcal{U} - A) \in j$, for some $j \in I\}$, or through the local function where $\Psi(A) = X - (X - A)^*$,the two definitions are equivalent. In the case of merge the fuzzy, the relationship between the two definitions is not achieved for all fuzzy points since the two definitions cannot be considered equivalent except after the exclusion of fuzzy points equal to the memberships. While our results suggested that the two definitions are not equivalent, more investigations in future work are necessary to confirm our results.

We also noticed a correlation between the four types of fuzzy Ψ_i-operators where the fuzzy Ψ_3-operator is considered the general condition of fuzzy Ψ_1-operator, fuzzy Ψ_2-operator, and Ψ_4-operator. However, there is no relationship between the fuzzy Ψ_4-operator and fuzzy Ψ_1-operator. We can use these types in the construction of fuzzy topological space.

5   Future Work

Although there is a great deal of research in fuzzy topological areas, further study is still needed. In [12, 13] the fuzzy set can be re-studied in relation to the creation of a fuzzy soft topology for use in various applied sciences to help solve some problems in different aspects of life.

References

1. Zadeh, L.A.: Fuzzy sets. Inform. Control **8**, 338–353 (1965)
2. Marudhachalam, R., Ilango, G.: Digital topological concepts applied to medical image processing. Int. J. Pure Appl. Math. **12**, 177–187 (2017)
3. Hazrati, K., Mittal, P., Chauhan, K.: Network topologies. Int. J. Inn. Resear. Tech. **12**, 263–266 (2015)
4. Chang, C.L.: Fuzzy topological spaces. J. Math. Anal. Appl. **24**, 182–190 (1968)
5. Sarkar, D.: Fuzzy ideal theory, fuzzy local function and generated fuzzy topology. Fuzzy Sets Systems. **87**, 117–123 (1997)
6. Al-Swidi, L.A., Al-Razzaq, A.S.A.: On Classification of Fuzzy Set Theory. J. Modern. Math. Stat. to Appear (2018)
7. Pao-Ming, P., Ying-Ming, L.: Fuzzy topology. I. neighborhood structure of a fuzzy point and moore- smith convergence. J. Math. Anal. Appl. **76**, 571–599 (1980)
8. Fars, F.E.M.: Fuzzy Al Gebra. Degree of doctor of philosophy. Zagazig University. Faculty of Engineering Department of Eng Physics and Math. 181 (2006)
9. Al-Swidi, L.A, Almohammed, R.: New Concepts of Fuzzy Local Function. Baghdad Sci. J. (2019). (to appear)
10. Hamlett, T.R., Jankovic, D.: Ideals in Topological Spaces and the Set Ψ –operator. Boll. Un. Mat. Ital. **7**, 863– 874 (1990)
11. Nasef, A.A., Mahmoud, R.A.: Some Topological Applications via Fuzzy Ideals. Chaos, Solitons Fractals **13**, 825–831 (2002)
12. Al-Swidi, L. A., Awad, F.S.S.: Analysis on the Soft Bench Point. J. Int. Conf. Adv. Sci. Eng. 330–335 (2018)
13. Al-Swidi, L.A., AL-Rubaye, M.S.: New classes of separation axiom via special case of local function. Int. J. Math. Anal. **23**, 1119–1131 (2014)

Programs Features Clustering to Find Optimization Sequence Using Genetic Algorithm

Manal H. Almohammed$^{(\boxtimes)}$, Esraa H. Alwan,
and Ahmed B. M. Fanfakh

Department of Computer Science, College of Science for Women,
University of Babylon, Hillah, Iraq
Manalalmohamed.h@gmail.com, {wsci.israa.hadi,
wsci.ahmed.badri}@uobabylon.edu.iq

Abstract. Finding the best optimization sequence order that can improve the performance for even a simple program is not an easy task. However, the modern compilers provide dozens of optimizations, making it not a practical solution to try all the optimization sequences manually to find the optimal one. In this paper, genetic algorithm is proposed to select the best optimization sequence for a cluster of similar programs. However, wide set of programs are elected to cover as much as possible all the features. The set of the programs are classified into three clusters depending on them features. Thus, the genetic algorithm in this work is learning method. This means any new program, unseen program, can take the optimization sequence of the cluster that has similar features to it. Moreover, two scenarios are proposed using genetic algorithm to find the best optimization sequence for each cluster. In the first scenario, programs are classified into three clusters according to program dynamic features. The genetic algorithm with Tournament selection method is applied on each cluster independently to obtain a good optimization sequence for a cluster. Moreover, the proposed method improved the execution time on average by 77% compared with the O2. The second scenario was exactly similar to the first one. While, different selection methods are used for each cluster. The improved average execution time for this scenario was 78% compared with the O2. LLVM framework is used to validate and execute the proposed method. In addition, Bolybench, Standerford, Shootout benchmarks are used to verify the effectiveness of the proposed method.

Keywords: Genetic algorithm · LLVM · Phase-ordering · Optimizations

1 Introduction

1.1 Motivation and Background

Compiler have a large number of optimization passes and choosing the right set of optimization have a significant impact on the program performance [1–3]. The goal of these optimization passes is giving a code similar to the source code in syntax and semantic with better performance. Therefore selecting a good set of optimization passes

© Springer Nature Switzerland AG 2020
L. C. Jain et al. (Eds.): ICICCT 2019, LAIS 9, pp. 40–50, 2020.
https://doi.org/10.1007/978-3-030-38501-9_4

manually is very difficult task because the search space is very large [1]. Compiler developers' offer stander optimization levels (e.g. O1, O2, O3) which consist of special passes, for example, there are more than 100 passes in the LLVM while the GCC has approximately (250) passes [5]. However, the problem with these flags are not comprehensive for all programs. Sometimes they might be good for a particular program however their performance is bad for another program or may produce a wrong code [4]. Using the performance counters to find the similarity between programs was first introduced by Cavazos et al. [3]. According to assumption that if the similar programs compiled by the same optimization sequence, they react approximately in the same way.

The information extracted from the performance counter called performance event, which represents the dynamic behavior of the programs. These events consider an important aspects of program improvement, for example, instructions, cache_references, emulation_faults, and others [2, 3]. A set of benchmark suites such us Bolybench, Standerford, Shootout benchmarks have been used to show the effectiveness of the proposed method. Moreover, genetic algorithm used to solve the problem of phase-ordering [6, 7], because of its authoritative capabilities in solving complex problems, including the large size search space [8].

In this paper, clustering approach for programs based on program features was proposed. The search space is divided into three clusters where GA works on each cluster independently to find the best sequence. Then, any unseen program can acquire an optimal sequence after determinate to which clusters it belongs by computing its similarity. The experiments use LLVM (Low Level Virtual Machine) compiler framework to implement the proposed method [4].

The remainder of the paper is organized as follows: the next Sect. 2 describes the proposed method where the implementation of Genetic Algorithm (GA) has been explained. Section 3 presents some experimental results from the application of the proposed method. Finally, in Sect. 4, some of related works have been introduced and the last section, Sect. 5, displays the conclusions and future work.

2 Proposed Method

In this paper, clustering approach for programs based on programs' features was proposed. The similarity between the programs' features in the search space and the one chosen out of them is computed. According to similarity result, the programs are divided into three classes. Moreover, genetic algorithm (GA) is applied over each cluster to find the best sequence. The proposed genetic algorithm works offline to optimized each cluster independently. Thus, the accuracy of the proposed method depends significantly on the collected programs' features of each cluster. Therefore, the higher diversity selected set of programs for clustering is an important factor. After computing the best optimized sequence for each cluster, any new unseen program can be matched with the most similar cluster. The similarity of the new unseen program is determined by extracting its features firstly. Then the similarity law is computed between the new unseen program features and the average programs' features for each cluster.

Next, the outline of the proposed method:

- Collecting feature vector f1 for each program, using compiler level O0
- Divided the program space into three clusters according to specific similarity threshold.
- Applying genetic algorithm over each cluster to find the best optimization sequence for each cluster.
- For unseen program, extract its feature in vector f2, using compiler level O0 and computing its similarity with other clusters.
- Using the optimization sequence of the most similar cluster to optimize the unseen program.

Where PS_i is the program i similarity value, $i = 1$ to N and N is the number of programs selected. T1, T2 and T3 are the threshold values.

The events extracted from performance counters is a compact summary of the program's dynamic behavior. In particular, they summarize the important aspects of a program's performance, e.g., context switches, cache_refrence, cpu_migrations, etc. Our approach uses this information to classify the programs into three clusters. More details about the proposed method is presented as follows:

2.1 Extracting the Features of the Program

This method can be summarized as follow:

1- Each program runs many times (three times) and the average values of 52 events are computed, its main goal to get more accuracy.

The similarity of each program is computed between its features and the features of other programs. Depending on this process, the clustering methods can be applied in step 3. In the literature, there are many laws that measure the similarity, like Jaccard, cosine, Euclidean, etc. In the proposed method, cosine law is used. The resultant of large value means low similarity. However, if the similarity between the base program and the other programs are equal to one.

2-means these programs are identical. Equation (1) illustrates the cosine low where p and pi represent the base program and other programs respectively.

$$\text{Sim}(p, pi) = \frac{\sum_{w=1}^{m} Pw \times Piw}{\sqrt{\sum_{w=1}^{m} (Pw)^2} \times \sqrt{\sum_{w=1}^{m} (Piw)^2}} \tag{1}$$

3-According to the result that obtains from the above step, clustering method is applied by dividing the programs into three clusters depending on them similarity value, PS_i, obtained from Eq. 1. Two specific thresholds value, $T1$ and $T2$, are used to classify all programs to three clusters as shown in the Fig. 1.

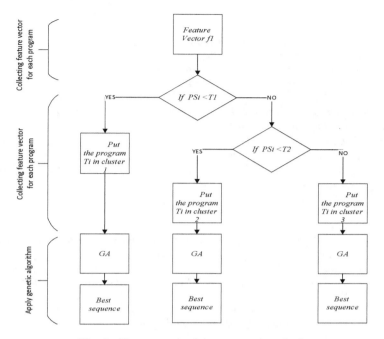

Fig. 1. The structure of the proposed method

2.2   The Genetic Algorithm for Sequence Optimization

It is a search method to obtain the approximate solutions to search problems and optimization. GA handles a population of possible solutions where; each solution is represented through a chromosome [10]. Based on the principles of genetics and natural selection. The genetic algorithm (GA) is an optimization and search technique [8]. In the proposed method, genetic algorithm is used as an optimizer to find the best compiler optimization sequence of passes. The best sequence can give smaller execution time when apply it to a program or a set of programs compared to other generated sequences. Algorithm 1, summarizes the steps used in the proposed method.

Algorithm 1 Genetic algorithm for sequence optimization
Require:
  Pop_size: The population size of the algorithm.
  Opt_passes: The vector of the optimization passes.

Chrom_length: The length of the chromosome in the algorithm.
Pn: Number of the program in the cluster.
Iterations: is the iteration index of the algorithm.
Maxgen: is the maximum number of iterations.
Fit: Fitness value.
Output: *The best optimization sequence of passes for the cluster of programs.*
1: *Generating the initial population.*
2: *for i=1 to pop_size do*
3:      *for j=1 to chrom_length do*
4:            *pop[i].sequence[j] ← select randomly a pass from vector Opt_passes*
5:            *Flip ←select random number from 0 to 8*
6:            *If flip=1 then*
7:                 *pop[i].sequence[j]=0 .*
8:            *else*
 9:                  *Pop[i].sequence[j]=select random pass from vector Opt_passes.*
10:          *end if.*
11:    *end for.*
12:            *Compute the average of execution time for the sequence Pop[i].*
sequence[j]
            during three runs applied over Pn programs.
13:          *Pop[i].fit ← average of the execution time for the three runs.*
14: end for.
15: *repeat*
        Iterations ← Iterations +1
16:      *Selecting a chromosome using a selection method.*
17:      *Applying the uniform crossover over the selected parents.*
18:      *Applying one-point mutation over the two new offspring.*
19:      *Evaluating the new offspring using steps 12 and 13.*
20:      *Replace the best individual from the new produces offspring*
            instead of the worse individual of a randomly selected group.
21:      *__Until__ the standard deviation between the last two generations*
            reaching to the less fitness function __or Iterations__>maxgen.

3 Experiment and Discussion

3.1 The Experimental Setup

This section presents an evaluation of the implementation of the proposed method that has been presented in Sect. 2. It discusses the results obtained from applying the proposed method to programs selected from three benchmarks. Before starting with evaluating the results. LLVM compiler framework is used to implement the proposed method. The clang in c language frontend can transform the source of c code of each program into IR code which can be saved in bitcode of machine-readable form. O2 optimization level is used to measure the effectiveness of the resulted sequences.

Our tests use a collection of LLVM (64) optimizations passes to find a sequence that will give best or close optimal execution time for each cluster. In the proposed method, the start parameters have been set to the following values, (population size equal to 100, the probability of crossover is 0.5, mutation is 0.01, the ratio of occurrence gene 0.4, stop criteria of the standard deviation is 0.01. Then the speedup has been computed in the Eq. (2). The improvement ratio of the execution time is computed in the Eq. (3).

$$Speedup = \frac{baseline\_runtime}{new\_runtime} \tag{2}$$

$$Improvemnent\ ratio = ((speedup) - 1) * 100 \tag{3}$$

Where the (*baseline-runtime*) refers to the execution time of optimized program with O2 flag [6].

3.2 The Experimental Results

In this part, we will discuss the results we have achieved in implementing the proposed method. We use 60 programs from deferent benchmarks have been used as case studies, and these benchmarks cover a different type of programs which are involving programs containing tail recursive, floating point arithmetic operation programs, image processing, linear algebra.

The script file is used to extract a program from a benchmark and execute it to construct an optimization sequence for that program and we used to compute the execution time.

After applying the clustering algorithm 1 to classify the programs into three clusters, steady state genetic algorithm is executed over each cluster. In other word, three genetic algorithms are running each one a cluster. The main goal of the genetic algorithm is to find the best optimized sequence that gives less execution time when it applies to the program.

However, GA is coding all the possible solutions into 100 chromosomes. While the maximum length of each chromosome was 60. Each pass is selected from a vector composed from 64 passes that available in the LLVM.

The encoded values for each pass are an integer value from 0 to 64. Where zero value is indicated no pass used, this will help in generating variable chromosome lengths. Execution time is subject to be the fitness function. In the experiments, two different scenarios are used. The first scenario used similar selection methods for all the three clusters. It uses tournament selection method.

While, the second scenario uses three different selection methods. These methods are tournament, stochastic universal sampling and roulette wheel. The three-selection methods were applied to each cluster and the one that gives the best is selected to the cluster. Therefore, clusters 1, 2, 3 are used, stochastic universal sampling, roulette wheel and tournament selection respectively.

A. First Scenario Results

As mention earlier, tournament triple selection method and UX crossover has been used for all clusters. After applying of the proposed genetic algorithm of this scenario, the best sequence for each cluster is selected. Number of unseen programs is selected to evaluate the proposed method. The feature of each unseen program is computed using perf tools. The similarity of each unseen program is computed and thus, each one can be classified to its similar cluster. However, the results of this scenario are the comparison of the execution time of the best sequence applied to the unseen program from each cluster and the sequence of the -O2 flag as in Figs. 2, 3, 4. As shown from these figures, the Multi-clustering genetic algorithm (MCGA) outperforms the O2 performance.

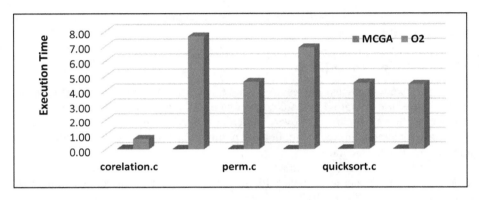

Fig. 2. Illustrates the comparison of the execution time between MCGA and the O2 in the first cluster

The comparison between the average of execution time for three clusters between O2 and MCGA is illustrated in Table 1.

B. Second Scenario Results

In this scenario, three different selection methods are picked to obtain high diversity among individuals in each population. The Stochastic universal sampling, Roulette wheel, and Tournament selections methods are used for each cluster respectively. Many experiments are conducted to select the best selection method for each cluster. The comparison between the MCGA of the second scenario with O2 are made in Figs. 5, 6, 7. They illustrate that the performance of MCGA method outperformed O2 in all clusters.

This scenario also gives three best Optimization Sequences one for each cluster. For any unseen program chosen, its similarity is computed. However, the sequence of the most similar cluster is selected to optimized the unseen program. Table 2 shows the average of the execution time of the second scenario for O2 and MCGA method. According to this table the MCGA method outperformer the O2 for all clusters.

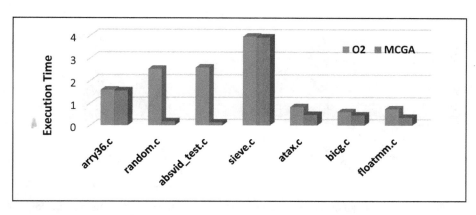

Fig. 3. Illustrates the comparison of the execution time between this method and the O2 in the second cluster

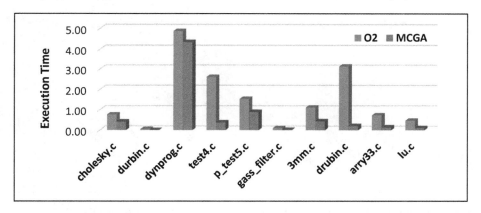

Fig. 4. The comparison of the execution time between this method and the O2 in the third cluster

Table 1. The comparison of the execution time between O2 and MCGA

Cluster name	O2	Proposed method
Cluster 1	4.56	0.03
Cluster 2	1.90	1.06
Cluster 3	1.53	0.73

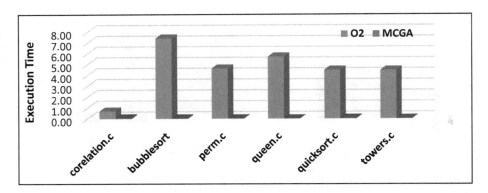

Fig. 5. Illustrates the comparison in the execution time between MCGA and the O2 for first cluster using Stochastic universal sampling selection.

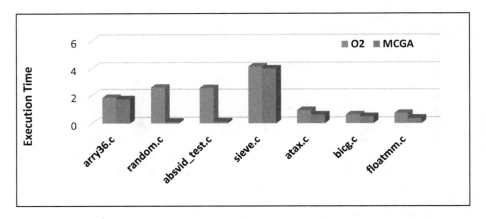

Fig. 6. Illustrates the comparison in the execution time between MCGA and the O2 for second cluster using Roulette wheel selection

Table 2. The average of the execution time for the best optimization sequence of the second scenario for each cluster

Cluster name	O2	Proposed method
Cluster 1	4.79	0.02
Cluster 2	1.83	1.01
Cluster 3	1.53	0.73

4   Related Work

Ashouri [1], introduced a Mitigates the Compiler Phase-ordering problem called MiCOMP which is an automatic optimization framework. They clustering the highest optimization level of LLVM's O3 into different clusters in order to predict a new sequence that can give better performance than old one. To tackle the problem, the predictive model uses dynamic features, an encoded version of the compiler sequence, and an exploration heuristic can be introduced.

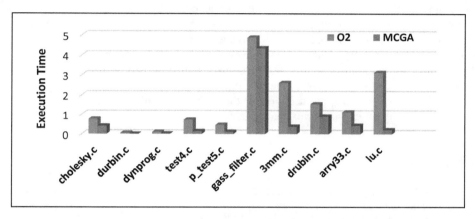

Fig. 7. The comparison of the execution time between MCGA and the O2 for third cluster using Roulette wheel selection.

Genetic and Random algorithms used by Purini and Jain in [6] found a few sets of optimization sequence. Then the optimization passes that have bad effect or the one that has not any improvement to the program performance are removed. Then the Best-10 algorithm will be executed to find the most affected 10 sequences among other.

In Kulkarni et al. [9] Markov process and the characterization of current state used to generate the good solution for the phase ordering problem. They constructed artificial neural network using neuro-evolution that has ability to predict a valuable optimization order for a piece of code [11]. They presented a program dynamic analysis for optimizing a program on complex processor architecture. According to this manual but systematic approach, efficient process can be resulted from gathering a small set of run time information. Decision tree is used to guide the optimization process. Pan and Eigenmann [12] proposed an algorithm called Combined Elimination that based on two developed algorithms called Batch Elimination and Iterative Elimination.

Cavazos, [3] used the supervised learning method to predict optimal sequence by using the performance counter. In this method, we will not use manual experimentation. In addition, it is possible to give a good result to the unseen program. Kulkarni [9] attempted to identify and understand the important properties of some commonly employed heuristic search methods to find near-optimal solutions quickly by using information collected during an exhaustive exploration of the phase order search space.

5 Conclusion

In this work, multi clustering genetic algorithm (MCGA) has been introduced to discover the best optimization sequence. It has the ability to determine an optimization sequence for each cluster depending on programs' features. The proposed method has been evaluated on different instances of real-world benchmarks distributed on three clusters. Each cluster contains similar programs that have been obtained by extracting event feature using the performance counter.

The proposed MCGA applied in two scenarios, one that used similar genetic operators and the second one used suitable selection method for each cluster. LLVM infrastructure has been used to validate the proposed method. The experiment results of the proposed method showed good effectiveness when compare them with the results of the O2 optimization sequence. Moreover, each obtained sequence for a cluster of programs can be used as a guided sequence for the new unseen program, which is the most similar one to the cluster. In the future, more programs can be used to expand the number of clusters. Thus, the accuracy of the computed similarity between programs' clusters and the unseen program can be increased.

References

1. Shouri, A.H., Bignoli, A., Palermo, G., Silvano, C., Kulkarni, S., Cavazos, J.: Mitigating the compiler phase-ordering problem using optimization sub-sequences and machine learning. ACM Trans. Archit. Code Optim. **14**(3), 1–28 (2017)
2. De Lima, E.D., de Souza Xavier, T.C., da Silva, A.F., Ruiz, L.B.: Compiling for Performance and Power Efficiency (2013)
3. Cavazos, J., Fursin, G., Agakov, F., Bonilla, E., O'Boyle, M.F., Temam, O.: Rapidly Selecting Good Compiler Optimizations using Performance Counters (2007)
4. Alkaaby, Z.S., Alwan, E.H., Fanfakh, A.B.M.: Finding a good global sequence using multi-level genetic algorithm. J. Eng. Appl. Sci. **13**(22), 9777–9783 (2018)
5. Cavazos, J.: Mitigating the compiler optimization phase-ordering problem using machine learning. ACM SIGPLAN Not. **47**(10), 147–162 (2012)
6. Purini, S., Jain, L.: Finding good optimization sequences covering program space. ACM Trans. Archit. Code Opt. **9**(4), 56 (2013)
7. Majumder, A., Ekbal, A., Naskar, S. K.: Feature Selection and Class-Weight Tuning Using Genetic Algorithm for Biomolecular Major Steps for Event Extraction (2017)
8. Trivedi, A., Srinivasan, D., Biswas, S., Reindl, T.: Hybridizing genetic algorithm with differential evolution for solving the unit commitment scheduling problem, Swarm Evol. Comput. 1–15 (2015)
9. Davidson, J.W., Tyson, G.S., Whalley, D.B., Kulkarni, P.A.: Evaluating Heuristic Optimization Phase Order Search Algorithms (2007)
10. Muslim, A.B., Ali, A.K.M.: The combination of genetic programming and genetic algorithm for neural networks design and training. J. Univ. Babylon **18**(1–2), 350–359 (2011)
11. Lattner, C., Adve, V.: LLVM: A compilation framework for lifelong program analysis & transformation. In: Proceedings of the International Symposium on Code Generation and Optimization IEEE Computer Society (2004)
12. Pan, Z., Eigenmann, R.: Fast and effective orchestration of compiler optimizations for automatic performance tuning. In: Proceedings of the International Symposium on Code Generation and Optimization IEEE Computer Society, 319–332 (2006)

Complex Event Processing Based Analysis of Cassini–Huygens Interplanetary Dataset

Ashraf ALDabbas$^{(\boxtimes)}$ and Zoltán Gál

Faculty of Informatics, University of Debrecen, Debrecen 4028, Hungary
{Ashraf.Dabbas,Gal.Zoltan}@inf.unideb.hu

Abstract. A complex event is an assortment of data observations that chimed to captivating of remarkable patterns in the implicit incident that are captured via the sensing devices. An innovative method is introduced in this research for the purpose of detecting complex events by analysing a huge dataset of images and other metadata that have been collected using remote sensing approach in Cassini–Huygens mission interplanetary expedition. The proposed method is competent to explore a big volume of data that is influenced by the indexed time-series manner using score array and the variation score of the target time intervals via the Weighted Complex Event Level (WCEL), allowing to convert the sensed data and images into spectacular detected events. It has been tested by applying it on all the batches which forming the included dataset and the results showed that almost in every batch there was more than a single complex event, within the 5 analysed phases of the mission, which include: Approach science phase, Extended phase, Extended-Extended phase, Tour, and Tour pre-Huygens phase.

Keywords: Special event detection · Extreme event · Weighted complex event processing · Classification · Sensory data · Big data · Pattern processing

1 Introduction

Enormous data gained from several sources such as home appliances, smartphones and smart devices, remote sensing data, and tracking system offer us the right set of circumstances to explore, establish a series of processing actions and to carry out the needed analysis. Contemporary progress in the Artificial Intelligence (AI), Big Data, and hosted services online on the internet all of these frameworks empower saving, processing, inspect and evaluate semantic data just like the data collected from remote sensing devices. In the role of the fact that sensors and hard disks storage equipment obtain outstanding abilities and efficiency with reasonable budget, the gathered data raise which produce very huge databases sooner or later it will be unattainable for researchers and analysts to monitor the whole gathered data within a sensible time interval. This is the reason that makes increasing need to provide approaches that help when processing and screening via metadata become unrealistic, the exclusive wise decision to do is to screen through image content.

During spacecraft missions a persistent recording will produce an enormous volume of images to examine and appraise. The images need to be recognized and

L. C. Jain et al. (Eds.): ICICCT 2019, LAIS 9, pp. 51–66, 2020.
https://doi.org/10.1007/978-3-030-38501-9_5

analysed by ground teams to detect their scientific value, as it may include fascinating events, the images can be subjected to various interpretations. To simplify the previously mentioned process the need for research such we are providing is raising.

The orientation of Cassini was guided via reaction wheels, which yield a remarkably balanced camera sighting. In the wide and narrow-angle cameras, filter wheels were put neatly in order to acquire more elasticity, also a detector was arranged on each camera. The process of event imaging starts with shutter reassign or reset from a former exposure, subsequently comes the changing the filter wheels position to their instructed locations, primarily the narrow camera then the wide angle camera. Afterward the light overflow and blot out, which includes flooding the light with a ratio the reaches around 50 times of the overabundance or saturation level, an instant readout process come behind. Such phase is required to eradicate variation in the image, also assists to guarantee regular, repeatable initiating condition with regard to the detector. Then the exposure takes place, at the time the exposure is accomplished, there are ordered readout process. The vast diversity of celestial bodies in the outer space are reminiscent of varied phases of maturation and evolution in which the celestial and planetary satellite in the whole solar system has subjected to acquire elaborated perception about each of these is priceless in order to consolidate our awareness concerning the universe. This enforces us to look deep and search hard and introduce the method of Complex Event Detection (CED). Remote sensing in several domains has provided great support in incubating appropriate evolutionary approaches by exploiting imaging instruments in the outer space missions such as Cassini mission.

Usually, events could be judged as a solitary incident that takes place in time, while a complex event can be seen as a set of circumstances or modality that encompass a certain coalition significance for the system. Every event is allotted with a type depending on two factors: content and its semantics. Primordial event is simple or has the feature as the non- decomposable, however complex event is pinpointed and drawn out by the Complex Event Processing (CEP) framework counting on a specified modality or regulations [1]. CEP intents to deduce consequential great extent-scale event (complex event) from a series of interconnected minimal level events [2].

Contemporary technologies establish the capability and potency to trace and notice significant events that are taking place in the outer space [3]. CED emanated as a neoteric technology that has the motive to distinguish complex events via classifying, analysing, extracting features, and events matching. The major notion on the far side of CEP is pinpointing out the occurrence of an incident or event by examining the reason/effect interconnection to each of the plain events that hold on particular facts or information learned about something. The outcome of the previously mentioned analysis is putting forth an appraisal concerning CED, which facilitates the process for the experts in related fields the power to influence or direct the situation or the course of events in order to proactively confronting difficult issues. The process of CED can be seen as a sophisticated identification of the data pattern, where the term pattern is investigated in order to recognize a specific event modality. Placement of the rule modality or methodology for matching normal events depending on semantic, provisional or the spatial distribution and relations are the main duty of CED framework. An extreme event is a category of collective conduct within a complex framework, frequently holds noticeable results.

Extreme events take place on a considerable assortment of a complex systematic scheme, an epitome of an extreme event may include surprising packet flow on a network, network jamming, huge increase on the number of website hits, or calamitous state such as immense level of disintegration produced by casual collapse of nodes [4].

The general precept is that extreme events have a significant role in the path of detecting complex events, through a systematic computational analysis of Big Data, by considering data as an unarranged spectrum of events arriving from diverse sources and sensing devices. Motivated by abnormality detection that exists in several fields around us, we contemplated applying queries to these elements in order to be able to provide the adequate response for the detected faults and ameliorate the pliability that is required to perform solutions for issues related to data monitoring and management.

The structure of the paper is following: In Sect. 2 works related to the current topic are listed. In Sect. 3 description of the analyzed data set and the methodology is given. Section 4 introduces special event detection algorithm and weighted complex event metric. In Sect. 5 Detected Complex Events of the Cassini-Huygens Mission are explained. Conclusion and possibilities of the continuation of this work is given in Sect. 6.

2   Related Work

Literature inspection and detailed analysis uncovers wide exploitation of sensory data in varied fields spanning from environmental conditions, healthcare applications, industrial to civilian and military implementation and social behavior recognizing, that are all founded on tracing and analysis of information [5].

In [6] were evolved four classifiers which can be exploited onboard of the space ship to pinpoint high primacy data to be sent to Earth, to enlarge the bandwidth utilization of bounded downlink, these classifiers used to recognize events on the cryosphere by hyperspectral imaging. In [7] was introduced and appraised communication methods that could impressively implement CED along distributed event provenance.

The authors of [8] proposed a method for determining whether and in what way or manner the evaluation process must be reinforced. This approach is counting on a few sets of restrictions to be accomplished via the observed values, delineated in a way to get improved evaluation process is ensured in case of any predefined principles is breached. The presented method intended to obviate false positives regarding the most advantageous resolution.

Authors of [9] depict a pattern-based approach querying methods that are related to remotely sensed data networks, scheme based techniques to data gathering has been taken into account to acquire more productivity. Additionally, they debated dependability based outcomes and contemplate the reliance among the sensors of network readings. Authors of [10] intend in their work to concentrate on the observation of abnormal events in an environment where complex activities are accomplished and potential abnormal situation differ in various phases, in the given model the abnormal situation among any phase could be converted into event modality termed as abnormal event patterns. The contexts within various times measured by sensors, a state transition is exploited to pattern each phase in order to detect the normal transition interval at the starting and the ending of every phase with the purpose of finding an abnormal situation. Another paper purpose is to put forward an approach to increase the realization regarding incidents, which could be characterized as events, they adopted an anomaly detection technique depending on rapid harmonic analysis method and decision-making regulations [11].

In [12] are presented CEP techniques that requires long time to accomplish their tasks, as they do not take in account resemblance and operators redundancy, that what Bok, Kyoungsoo et al. mentioned and put forward in their work an approach that contemplate resemblance and redundancy of operator for sensors data streaming. The focus of work [13] is to provide a framework concerning vehicle interaction, whereas uncertainty is occupying a considerable amount of their proposed technique as no methodical procedure can completely depict the scenario of the complex event. Markov logic network is utilized among the terrestrial domain in which dynamic characteristics and relevance between vehicles are grasped via several sources and sensors. Computations to address the concept of complex events within big streaming of data have proliferated remarkably, also consummation time is overdue owing to the fact that events could not be processed in a prescribed interval of time. Consequently, research on mitigation of the aggregate computational encumbrance is required and must be included in a situation of events which are primitive which lead to having complex events that possess redundant and resembling operations.

Feature extraction contemplated as a decisive aspect in the domain of pattern modelling, in addition to an important role in the remote sensing field, as it simplifies the posterior data handling, representation and classification. Features could be commonly ranged from bottom to top scale features, despite the fact that there is no apparent separation between them [14]. An essential part of the event comprehension is to perceive what triggered it and in which time it is taking place [15].

3 Description of the Data Set and the Methodology

Since the Cassini–Huygens spacecraft arrived at its destination it has provided us with images and a huge volume of data that has not gained at any time before. Almost 13 years spent by Cassini mission in the orbit of Saturn were the headmost chance to study this anonymous planet charm, which seduces us for a closer look to it, throughout the spacecraft's passing near Saturn. By this stage within the approach gradation, the planet Saturn was huge quite so that dual narrow-angle camera was needed to cover the whole size of planet with its rings and some of its icy moons such as Pandora, Epimetheus, Mimas, Janus, Prometheus and Enceladus. Imaging Science Subsystem (ISS) was the device of remote sensing that took images of Saturn system with its moons and rings. 407,302 samples have been included in our analysis which are collected from 106 volumes of the ISS data source among 13 phases of the mission. The duration of these phases are very different between just half dozen of samplings to more than one thousands of samples. This made us to concentrate to just five mission phases having considerable amount of samples (see Table 1).

Table 1. Analysed mission phases

Phase name	Start sample ID	Start time	End sample ID	End time
Approach science	1	2004-037T02:07:06.498	10,675	2004-162T14:47:05.854
Tour pre-huygens	10,940	2004-164T02:33:51.000	31,640	2004-358T13:47:22.548
Tour	32,032	2005-015T18:28:29.491	166,187	2008-183T09:17:06.323
Extended mission	166,188	2008-183T21:04:09.008	235,887	2010-283T14:14:20.741
Extended-extended mission	235,888	2010-285T05:23:32.745	407,303	2017-257T19:59:04.075

A vigorous approach for recognizing variation has applied to find out if an event has happened in a specific image, based on difference assessment made with the previous and the following taken images within a 3 groups of parameters (Group 1: Timing; Group 2: Temperature; Group 3: communication) and its related 15 variables of metadata. The analysed variables sets are given in Table 2.

Table 2. Groups of analysed variables

Variable group	Variable name	Description
A. Timing	IMAGE_NUMBER	Mean time of image
	START_TIME	Capture start time of the sample
	END_TIME	Capture end time of the sample
	EARTH_RECEIVED_START_TIME	Start time of the receiving msg.
	EARTH_RECEIVED_END_TIME	End time of the receiving msg.
B. Temperature	FILTER_TEMPERATURE	Temp. of the optics filter [°C]
	OPTICS_TEMPERATURE_FRONT	Temp. of the front optics [°C]
	OPTICS_TEMPERATURE_REAR	Temp. of the rear optics [°C]
	DETECTOR_TEMPERATURE	Temp. of the detector [°C]
	SENSOR_HEAD_ELEC_TEMPERAT	Temp. of the head electronics [°C]
C. Communication and scale	EXPECTED_PACKETS	No. of expected radio packets
	INSTRUMENT_DATA_RATE	Data rate of the transmitter [kb/s]
	INST_CMPRS_RATIO	Compression ratio of the msg.
	RECEIVED_PACKETS	No. of received radio packets
	PIXEL_SCALE	Pixel scale [km/pixel]

The categories are specified by a commonly unanimous convention that the event is recognized based on the amount or the presence of significant attributes, e.g. a change in size or the colour and even the previous known shape. The connection between events within the scope of conditional probability commonly holds a conditional likelihood provided that an event conditional distribution continues to be fixed with the given measures. The nominated approach is an acclimation of supervised learning as the method must impart visual significances pertinent to a particular event class by extracting the whole related visual motif and patterns for an event from images, the approach includes constructing a weighted extreme pattern score (WEPS) array for the related target metadata time series. A set of weighted score for each submission and a sensitivity threshold value are detected, while the correctness of the method is specified by the consecutive window size via the difference of the time moments and intervals. After that, an alteration score for each timestamp is computed based on the WEPS score array, a time series can be defined as data concatenation of marks gathered over a period of time.

4   Special Event Detection (SED) Algorithm and Calculation of Weighted Complex Event Level (WCEL) Metric

The concept of processing a complex event is concerned with the issue of processing several events with the attention of pointing out the significate events among a timely stream of events sequence. Introducing the label of weight with event is to provide a numeric value that quantifies the degree of compactness of an event, this weight spans with a range of 0 till 1. The weight is like a temporal dialectic with sensible way concerning events and their impacts. Aiming to promote the competence of the detection process with a scale modality we have introduced the concept of wCEL, which is an extension of a weighted complex event (Fig. 1).

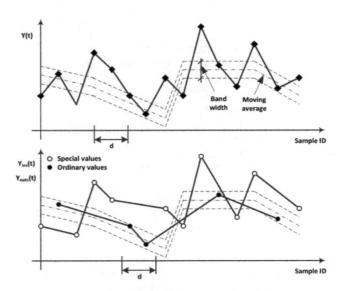

Fig. 1. Special Event Detection (SED) mechanism of variable Y(t). SED decomposes the data set into two disjunctive subsets: *special values* and *ordinary values*. For a window of size *d* is determined the moving average. The width of band is given by $2 \cdot \rho \cdot \sigma$ (ρ: SED sensitivity parameter, σ: standard deviation of the time series in the actual window, d). Samples outside of band are special values and the rest of samples are ordinary values.

4.1   Algorithm_A (Special Event Detector: SED)

Let us have variable $Y(t)$ with $t \in (1, N)$, where N is the number of samples collected by sensors. Outliers of the time series $Y(t)$ can be detected with the following algorithm:

(i)   Let $t = 0$, $V(1{:}N) = 0$.
(ii)   We use a number d of consecutive samples generating vector $Y(t : t + d)$ and we determine the average value $mY(t, d) = mean(Y(t{:} t + d))$ and standard deviation $\sigma(t, d) = STD(Y(t : t + d))$, respectively.
(ii)   Set element $V(t + d)$ corresponding to element $Y(t + d)$ of the time series to be outlier (i.e. $V(t + d) = 1$), if following relation hold:

$$|Y(t + d) - mY(t, d)| \geq \rho \cdot \sigma(t, d). \tag{1}$$

(iii)   Repeat $(t = t + 1)$ and (Go to step ii) Until $t > N$.

Parameter ρ is the sensitivity and d is the memory of the SED mechanism. Both parameters are given by the service specialist. It can be observed that for $\rho = 0$ each sample $Y(t)$ is outlier and for $\rho \geq \rho^*$ no outlier will exist. We name ρ^* cut value of the SED sensitivity. Special events of variable $Y(t)$ are considered outliers of $Y(t)$ and are stored in corresponding variable, $V(t)$. Conform to relation (1) $V(t)$ has following property for each $t = 1, \ldots, N$:

$$V(t) = \begin{cases} 0 & \text{if } Y(t) \text{ is } ordinary\ event \\ 1 & \text{if } Y(t) \text{ is } special\ event \end{cases} \tag{2}$$

First d elements of the variable are ordinary events.

4.2   Algorithm_B (Processing of Weighted Complex Event Level: wCEL)

For a group of values $G_k(t : t + d) = \{Y_1(t : t + d), Y_2(t : t + d), \ldots, \{Y_k(t : t + d)\}$ can be determined the weighted Complex Event Level conform to following algorithm:

(i)   Let $t = 0$.
(ii)   Let $i = 1$.
(iii)   Calculate weights:

$$w_i(t + d) = \sum_{j=t}^{t+d} V_i(j). \tag{3}$$

(iv)   Calculate cumulative weights:

$$w(t + d) = \sum_{j=1}^{k} w_i(t + d). \tag{4}$$

(v)   Repeat $(i = i + 1)$ and (Go to iii) Until $i > k$.

(vi) Calculate weighted Complex Event Level:

$$wCEL(G_k, t+d) = \sum_{i=1}^{k} \frac{w_i(t+d)}{w(t+d)} V(i). \tag{5}$$

(vii) Repeat $(t = t+1)$ and (Goto step ii) Until $(t > N)$.

It is stated that $wCEL(G_k, t+d)$ is a metric of event complexity and has limited values:

$$wCEL(G_k, t+d) \in [0, 1] \tag{6}$$

Based on the formula (4) we have:

$$\sum_{i=1}^{k} \frac{w_i(t+d)}{w(t+d)} = 1 \tag{7}$$

Conform to formula (2) values of $V(t)$ are special: $V(t) \in [0, 1]$. This feature together with relation (8) implies relation (6). The higher is the number of special events in the group $G_k(t : t+d)$, the greater is the $wCEL(G_k, t+d)$. The higher is the special events in group G_k in sampling moment $(t+d)$, the greater is the metric $wCEL(G_k, t+d)$.

We want to characterize a given group G_k of variables globally, independently of the time. For this we determine the mean of weighted Complex Event Level by following formula:

$$mwCEL(G_k) = \frac{\sum_{t=d}^{N} wCEL(G_k, t)}{N - d + 1} \tag{8}$$

This metric depends on the sensitivity, ρ and memory, d of the SED mechanism and globally characterize the variable group G_k.

4.3   Evaluation of the Variables

The arrangement of the evaluation process has been positioned according to the groups sequence below:

4.3.1   Timing Variables (*Group_A*)

The previously mentioned timing group contains the time moments for the images that include mean, starting and ending capture time of the sample at the Cassini and the starting and ending time of the sample at the Earth. Figure 2 expounds the sampling moments among time intervals and Fig. 3 shows the number of samples in each year of the project. It should be mentioned that the inhomogeneity of the sampling intervals producing nonlinear curve on Fig. 2. Majority of the samplings were captured in the first half of the expedition and smaller number of capturing was done in the seven years long phase named Extended-Extended Mission.

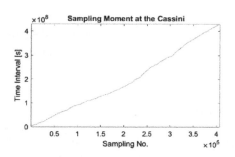

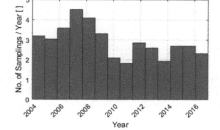

Fig. 2. Sampling time vs. sampling ID

Fig. 3. Number of samples per year

4.3.2 Temperature Variables (*Group_B*)

The temperature of 5 metadata variables is analysed (filter temperature, detector temperature, front and rear optic temperature, sensor head electronics temperature), the filter temperature indicates the real temperature of the filter wheels, the detector operating temperature (−90 °C), while the optimal operating temperature spans between 0 to 10 °C for both the narrow-angle camera and the wide-angle camera. (Figs. 4 and 5) shows the filter and detector temperatures over sampling.

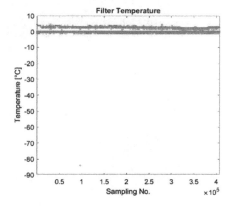

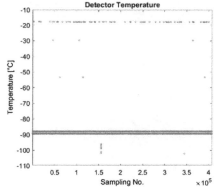

Fig. 4. Filter temperature over sampling

Fig. 5. Detector temperature over sampling

Throughout the flight mission the devices, excluding the detector (−90 °C approximately) shall not experience temperatures higher than +35 °C or lower than −25 °C. The average of temperature variation will not transcend the average of 8 °C per hour. Regarding the detector, the operating temperature is required for the repression of dim current also to reduce the neutron and gamma proton emission impact. The temperature is preserved through passive/active warming device, beside decontamination heaters with the purpose of taking off outgassed chemical elements volatiles that may gathered on the charge-coupled device window, optics temperature has two valued array, the 1st one is related to the front optic, but as there is no temperature for the rear optic of the wide-angle camera, thus if the instrument ID was equal to ISSWA, then the second array value shall permanently be −999.0 °C (−999.0 °C, −999.0 °C when the extended header does not exist).

4.3.3 Communication Variables (*Group_C*)

The communication group have 4 labels (expected packets, received packets, instrument data rate, instrument compression ratio) The label of the expected packets demonstrates the gross number of expected packets that will be stored on the spacecraft solid-state recorder for each image, in order to transform to volume measured in bits, we need to multiply the value of the expected packet by 7616 bits per packet. The concept of received packets signifies the veritable packets number which is received from the spacecraft solid-state recorder for each image, both of these keyword and their histograms are given in (see Fig. 6).

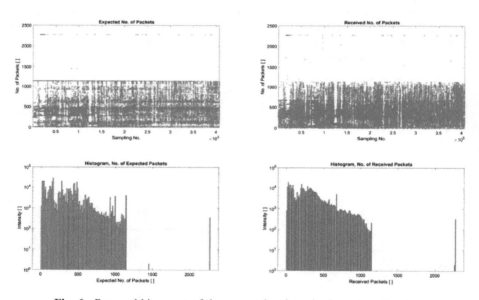

Fig. 6. Rate and histogram of the expected and received number of packets

Instrument data rate label is measured by (kilobits/second) and it represents the data rate that the data was transferred out, it has 5 modes the lowest rate is 60.9 Kbits/sec and the highest is 356.6 Kbits/sec. data rate can be also expressed by packet/sec with 8 as the lowest rate followed by 16, 24, 32 and 48 as the highest.

5   Detected Complex Events of the Cassini-Huygens Mission

The proposed method has detected a list of complex events through the metric of wCEL among the image batches of Cassini mission, among 5 analysed phases of the mission, which include: Approach science mission, Extended mission, Extended-Extended mission, Tour mission, Tour pre-Huygens mission. 15 variables of the metadata were involved in the analysis process. Selected group of the most remarkable events are listed below (Figs. 7, 8, 9, 10, 11 and 12). In numerous situations these distinctive events should be processed in order to understand the phenomenon that stands behind them. The selected list shows that as the values of (σ) decreases the wCEL increases.

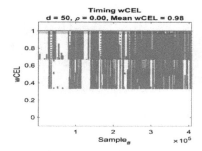

Fig. 7. wCEL of timing (d = 50, $\rho = 0$)

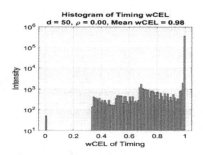

Fig. 8. Histogram of timing wCEL (d = 50, $\rho = 0$)

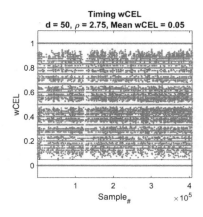

Fig. 9. wCEL of timing (d = 50, $\rho = 2.75$)

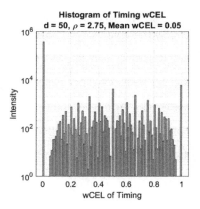

Fig. 10. Histogram of timing wCEL (d = 50, $\rho = 2.75$)

In Group_A (Timing) of variables three components were used for the special event detection: (i) Sampling moment at the Cassini; (ii) Shutting time at the Cassini (being the difference between starting and ending moments); (iii) Sampling moment at the Earth.

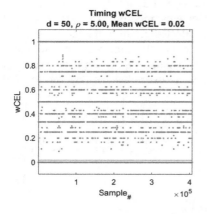

Fig. 11. wCEL of timing (d = 50, ρ = 5.00)

Fig. 12. Histogram of timing wCEL (d = 50, ρ = 5.00)

Transmission of the samples to the Earth was influenced by the relative position of the Saturn, Cassini and Earth causing variable temporal storage durations of the samples at the spacecraft (Figs. 13, 14, 15 and 16).

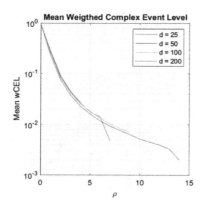

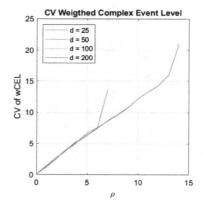

Fig. 13. Mean wCEL (Group A) for different parameters, d (sampling window sizes)

Fig. 14. Coefficient of variance of wCEL (Group A) for different parameters, d (sampling window sizes)

Based on the Fig. 13 can be concluded the cut values of the complex event level detector mechanism for group A of the variables (see Table 3).

Table 3. Estimated cut value of the sensitivity of the average wCEL (Group A: timing)

Sample window size (d)	ρ^*
25	5
50	8
100	10
200	15

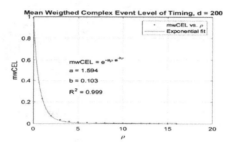

Fig. 15. Estimation of the mean wCEL (Group A) of timing, sampling window, d = 200)

It was found that the mean weighted Complex Event Level metric for the timing is exponential function of an exponential function and has following formula:

$$mwCEL(\rho) = \begin{cases} 1 & \text{if } \rho = 0 \\ \exp(-a \cdot \rho \cdot \exp(-b \cdot \rho)) & \text{if } \rho \in (0, \rho^*) \\ 0 & \text{if } \rho \geq \rho^* \end{cases} \tag{10}$$

where *mwCEL* is the mean of weighted Complex Event Level, ρ is the sensitivity of the Special Event Detector, *a* and *b* are parameters depending weakly on the sample window size (d). Parameter ρ^* is the cut value of the SED sensitivity and based on Fig. 17 depends on the sample window size: the greater is *d*, the greater becomes ρ^*. This phenomenon is caused by the smoothed property of the SED influenced by the moving average function. Estimated fitting parameters *a* and *b* are given in Table 4 with the coefficient of determination, $R^2 > 0.99$.

Table 4. Estimated parameters of the average wCEL (Group A: timing)

Sample window size (d)	a	b	R^2
25	1.353	0.070	0.999
50	1.413	0.086	0.999
100	1.483	0.085	0.999
200	1.594	0.103	0.999

In Group_B (Temperature) of variables five temperature components were used for the special event detection: (i) Temp. of the optics filter; (ii) Temp. of the front optics; (iii) Temp. of the rear optics; (iv) Temp. of the detector; (v) Temp. of the head electronics.

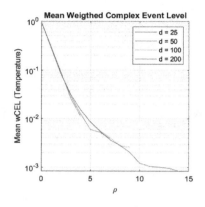

Fig. 16. Mean wCEL (Group B) for different parameters, d (sampling window sizes)

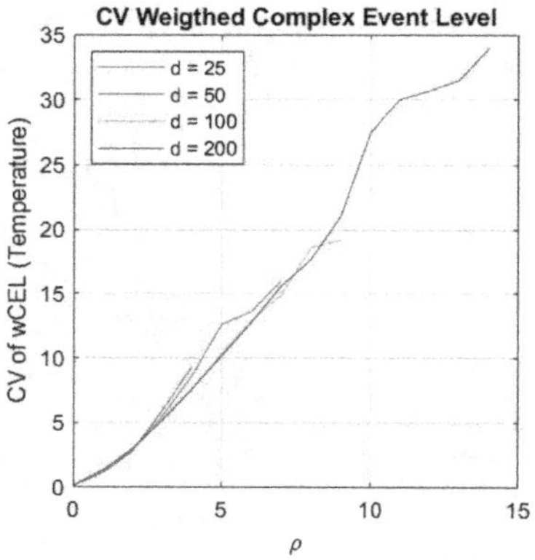

Fig. 17. Coefficient of Variance of wCEL (Group B) for different parameters, d (sampling window sizes)

In Group_C (Communication) of variables five components were used for the special event detection: (i) No. of expected radio packets; (ii) Data rate of the transmitter [kb/s]; (iii) Compression ratio of the msg; (iv) No. of received radio packets; (v) Pixel scale [km/pixel] (Fig. 18).

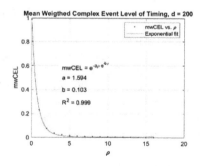

Fig. 18. Estimation of the mean wCEL (Group C, sampling window, d = 200)

6 Conclusions and Remarks

With the current advances in our world and huge amounts of data that is available everywhere in the surrounding environment, there is need for the enforcement of sophisticated mechanism in the field of complex event processing and detection. It has been examined by applying it on the adopted 5 phase batches, the outcomes confirmed that almost in every window of time interval there was more than a single complex event. In this work we aim to detect complex event in a huge images dataset related to Cassini-Huygens outer space mission, processing semantic data have always been a presumed defy to defeat. Nevertheless, just few research studies have been undertaken to declaim the topic of data inclusion and diversity.

CED is an affluent research field accompanied by several aspects to be investigated the exploitation of learning techniques to classify and detect special events offers a considerable prospective to be utilized on the incoming outer space expeditions, as the huge distance between the spacecraft and the Earth stations requires relatively a lot of crucial time, so the response to a certain event may not be accomplished within the needed time interval. Our proposed research work, which is based on remotely sensed data has shown a boost in detection a complex event over the consecutive learning process that relies on both real data and learning. The analysis of remote sensed images always concentrated on the extreme change detection (signature) of the pattern, fewer interest is given to the spatial modality existed in these images. Whereas, the patterns of these spatial orientation have constantly provided the essential part to map, create and stimulate ideal disciplines in many aspects of our life.

As complex event detection is an integral aspect of this paper we aim to keep improving our methodology exactitude and extending it. Also, we are concerned in the establishment of a user interaction interface, which would promote the user cordial navigation possibilities to have user friendly access to the extracted information from the dataset collected by Cassini-Huygens interplanetary mission. As a continuation of this work can be the classification of the sampled records based on learned data set.

References

1. Flouris, I., Giatrakos, N., Deligiannakis, A., Garofalakis, M., Kamp, M., Mock, M.: Issues in complex event processing: status and prospects in the big data era. J. Syst. Softw. **127**, 217–236 (2017)
2. Agrawal, J., Diao, Y., Gyllstrom, D., Immerman, N.: Efficient pattern matching over event streams. In: SIGMOD 2008, New York, NY, USA (2008)
3. ALDabbas, A., Gál, Z.: Getting facts about interplanetary mission of Cassini-Huygens spacecraft. In: 10th Hungarian GIS Conference and Exhibition, Debrecen, Hungary (2019)
4. Chen, Y.-Z., Huang, Z.-G., Lai, Y.-C.: Controlling extreme events on complex networks. Sci. Rep. **4**, 6121 (2014)
5. Aggarwal, C.C.: An introduction to sensor data analytics. In: Aggarwal, C.C. (ed.) Managing and Mining Sensor Data, pp. 1–8. Springer, Boston (2013)

6. Castano, R., Mazzoni, D., Tang, N., Greeley, R., Doggett, T., Cichy, B., Chien, S., Davies, A.: Onboard classifiers for science event detection on a remote sensing spacecraft. In: Proceedings of the 12th ACM SIGKDD International Conference on Knowledge Discovery and Data Mining, Philadelphia, PA, USA, 20–23 August, pp. 845–851 (2006)

7. Akdere, M., Çetintemel, U., Tatbul, N.: Plan-based complex event detection across distributed sources. Proc. VLDB Endow. 1(1), 66–77 (2008)

8. Kolchinsky, I., Schuster, A.: Efficient adaptive detection of complex event patterns. In: The 44th International Conference on Very Large Data Bases, Rio de Janeiro, Brazil, August 2018 (2018). Proceedings of the VLDB Endowment, vol. 11, no. 11 (2018)

9. Deshpande, A., et al.: Model-based approximate querying in sensor networks. VLDB J. 14 (4), 417–443 (2005)

10. Lu, T., Zha, X., Zhao, X.: Multi-stage monitoring of abnormal situation based on complex event processing. Procedia Comput. Sci. 96, 1361–1370 (2016)

11. Butakova, M.A., Chernov, A.V., Shevchuk, P.S., Vereskun, V.D.: Complex event processing for network anomaly detection in digital railway communication services. In: Telecommunication-Forum (TELFOR), Belgrade, pp. 1–4 (2017)

12. Bok, K., Kim, D., Yoo, J.: Complex event processing for sensor stream data. Sensors (Basel) 18(9), 3084 (2018)

13. Lu, J., Jia, B., Chen, G., Chen, H., Sullivan, N., Pham, K., Blasch, E.: Markov logic network based complex event detection under uncertainty. In: Proceedings of SPIE, Sensors and Systems for Space Applications XI, vol. 10641, 106410C (2018)

14. Elnemr, H.A., Zayed, N.M., Fakhreldein, M.A.: Feature extraction techniques: fundamental concepts and survey. In: Kamila, N.K. (ed.) Handbook of Research on Emerging Perspectives in Intelligent Pattern Recognition, Analysis, and Image Processing, pp. 264–294. IGI Global, Hershey (2016)

15. Luckham, D.: The Power of Events, vol. 204. Addison-Wesley, Reading (2002)

Integrating OpenMTC Framework into OneM2M Architecture for a Secure IoT Environment

Mohammed Misbahuddin[1(✉)], Aditi Agrawal[2], Aishani Basu[2], and Ankita Jain[2]

[1] C-DAC, Bangalore, India
misbah@cdac.in
[2] PESIT, Bangalore, India
mdmisbahuddin@gmail.com, aditiagrawal3050@gmail.com,
aishani.basu691@gmail.com, jaina865@gmail.com

Abstract. As the current era of Internet of Things (IoT) and M2M communication advances, intercommunication of devices becomes instrumental for establishing a smart and secure IoT environment. To enable such a smart and secure system, which seamlessly integrates devices like sensors and actuators independent of the underlying legacy networks, a set of standards called oneM2M was devised. Thus, this set of standards has successfully facilitated communication between any two applications which may have different underlying communication protocols. A common service layer implements these set of standards which comes in between the application and the underlying network entity. The overall architecture comprises of the open-source integration middleware platform OpenMTC, which is based on oneM2M standards and is implemented using Python. This paper presents the results of the successful intercommunication between three sensors and actuators each relying on a different communication protocol, namely Wi-Fi, Bluetooth, Serial Communication and Ethernet in a secure IoT environment and the results obtained have successfully proved the efficacy of the proposed system.

Keywords: oneM2M · OpenMTC · SSL · M2M · IoT · Python

1 Introduction

Recent challenges in the field of IoT include providing interoperability among IoT data, interpreting data generated by IoT devices and today, most M2M solutions use proprietary systems that often comprise all layers, from application to physical, to provide their specialized M2M services to the customers which makes it difficult to extend these systems to support new services, integrate new data, and interoperate with other M2M systems [1].

In the context of M2M and IoT, it is commonly observed that the terms, M2M and IoT are used interchangeably. M2M communication refers to automated data transfer between any two devices. [2] For example, a sensor which measures any change in a physical property may regulate a particular variable which then controls the output of

© Springer Nature Switzerland AG 2020
L. C. Jain et al. (Eds.): ICICCT 2019, LAIS 9, pp. 67–76, 2020.
https://doi.org/10.1007/978-3-030-38501-9_6

an actuator, like LEDs, DC motors, etc. without any human intervention. IoT, on the other hand, involves M2M communication, integration of web applications with it, and sending all the data to the cloud for processing or storage [3, 9].

These issues involving interoperability of M2M systems with different connectivity motivated various standard organizations to establish a new partnership project, "the oneM2M Global Initiative," to standardize a common horizontal M2M service layer platform for globally applicable and access-independent M2M and IoT services. This international partnership project was established in July 2012 between the eight most important SDOs of the world: Association of Radio Industries and Businesses (ARIB) and Telecommunication Technology Committee (TTC), Japan; the Alliance for Telecommunications Industry Solutions (ATIS) and Telecommunications Industry Association (TIA), United States; the China Communications Standards Association (CCSA), China; the European Telecommunications Standards Institute (ETSI), Europe; the Telecommunications Technology Association (TTA), Korea and Telecommunications Standards Development Society (TSDSI), India with the goal to work towards the global unification of the M2M and IoT community. In January 2015 the first release (Release 1) of oneM2M was published and since then, several Open Source foundations and projects have been using oneM2M architecture and standards for developing new M2M and IoT applications and one such platform is OpenMTC which became open source on the eve of Berlin5GWeek [4, 10].

1.1 oneM2M

The global oneM2M standard successfully defines a horizontal service layer that fits different verticals and enables seamless communications between various heterogeneous entities independent of their underlying network and vendor-specific device technologies. Thus, making it possible to reach and deliver a message to any entity in the system. Some of the main goals are: interoperability between various heterogeneous devices, minimizing market fragmentation, easier deployment and boosting M2M economy by minimizing the cost of production [8, 14].

The common service layer is achieved with the help of oneM2M architecture which is composed of four functional entities called nodes, namely the application dedicated node (ADN), the application service node (ASN), the middle node (MN) and the infrastructure node (IN) and each node contains a common services entity (CSE), an application entity (AE), or both. An AE provides application logic, such as remote power monitoring, for end-to-end M2M solutions. A CSE is a logical entity that is instantiated in an M2M node and comprises of a set of service functions called common services functions like, registration, security, application, service, data and data management, etc. that can be used by applications and other CSEs. In addition to CSFs, a CSE includes a service enabler to ensure the extensibility of services. It currently defines three reference points (i.e., Mca, Mcc, and Mcn) as shown in Fig. 1. The Mca reference point enables AEs to use the services provided by the CSE. The Mcc reference point enables inter-CSE communications. The Mcc' reference is like Mcc but provides an interface to another oneM2M system. The Mcn reference point is between a CSE and the service entities in the underlying networks.

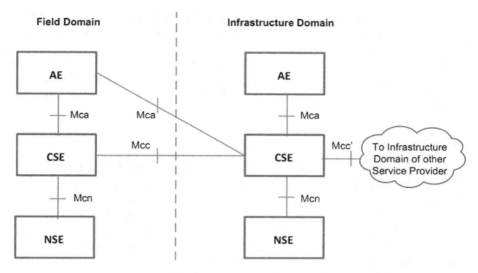

Fig. 1. oneM2m architecture

oneM2M adopted a resource-based data model where all services and data are represented as a resource which can be uniquely addressed by a uniform resource identifier and data can be modified by performing operations like CREATE, RETRIEVE, UPDATE and DELETE (CRUD) [4].

1.2 OpenMTC

OpenMTC is a reference prototype implementation of the international oneM2M standard which interconnects various sensors and actuators from different vertical domains and the aggregated data is forwarded to applications which initiate instructions to end devices for event–driven actions. Through REST application programming interfaces data can be accessed from any device irrespective of the underlying technology or hardware. In an OpenMTC system, all entities are represented as resources like oneM2M standards and these resources are either Containers consisting of sub-containers or content instances and Content Instances holding actual values.

The integral parts of OpenMTC are:

- Application Entity: It consists of an application provided by the application developer to access data within the OpenMTC-based IoT system. An example would be a simple GUI used for visualizing data.
- Interworking Proxy (IPE): It allows translating data from a non-oneM2M domain to a oneM2M domain. For example, an application that reads out sensor values and translates it to oneM2M. This block helps to integrate legacy devices into the OpenMTC architecture for communication using the oneM2M standard.
- OpenMTC Gateway: It can be defined as a software node that is central to every field domain that collects data from various IPEs. An application can access the resources via the gateway. These gateways are located close to resources or devices

which generates data. The responsibilities of the gateway are to send the aggregated data, retrieving sensor data and other local services like caching of data, device management and authentication.

- OpenMTC Backend: It is the central software node within the infrastructure domain. Often, it acts as the root node within a hierarchy of OpenMTC Gateways. Backend integrates multiple gateways and provides unique oneM2M compiled APIs [5].

1.3 SSL

The security of the communication is enabled by SSL or Secure Socket Layer. Secure Socket Layer is an industry standard security protocol for creating an encrypted connection between a server and client-typically a web server and a browser. SSL enables all information in transit between the client and server to be encrypted such that it cannot be intercepted by any third party who wants to eavesdrop. This protocol is extremely necessary to ensure sensitive information like social security numbers, credit card numbers, and login credentials remain unaffected to attacks by any third party. SSL being a security protocol, dictates how the encryption algorithm is to be executed [6].

OpenMTC uses SSL for security of the implemented architecture. When any device wants to send data to or access data from the OpenMTC Gateway, it requests the Gateway to identify itself. The Gateway sends its SSL certificate to the device, and the device matches the certificates' Root CA with a list of trusted CAs. Once it is satisfied, it sends a symmetric session key, encrypted using the Gateway's public key. The Gateway decrypts this message using its own private key and sends back an acknowledgement encrypted using the session key for starting the data transfer. Henceforth, all communication between the device and Gateway for the current session is encrypted using this session key. The OpenMTC Backend also uses SSL for security in a similar manner [7].

2 Proposed Architecture

In an IoT environment, corresponding to the proposed architecture shown in Fig. 2, data can be aggregated from 'n' number of sensors and devices operating in different or same field domains, and distributed to central nodes of multiple field domains. With the help of OpenMTC which is described in the previous section, an OpenMTC Gateway is setup for every field domain. All the oneM2M devices within this field domain would register themselves with the gateway directly and the non-oneM2M devices would need an Interworking Proxy Entity (IPE) in between to translate to oneM2M.

The gateways are registered to the OpenMTC Backend within the infrastructure domain, in a RESTful architecture. The OpenMTC Applications could either connect directly to the gateways or the backend to provide or collect data. Through the oneM2M framework the 'n' devices can communicate through other 'n' devices irrespective of their underlying communication protocols.

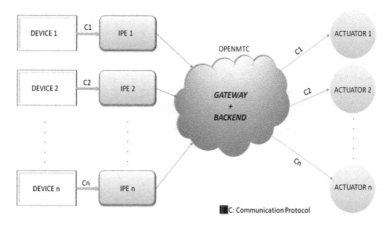

Fig. 2. Proposed architecture

Say, we have a device which uses only Bluetooth protocol to communicate and wants to communicate to another device which can even be an actuator which has only in-built Wi-Fi and now, this can be achieved through oneM2M architecture which is implemented here using the OpenMTC framework.

oneM2M successfully converts vertical data communication into a horizontal communication with the help of a common services layer. In absence of oneM2M each AE (Application Entity) would communicate with the underlying NSE (Network Service Entity) and there would be no question of two different sensors or actuator or both communicating with each other, but with the help of oneM2M this communication is now possible. This architecture is successfully integrated into the OpenMTC framework using the steps described above.

2.1 Deployed Architecture

A system corresponding to the proposed system above is implemented as shown in Fig. 3. To build such a system the corresponding hardware [15, 16] and communication protocols have been used and in order to incorporate the oneM2M architecture OpenMTC was installed [5] and implemented.

(1) *Sensors*: LM35 (temperature sensor), DHT11 (temperature and humidity sensor), EM-18 RFID tag and reader
(2) *Microcontrollers*: Arduino UNO, NodeMCU ESP8266, Raspberry Pi 3
(3) *Actuators*: LED
(4) *Communication Protocols*: Wi-Fi, Bluetooth, UART, Ethernet

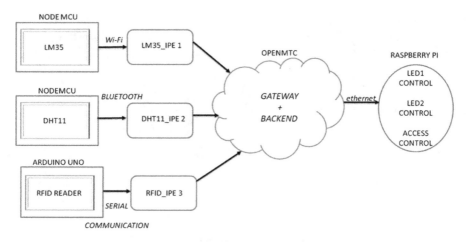

Fig. 3. Deployed architecture

2.2   Input

For the first input, the LM35 sensor is connected to the NodeMCU to obtain the temperature measurements. The temperature readings are obtained in the IPE (Inter-working Proxy Entity) block, by using Wi-Fi. This process is enabled by the Node MCU board which has an in-built Wi-Fi module. Thus, the temperature measurements get uploaded to the IP address assigned to the NodeMCU board in the private network being used. These values are procured by the IPE from this webpage using GET request [11] as shown in Fig. 4, where the IPE is connected to the same network.

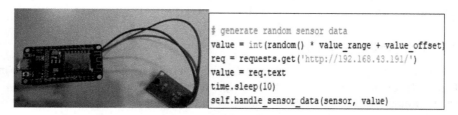

Fig. 4. Input 1 connections and IPE

For the second input, the DHT11 sensor is used, which is connected to another ESP8266 and the data is transferred to another IPE block with the help of Bluetooth. The Bluetooth client-server model is created using an RFCOMM socket in Python [12]. The Bluetooth server binds the Bluetooth port address of the host device to the socket created and waits for the clients to connect. As the client connects, it receives the temperature data sensed by the DHT11 sensor, and closes the connection when it is done. The data obtained by the server is stored in a file on the localhost. The second IPE obtains this data from the file and sends it to the OpenMTC Gateway as shown in Fig. 5.

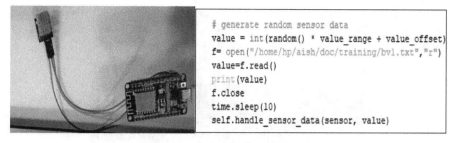

Fig. 5. Input 2 connections and IPE

For the third input, the UART (Universal Asynchronous Receiver and Transmitter) of the Arduino Uno is used. When the RFID tag is brought near the RFID reader, which is connected to the Arduino board, the tag data is read by the reader and displayed on the serial monitor of the Arduino IDE. This data is stored in a file and the third IPE block gets this data and sends it to the OpenMTC Gateway (Fig. 6).

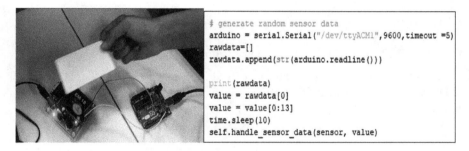

Fig. 6. Input 3 connections and IPE

OpenMTC Gateway and Backend: The OpenMTC Gateway takes data from all the three IPEs in the field domain. The Gateway is registered to the OpenMTC Backend and it is responsible for posting all the data obtained from all IPEs in each field domain to the Backend. The OpenMTC Backend is the central software node within the infrastructure domain. It is responsible for storing all the data obtained from the Gateway in a RESTful architecture. Any GUI (graphical user interface) which subscribes to the Backend can access resources from the resource tree [13].

2.3   Output

For displaying the output, we have used two LEDs connected to a Raspberry Pi board. The Raspberry Pi is connected to the system used via an Ethernet cable. The GUI displays the sensor data and stores the data it displays in a file, which can be accessed by the actuator or LED control code. Thus, depending on the temperature, a threshold is set, which is 21 for our demonstration, such that the first LED, LED1 is switched ON only when the temperature falls below 21 °C, and remains OFF otherwise. The second LED control is the opposite, which means LED2 is switched ON when temperature rises below 21 °C else switched OFF otherwise (Fig. 7).

Fig. 7. Output connection

3 Results

1. LM35 (Wi-Fi) and DHT11 (Bluetooth) Output

In the first case the temperature read by LM35 and DHT11 is 22.11 °C. Thus, LED1 is turned OFF and LED2 is turned ON as it is greater than 21 °C which is set as our threshold value (Fig. 8).

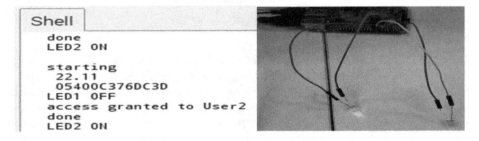

Fig. 8. Case 1 results

In the second case the temperature read by LM35 and DHT11 is 20.44 °C. Thus, LED1 is turned ON and LED2 is turned OFF as it is less than 21 °C (Fig. 9).

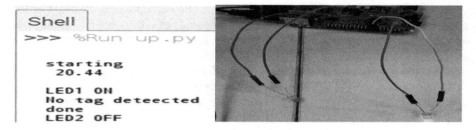

Fig. 9. Case 2 results

2. RFID (Serial Communication) Output

In the third case the RFID tag ID read by the RFID reader is 05400C376DC3D and hence, access is granted to USER1 and similarly, in the second case the RFID tag ID read by the RFID reader is 05400C3961110 and hence, access is granted to USER2 and In this last case no RFID tag is in the range of the RFID reader and hence, it doesn't read any value and the error message "No Tag Detected" is displayed on the console (Figs. 10 and 11).

```
Shell
    done
    LED2 ON

    starting
     22.11
     05400C376DC3D
    LED1 OFF
    access granted to User2
    done
    LED2 ON

    starting
     22.11
     05400C3961110
    LED1 OFF
    access granted to User1
    done
    LED2 ON
```

```
Shell
>>>  %Run up.py

    starting
     20.44

    LED1 ON
    No tag deteected
    done
    LED2 OFF

    starting
     21.00

    LED1 ON
    No tag deteected
    done
    LED2 OFF
```

Fig. 10. Case 3 results   **Fig. 11.** Case 4 results

4   Conclusion and Future Work

In a world, where the field of IoT is rapidly advancing, the need for device inter-communication is very essential. IoT and M2M communication is used in every domain, in vehicles, hospitals, homes and much more. In a few years, the amount of data being sent over to the middleware devices such as the gateway and backend will increase greatly. To facilitate such huge data traffic, we need to integrate the underlying communication protocols to form one single system, instead of multiple fragmented systems, incapable of withholding huge amounts of data. This is what has been achieved through this paper. We have successfully implemented an IoT system that involves M2M communication between sensors using four different communication protocols, namely, Wi-Fi, Bluetooth, Ethernet and serial communication via UART.

Any kind of communication over the Internet is however prone to attacks, like Dos (Denial of Service) attacks, phishing, eavesdropping, etc. Thus, there is a need for additional security along with the security protocol that we have used. SSL allows us to encrypt the data and send, while when any genuine resource wants to access it, can subscribe to the Backend to gain access to it. However, a third party might be able to deny the services to any such resource by theft or misuse of the SSL certificate that the web browser owns. It may also gain access to sensitive data by pretending to be a genuine resource and the results can be alarming. In such a case, by the time any action is taken, the harm might already be done. Thus, further improvements can be made on the implemented system to secure the data communication by issuing new SSL certificates, issued through a trusted certificate chain, to any verified device that wants to communicate with any device existing in the system.

References

1. https://www.finoit.com/blog/enterprise-challenges-in-iot/
2. https://www.gartner.com/it-glossary/machine-to-machine-m2m-communications
3. https://www.peerbits.com/blog/difference-between-m2m-and-iot.html
4. http://www.onem2m.org/
5. https://www.openmtc.org/
6. http://info.ssl.com/article.aspx?id=10241
7. https://www.digicert.com/ssl/
8. Swetina, J., Lu, G., Jacobs, P., Ennesser, F., Song, J.: Toward a standardized common M2M layer platform: introduction to oneM2M. IEEE Wirel. Commun. 21(3), 20–26 (2014)
9. Wu, C.-W., Lin, F.J., Wang, C.-H., Chang, N.: OneM2M-based IoT protocol integration. In: 2017 IEEE Conference on Standards for Communications and Networking (CSCN) (2017)
10. Abdurohman, M., Sasongko, A., Herutomo, A.: M2M middleware based on OpenMTC platform for enabling smart cities solution. In: Lecture Notes of the Institute for Computer Sciences, Social Informatics and Telecommunications Engineering, pp. 239–249 (2015)
11. https://www.w3schools.com/tags/ref_httpmethods.asp
12. http://pages.iu.edu/~rwisman/c490/html/pythonandbluetooth.html
13. https://fiware-openmtc.readthedocs.io/en/latest/reference-doc/gateway-and-backend-configuration/index.html
14. http://www.onem2m.org/images/files/deliverables/TR-0009-Protocol_Analysis-V0_7_0_1.pdf
15. https://github.com/mbenalaya/onem2m-clientarduino/blob/master/src/onem2m-client-2.ino
16. https://www.hackster.io/onem2m/onem2m-demo-57022e

Deep Convolutional Neural Networks for Human Activity Classification

Hamid Aksasse$^{(\boxtimes)}$, Brahim Aksasse, and Mohammed Ouanan

ASIA Team, M2I Laboratory, Department of Computer Science,
Faculty of Sciences and Techniques, Moulay Ismail University of Meknes,
BP 509 Boutalamine, 52000 Errachidia, Morocco
haksasse@gmail.com, b.akssasse@fste.umi.ac.ma,
ouanan_mohammed@yahoo.fr

Abstract. In this work, we investigate the potential of applying deep learning techniques in the field of Visual Lifelogging (VL), which means use a wearable camera to acquire information of an individual. Visual Lifelogging could be interpreted, as a complete and comprehensive black box of human's daily activities. This black box will offer great potential to extract accurate and opportune knowledge on how people live their lives. The sensing technology advent that allowing efficient sensing of personal activities, had led to huge collections of data that are available. The ability to process this data had increased as well. This is well seen in the popularity and growing interest given by the scientific community to the hot field of lifelogging. Using features that separate activities are vital for human behavior understanding and characterization. In this paper, we emphasize more particularly on human activity recognition (HAR) captured by a low temporal resolution wearable camera. To achieve this goal, we have used an already trained Deep Convolutional Neural Network (DCNN). The training is done on the large Dataset ImageNet, which contains millions of images and transfer this knowledge to recognize and classify automatically the daily human activities into one of the categorized activities. The numerical results of the proposed approach are very encouraging with an accuracy of 98.78%.

Keywords: Lifelogging · Daily activities · CNNs · Transfer learning · Activity recognition

1 Introduction

Record our everyday activities is not something new for us. The diary of writing is a way that people used for a long period to record their activities, and this tool has been transmitted from one generation to another for many centuries. With the arrival of information technology everywhere, the model that people currently use to record their experiences and activities begins to change. The digital blogs are one alternative and a new form of the writing journal that became very popular recently. While the traditional writing journal with a pen on a notebook is generally private and intended for one's use, blogs are opposite in the sense that they are generally open to a general public and used to share our experiences, feelings, opinions, comments etc. The

© Springer Nature Switzerland AG 2020
L. C. Jain et al. (Eds.): ICICCT 2019, LAIS 9, pp. 77–87, 2020.
https://doi.org/10.1007/978-3-030-38501-9_7

problem with the writing journal and blogs is that we could record only a small part of our daily experiences and activities by selecting and manually editing the content. The idea of the automatic recording of a life is to record every detail of our daily experiences and activities. The current research area of lifelogging in which the present work is situated tries to answer questions such as: How can we record efficiently every aspect of our daily activities with the advanced devices of recording and sensing? How can we efficiently access the content of such recordings and find or extract useful knowledge from this large volume of information?

Lifelogging is the activity consisting on carrying and wearing over an extended period, of time, a camera (Fig. 1) (or another recording device) in order to capture automatically daily human experiences from the egocentric point of view. When the data captured are only visual signals (images or videos), it is called the visual lifelogging. This idea of the generation of personal digital archives began since 1945 when Bush [1] expressed his vision that our daily activities can be captured and recorded using the technology tools.

We name a person who has chosen to capture, using a wearable camera, his/her daily activities, and experiences so that to create a digital archive of his/her life a Lifelogger.

The vulgarization of cameras used in Lifelogging has expanded the need for the automatic treatment of that large number of images and videos those cameras could produce. These cameras make lifeloggers' daily experiences easily shareable with the rest of the world by using social networks. By discreetly cutting onto the clothes, such a camera will capture photos of our daily life, or record on video what we see in front of us, which can then be connected to the cloud or shared among our favorite social networks or personal blog. The amount of data that these cameras generate has created a huge demand for human activity recognition (HAR). Lifelogging is one of many applications that can benefit from HAR. As a result, we noticed recently that many approaches were proposed for the activities classification task.

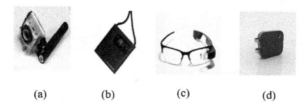

(a)   (b)   (c)   (d)

Fig. 1. A variety of lifelogging cameras

In Fig. 2 below, we show some examples of images taken by a SenseCam camera device.

Fig. 2. Example of images taken by a SenseCam

During a day, a person has 16 h awake on average, which means approximately 960 min a day. Lifelogging cameras (Fig. 1) are usually programmed to grab a picture meanwhile 30 or 60 s. This capturing speed leads to thousands of images captured every week. This huge amount of data gave birth to an imperative need of automation and personalization to the process of images analysis and categorization

Lifelog is referred to the digital archive of data collected, over a long period, using a lifelogging camera. This digital archive provides great potential for digging or/and deducting knowledge about the way human beings experience their lives. [2], hence allowing a large number of utilizations that have emerged. In fact, a group of studies published in the American Journal of Preventive Medicine in a special issue [3] has shown that a huge potential exists within the use of digital archive (lifelog) captured by a wearable camera like SenseCam from several points of view. In particular, it has been proved that the lifelog can be of the most value when used as a tool to comprehend the behavior or the lifestyle of a person. It has also been shown that VL could be employed as a tool for do the re-memory cognitive training, so to speak, that visual lifelogs would enable to prevent the cognitive and functional decrease in aged human beings [4, 5]. In this paper, we propose the use of one deep learning approach based mainly on ConvNets(CNN) [6–8] to recognize activities in the lifelogging applications domain. There are two key advantages when applying CNN to HAR:

Local Dependency: CNN captures local dependencies of an activity image. In image classification problems, typically all the nearby pixels have a powerful relationship with each other. Likewise, in HAR given one activity, the nearby acceleration readings are chances are to be matched.

Scale Invariance: CNN preserves feature scale invariant. In image classification, the training images might have different scales. In HAR, a person may walk or run with different paces (e.g., with different motion intensity).

For the purposes of clarity, we have divided the present paper into six sections as follows: In Sect. 2, we present the related works; Sect. 3 describes our CNN-based method for human activity recognition; Sect. 4 presents the experimental results of the proposed approach to demonstrate its applications, Sect. 5 presents the experiment results of our method. Ultimately, Sect. 6 concludes the paper.

2   Related Works

In this section, we review and explore the related methods on the HAR problem. Specifically, we are about to present human activity recognition based on traditional and deep learning methods. Many interesting applications for lifelogging have recently emerged and then are actively researched. The most important application is the use of VL as a human activity analysis and behavior understanding.

Extracting the features for the HAR is a very important task; this has been studied for many years before. The most commonly used features in literature are statistical like standard deviation, mean, entropy, correlation coefficients, etc.

Since the input data are images and the output is a label of one activity class, we consider the HAR task as an image classification problem.

The literature review in scientific research regarding the subject of HAR using deep learning is very few. Among the first works that ventured in the literature are [9], which have used the restricted Boltzmann machines technique (RBM), and [10, 11], which have both used the slightly different sparse-coding techniques. In fact, all the above mentioned deep learning techniques automatically extract features from time-series sensor data, but all of them use a fully-connected architecture that does not catch the local dependency characteristics of sensor readings [6]. Convolutional Neural Networks, sometimes referred to as ConvNets, were ultimately combined with accelerometer and gyroscope data to address the gesture recognition problem as described in the work by Duffner et al. [12], which have concluded that ConvNets could perform better than other methods of the literature in regard with gesture recognition including Hidden Markov Model (HMM) and Dynamic Time Warping (DTW). The work [13] of Zeng et al., treats one method to automatically achieve the extraction of discriminative features to recognize an activity. More specifically, the authors developed an approach based on ConvNets (CNN). This method can be used to capture local dependency and ensure the scale invariance property of a signal likewise; it has been shown in image recognition and speech recognition domains.

Zheng et al. [14] have applied the technique of convolutional neural networks to deal with the Human Activity Recognition (HAR) task, but the first-mentioned has assessed the problem from the time-series point of view in a general manner. The latter mentioned have used a CNN with only one layer; thus, it ignores the high potential benefit of hierarchical feature extraction. On the other hand, Yang et al. [15] made the application of convolutional neural networks to recognize the daily human activity using a hierarchical model to affirm the eminence in many benchmark problems.

Talavera et al. [16] used the AlexNet [17] CNN as a fixed tool to extract features for image representation. In this work, the authors have presented a new methodology for automatic egocentric video segmentation. The presented method is able to segment low temporal resolution data by global low-level processing. This proposed method uses a graph cut algorithm to separate tentatively the photo streams and includes an agglomerative clustering approach with concept drifting methodology, called ADWIN.

In [18], Daniel Castro et al. presented a technique to analyze images collected by a wearable camera to do the prediction task of daily activities. The authors have used a ConvNet to conduct the classification task where they use a new method called the late fusion ensemble. In this article [18], the authors have treated the problem of classification and prediction of human activities, thus the classification task is performed using a combination of the CNN approach and a random decision forest tree.

The paper presented by Medjahed et al. [19], illustrates a system based on the fuzzy logic applied in activity recognition task by the use of a set of sensors such: microphones, debit sensors, state-change sensors, infrared sensors, and physiological sensors.

3 Activity Recognition

3.1 Convolutional Neural Networks (CNN)

Convolutional Neural Networks (CNNs) [6–8], referred to more commonly, as Deep Learning (DL) techniques are deep neural network architectures used to automatically extract low-level features from input volume data without the need for the traditional machine learning feature engineering. These low-level features are automatically extracted from the input data through the main layers of a CNN that are convolutional layers and pooling layers and then these features are put into a multi-layer perceptron (MLP). This kind of neural network have shown recently that they are now capable of outperforming human capabilities in some computer vision tasks, such as in image classification task. CNNs are inspired by human visual areas and can be described as follows:

- Segmenting the input data into several domains. This process is equivalent to a human's receptive fields.
- Extracting the features from the respective domains, such as edges and position aberrations.

In this present work, we use a famous very deep CNN named VGG16 [20] already pre-trained on a large dataset called ImageNet. Once this is done, we reuse the trained weights to construct our new model to recognize and classify activities. The last layer is implemented using the softmax activation function to output the prediction probability to recognize and determine which activity class belongs to each input image (activity). The graphical model of a CNN is not similar to that of an ordinary neural network. In general, a strong ConvNet contains various feature extraction blocks; more particularly, we find in each block a layer of convolution followed by a non-linear function layer, and then a pooling layer. Here (Fig. 3) is an example of a CNN architecture with four stages of Convolution + Relu + Max-Pooling followed by two fully connected layers and ultimately the output layer:

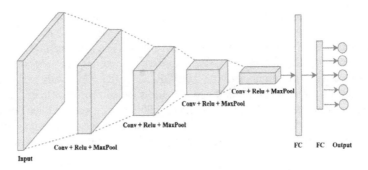

Fig. 3. Example of CNN architecture

Convolutional Layer: This kind of layer forms the vital piece of a CNN, which is the convolution computations where the feature extraction takes place. In these layers, we apply several filters to the input image to extract features. These filters also called kernels, and the output-convolved images are called feature maps. First, the kernels are initialized randomly then they will be updated during the training process. Each output value $h_{i,j}$ is calculated using the following formula (1) where m represents the kernel size, w is the kernel or the filter and x is the input.

$$h_{i,j} = \sum_{k=1}^{m} \sum_{l=1}^{m} w_{k,l} x_{i+k-1,j+l-1} \tag{1}$$

Relu Layer: The main purpose of applying the Relu (Rectified Linear Unit) activation function is to introduce and increase the non-linearity in the model. We apply the function $f(x) = \max(\text{zero}, x)$ as activation function to all of the values in the input volume. In other terms, the Relu layer removes all the negative values and so then make the next computations easy.

Pooling Layer: Once the feature maps are computed during the previous convolution layers. The feature's exact location is no longer important. The max-pooling layers are stacked over convolutional layers. These layers do not train or learn by themselves but they just down sample and reduce images propagated from previous convolutional layers. The preservation of scale-invariant characteristic is considered as another crucial property of convolutional neural networks, which is also ensured by the max-pooling layer computation. In max-pooling layers, we chose the maximum of a matrix of 2×2 or 3×3 for each possible location. So mathematically (Eq. 2):

$$h_{i,j} = \max\{x_{i+k-1,j+l-1} \ \forall \ 1 \leq k \leq m \ and \ 1 \leq l \leq m\} \tag{2}$$

3.2   Transfer Learning

3.2.1   Motivation

In practice, train a CNN from scratch (with random initializations) is not a trivial task, because it is rare to have a large dataset and this becomes a handicap to do sufficient training of the model. To solve this problem, One of the most powerful ideas actually in deep learning is to take the knowledge a Neural Network that has already learned from one task and try to apply this knowledge to another separate but related task. Suppose we have trained a network on image classification using the supervised learning algorithm, so first, we take the neural network and train it on (X, Y) pairs where X represents the input images and Y are the corresponding labels class. During the first phase of training we are training on an image classification task, we train all of the usual parameters within the network all the weights all the layers and now we have something that learns to make image classification. Studies have shown that features learned from huge image dataset, like ImageNet dataset, are highly transferable and could be applied in a variety of image classification tasks. There are several ways that we can use to transfer the knowledge from one model to another. The easiest one is to chop off the top layer of the already trained model and replace it with a randomly initialized one.

3.2.2   Goal

The next step after having trained the network is to swap in a new dataset. Our ultimate purpose is to take the knowledge learned from image classification and apply it or transfer it to HAR. The reason why this process can be helpful is that many low-level features such as detecting edges or curves learned from a very large image classification dataset to recognize and classify the input image into one of a thousand categories might help our learning algorithm to do better in HAR.

First, the network we are using here is already trained using the ImageNet [21] dataset to recognize and classify each input image into one of a thousand possible classes. This will determine all the hyper-parameters of the ConvNet, such as the weights of the convolutional kernels, the stride, and the padding. For our activity recognition task, we will use the VGG-16 [20] network, the following Sect. (3.3) explains the VGG16 in details.

3.2.3   Data Augmentation (DA)

To build a deep learning model with higher accuracy, many training data are necessary. To cope with this limited data problem, we usually apply the data augmentation operation [17] to improve the generalization of CNN networks. Data augmentation occurs to create new data based on random transformations of the existing data. In our case, since the data are images so the DA operation would include operations like flipping the image horizontally or vertically, rotating the image, zooming in or out on the image, cropping the image and varying the color on the image. In general, in the deep learning domain, we apply DA for two purposes. First, when we need to obtain more data for training the model or to minimize the gap existing between the training and the validation error or what is so-called the overfitting problem.

3.3   VGG-16

VGG16 is a deeper CNN with 16 weight layers designed by the Visual Geometry Group [20] and presented in the scope of the ImageNet Large-Scale Visual Recognition Challenge (ILSVRC) [21], VGG16 model contains 16 learnable weights layers: 13 are convolutional layers and 3 are of type fully connected layers. VGG16 is a CNN that have been trained on a massive dataset with more than a million images and can achieve higher accuracy in image classification into 1000 object categories. This model not only achieves excellent accuracy on the ImageNet classification task [21] but it can also apply to other image recognition datasets. Very small matrix of size 3 × 3 (Kernels) are used in every convolutional layer to have the number of learnable parameters reduced as much as possible. We summarized the architecture of the VGG-16 network in Fig. 4.

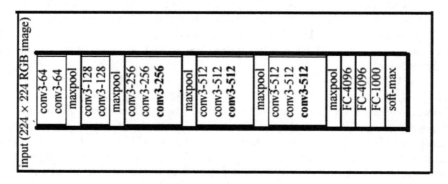

Fig. 4. VGG-16 architecture

4   Experiments and Dataset

We briefly introduce the dataset used in Sect. 4.1 and describe then in Sect. 4.2 the architecture for our CNN model. Then Sect. 5 presents briefly the experimental results for the HAR task.

4.1   Dataset Used

For the HAR purpose, in order to train our CNN model, we perform the experiments on the EDUB Dataset [22]. EDUB is a public Dataset composed of pictures and images captured using a lifelogging camera of type Narrative that takes images in a passive manner in the interval of 30 to 60 s; each image size is 384 × 512. Four persons have committed to acquire The EDUB Dataset. To validate our results, we used the same EDUB dataset. Note that the current paper is a continuity of our previous work cited in [23], which aimed to give an overview of some public egocentric datasets used in the lifelogging field.

4.2   Model Architecture

The architecture of our model is based on the VGG16 network previously trained on the ImageNet [21] dataset used to identify custom classes. More specifically, to apply the transfer learning paradigm, we have replaced the VGG16's fully connected layers by two layers of type fully connected of 256 size. The output layer, implemented using the softmax activation function, outputs whether the input image (activity) belongs to the class activity number one ("Talking_in_phone"), class number two ("Walking") class number three ("Watching_TV"), class number four ("Eating"), Talking and so on. 1600 different training images were used for training each class, and augmented using the data augmentation explained in the Sect. 3.2.3. The goal is to produce more images of each image from the original 1600. We have trained the network over 12 epochs and the results shown above with images completely unseen by the network are shown in the results section.

5   Results

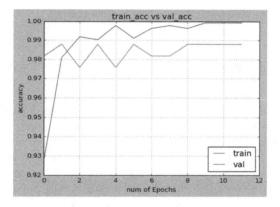

Fig. 5. Train accuracy vs validation accuracy

The Fig. 5 shows the training accuracy curves vs the validation accuracy. The classification accuracy of our Human Activity Recognition model is 98.78% over only 12 epochs. As known, we could increase our model's accuracy by getting bigger the number of epochs as well as the amount of the data used while the training phase, but this will also increase the time needed to train the model. We limited to 12 epochs because we judge that 98.78% is good enough to do the classification task (Fig. 5).

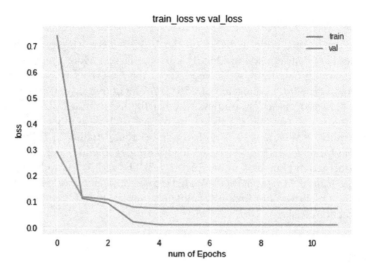

Fig. 6. Training loss vs validation loss

6 Conclusion

We have presented in this paper, a powerful model for human daily activity recognition using the VGG-16; which is a ConvNet already pre-trained using the ImageNet dataset. The accuracy of our model achieved 98.78%. We believe that, this work will be a tool and an evidence to demonstrate the potential of applying Deep Learning techniques on lifelogging in order to characterize and recognize a person's everyday activities. The next version of our system will focus on including more classes of activities as well as more powerful deep learning techniques. In addition, as another future work, we emphasize to apply those deep learning approaches to the human activity-captioning task. Moreover, we expect in the future that connecting the classification with a semantic interpretation will be an interesting achievement.

References

1. Bush - As We May Think (Life Magazine 9 October 1945)
2. Bolanos, M., Dimiccoli, M., Radeva, P.: Toward storytelling from visual lifelogging: an overview. IEEE Trans. Hum.-Mach. Syst. **47**(1), 77–90 (2017)
3. Doherty, R., Hodges, S.E., King, A.C., Smeaton, A.F., Berry, E., Moulin, C.J., Lindley, S., Kelly, P., Foster, C.: Wearable cameras in health. Am. J. Prev. Med. **44**(3), 320–323 (2013)
4. Hodges, S., Williams, L., Berry, E., Izadi, S., Srinivasan, J., Butler, A., Smyth, G., Kapur, N., Wood, K.: SenseCam: a retrospective memory aid. In: Dourish, P., Friday, A. (eds.) Ubiquitous Computing, UbiComp 2006, vol. 4206, pp. 177–193. Springer, Heidelberg (2006)
5. Lee, M.L., Dey, A.K.: Lifelogging memory appliance for people with episodic memory impairment, p. 44 (2008)

6. Lecun, Y.: Gradient-based learning applied to document recognition. Proc. IEEE **86**(11), 47 (1998)
7. Neubauer, C.: Evaluation of convolutional neural networks for visual recognition. IEEE Trans. Neural Netw. **9**(4), 12 (1998)
8. Oliver, N., Horvitz, E., Garg, A.: Layered representations for human activity recognition. In: Proceedings of the 4th IEEE International Conference on Multimodal Interfaces, p. 3 (2002)
9. Plötz, T., Hammerla, N.Y., Olivier, P.: Feature learning for activity recognition in ubiquitous computing. In: IJCAI Proceedings-International Joint Conference on Artificial Intelligence, vol. 22, p. 1729 (2011)
10. Bhattacharya, S., Nurmi, P., Hammerla, N., Plötz, T.: Using unlabeled data in a sparse-coding framework for human activity recognition. Pervasive Mob. Comput. **15**, 242–262 (2014)
11. Vollmer, C., Gross, H.-M., Eggert, J.P.: Learning features for activity recognition with shift-invariant sparse coding. In: International Conference on Artificial Neural Networks, pp. 367–374 (2013)
12. Duffner, S., Berlemont, S., Lefebvre, G., Garcia, C.: 3D gesture classification with convolutional neural networks, pp. 5432–5436 (2014)
13. Zeng, M., Nguyen, L.T., Yu, B., Mengshoel, O.J., Zhu, J., Wu, P., Zhang, J.: Convolutional neural networks for human activity recognition using mobile sensors (2014)
14. Zheng, Y., Liu, Q., Chen, E., Ge, Y., Zhao, J.L.: Time series classification using multi-channels deep convolutional neural networks. In: Li, F., Li, G., Hwang, S., Yao, B., Zhang, Z. (eds.) Web-Age Information Management, vol. 8485, pp. 298–310. Springer, Cham (2014)
15. Yang, J., Nguyen, M.N., San, P.P., Li, X., Krishnaswamy, S.: Deep convolutional neural networks on multichannel time series for human activity recognition. In: IJCAI, 2015, pp. 3995–4001 (2015)
16. Talavera, E., Dimiccoli, M., Bolaños, M., Aghaei, M., Radeva, P.: R-clustering for egocentric video segmentation. arXiv:1704.02809 Cs (2017)
17. Krizhevsky, A., Sutskever, I., Hinton, G.E.: ImageNet classification with deep convolutional neural networks. Commun. ACM **60**(6), 84–90 (2017)
18. Castro, D., Hickson, S., Bettadapura, V., Thomaz, E., Abowd, G., Christensen, H., Essa, I.: Predicting daily activities from egocentric images using deep learning. In: Proceedings of the 2015 ACM International symposium on Wearable Computers, pp. 75–82 (2015)
19. Medjahed, H., Istrate, D., Boudy, J., Dorizzi, B.: Human activities of daily living recognition using fuzzy logic for elderly home monitoring. In: IEEE International Conference on Fuzzy Systems, FUZZ-IEEE 2009, pp. 2001–2006 (2009)
20. Simonyan, K., Zisserman, A.: Very deep convolutional networks for large-scale image recognition. arXiv:1409.1556 Cs (2014)
21. Russakovsky, O., Deng, J., Su, H., Krause, J., Satheesh, S., Ma, S., Huang, Z., Karpathy, A., Khosla, A., Bernstein, M., Berg, A.C., Fei-Fei, L.: ImageNet large scale visual recognition challenge. arXiv:1409.0575 Cs (2014)
22. Bolaños, M., Radeva, P.: Ego-object discovery. arXiv preprint arXiv:1504.01639 (2015)
23. El Asnaoui, K., Aksasse, H., Aksasse, B., Ouanan, M.: A survey of activity recognition in egocentric lifelogging datasets. In: 2017 International Conference on Wireless Technologies, Embedded and Intelligent Systems (WITS), Fez, Morocco, pp. 1–8 (2017)

A Model to Forecast the Student's Grade and Course Recommendation: A Case Vietnamese Students

Minh Tu Nho[1], Hoa-Huy Nguyen[2], Ton Quang Cuong[3],
and Viet Anh Nguyen[1(✉)]

[1] VNU University of Engineering and Technology,
Vietnam National University, Hanoi, Vietnam
{15021038,vietanh}@vnu.edu.vn
[2] VNU Center for Education Accreditation,
Vietnam National University, Hanoi, Vietnam
huynguyen@vnu.edu.vn
[3] VNU University of Education, Vietnam National University,
Hanoi 144, Xuan Thuy, Cau Giay, Hanoi, Vietnam
cuongtq@vnu.edu.vn

Abstract. This paper presents a model for forecasting the learning outcomes and suggesting the recommend courses for undergraduate students. In this research, we propose methods based on machine learning techniques and recommender systems to answer three research questions of forecasting student's outcomes problem: (1) How to forecast the course's grade for the next subject. (2) How to predict the final grade point average basing on the subjects students have studied. (3) How to suggest a list of subjects that students should learn in the next semester. The model has been tested with the grade data in 22-course subjects with the participation of 580 students. The results, in the best case, for predicting the missing subject grades with the tested data set were 0.656, a good result for scores in the range [0, 10]. Assessing the user's satisfaction of the model through the survey, the results show that 68.2% of students think that the system is useful for them.

Keywords: Learning outcomes · Recommender systems · Learning analytics · Blended-learning

1 Introduction

Recently, one of the highly concerned issues of higher education is to personalize learning activities to suit each learner to achieve the highest efficiency. To gain an insight into the issue, this study was conducted at the University of Engineering and Technology, Vietnam National University, Hanoi, in Vietnam with a scale of about 3,500 full-time students of 4-year training, in which several blended-learning courses have been implemented. One of the problems causing difficulties for students participating in blended-learning is that they do not have the tools to support counseling, forecasting results, as well as assessments when they participate in the learning process.

© Springer Nature Switzerland AG 2020
L. C. Jain et al. (Eds.): ICICCT 2019, LAIS 9, pp. 88–97, 2020.
https://doi.org/10.1007/978-3-030-38501-9_8

According to some studies, early identification of students with unexpected academic results and the timely warning will help the learners to significantly improve results, especially when they are participating in online learning, the environment lacks close supervision compared to face training [1]. To predict the learning outcomes of the courses and give warnings to students who have not achieved good results so that the learners have an overview of the current situation. On the other hand, for students who are in the process of achieving good academic results, the GPA (Grade Point Average) prediction can help students adjust and have early preparing solutions.

In the training program, beside the compulsory subjects, there also includes a lot of selected subjects. To select the subjects that are suitable for each learner, which equip knowledge of specialized areas as well as are possible to achieve high results, is a problem for every student when enrolling for the course. Therefore, it is necessary to provide students enrolling in the next semester a list of suggested subjects, especially for those who do not have complete information about the content of the training program.

This study presents the results of building and testing the learning outcomes forecasting model and suggesting the learning path based on the educational data mining application [2] to answer the following three research questions: (1) How to forecast the course's grade for the next subject. (2) How to predict the final grade point average basing on the subjects students have studied. (3) How to suggest a list of subjects that students should learn in the next semester. In the next section, we review some related studies. The proposed model is described in detail in Sect. 3. Section 4 of the paper presents some results obtained. Then the last section is some discussions as well as implications and conclusions.

2 Literature Review

The application of machine learning techniques in predicting learning outcome problem is currently being considered. The research [3] proposes a system suggests students' grades and provides a list of subjects that they should register based on their competency, meeting their qualification requirements and the ability to acquire the subject well based on background knowledge. The results showed that the suggested subject based on kNN method is more accurate than the clustering method. The test results with different repeating conditions showed that the large data set (about 50,000 students) had the best root mean square error (RMSE) results.

Saarela et al. [4] used a neural network to analyze the students' learning results based on the sparse data sets during the period of 2009–2013, the study results showed the ability to study of learners was one of the factors to predict their learning outcomes. Sweeney et al. [5] conducting experiments with 33,000 students using a combination model of Factorization Machines and Personalized Multi-linear Regression to predict student learning results for the next semester.

In the research [6], Matrix Factorization (MF) technique is used in predicting students' performance in the e-learning system. The MF method allows searching for the hidden coefficients and models suitable for each learner. In addition, because students' performance changes over time, the author uses tensor factorization to re-calculate the coefficient of the matrix.

Iqbal and colleagues [7] applied machine learning techniques to predict students' grades. The techniques used are Collaborative Filtering (CF), MF, and Restricted Boltzmann Machines (RBM) with the actual point data set of the students at Information Technology University (ITU), Lahore, Pakistan. The results showed that the initial test score affects the student's cumulative GPA. The study also showed that the RBM method giving the best results. CF is a popular method to predict learning outcomes by its simplicity. However, because the methods depend on students' data in the past, they are not effective for sparse data sets.

Zimmermann et al. [8] conducted an analysis of the factors affecting the quality of graduate students. The authors used 81 indicators of 171 students from the master's and bachelor's degree in computer science programs. In their research, the authors studied regression models associated with variable selection and are embedded in a two-layer cross-validation round. The results showed that the performance of the learning process can determine 54% for the result of final graduation. In particular, the academic results of the third year had the greatest impact on the final grade. In addition, the research results also provided an additional assessment to help employers.

El Badawy [9] used various algorithms to predict students' learning performance, determine who is at risk of expulsion as well as must repeat to learn the course. The study used both the personalized multi-regression and the matrix factorization method to evaluate the learner performance outcomes in the future. The results showed that Factorization Machines gave the best results.

3   Methods

3.1   Participant and Data

Data is extracted from the learning management system, including the scores of 580 students who are studying information technology major for 4 years from 2015 to 2019. Some descriptive statistics showed in the following Tables 1 and 2:

Table 1. The number of students with the enrolled courses

Number of the enrolled courses	Number of students
Less than 5 courses	51 (8.8%)
From 5 subjects to 10 subjects	247 (42.6%)
More than 10 subjects	282 (66.2%)

Table 2. Percentage of the subjects is missing grade values

Percentage of the subjects is missing grade values	Number of the courses
<10%	2
≥ 10% and <50%	5
>50% and <79%	15

3.2   The Model for Forecasting the Student Learning Outcomes

The model of learning outcomes forecasting as well as the learning path suggestion is carried out through the following basic steps: (1) Extract data from the learning management system; (2) Normalize data; (3) Forecast the grade of the course; (4) Forecast the GPA. Next, we will present some steps of the learning outcome process in detail as well as suggestions of the learning path for students in the next semester.

3.2.1   Data Normalization

In a model, with a better - standardized data set, it improves the model results clearly. Therefore, it is necessary to standardize the data set to get the best results when implementing the forecasting model.

When observing the matrices (students, points) collected from the LMS, we see that there are many subjects that do not have scores (the subject with the most missing score is 80% missing values). So how much do we need to fill in the missing point value? In this model, we consider to choose one of the three methods: (1) The method of filling the missing value is '0', (2) The method of filling the missing value is the average value, (3) The method of subtracting the average.

The Method of Filling the Missing Value is '0': The method of filling missing values is '0' very popular. We can see that it means nothing, or the value to predict. On the other hand, considering the technical aspect, the matrix will not store cells that are worth '0', thus optimizing memory usage. Especially with large data sets, memory optimization is very important. However, for the actual data set collected in this research, there are many subjects have the score '0', the value '0' is the lowest score on the '10' scale or the learner who has learned but with no satisfaction (students do not like that subject). So filling '0' into the missing value leads to an unclear data set. Otherwise, the data after normalization does not clearly distinguish a subject '0' from which the student does not pass the subject (dislike) or the subject has not yet studied (equivalent to students who do not like or dislike). Therefore, a value of '0' is not reasonable.

The Method of Filling the Missing Value is the Average Value: With the scale used in this study, the average score is '5'. Should we fill in the missing value '5'? Firstly, we also have the same problem as when entering the value as '0', because the value of '5' does not indicate correctly that the subject is missing points. Secondly, the value of '5' does not clearly indicate whether it is good or not for each different student. Specifically, for some difficult subjects, to meet the requirements of the subject is the goal that students who have normal learning ability desire. In other words, the subject has scored '5' value is an acceptable point (it means almost like the learners are satisfied, they like that subject). However, it is not favored by other students who expect good results to get scholarships. For example, the requirement is that the subject grade point must be at least '7'. So, in this case, '5' point is not satisfactory to them (or equivalent to not like that subject). So it is not reasonable to fill the average value with this data set, it does not indicate that the subject has not been studied.

Average Deduction Method: We can see that the matrix obtained after subtracting the average value is a new matrix consisting of positive and negative values. Here the value is positive for students who like that subject and in contrary, the result is negative in the case of the students who are not interested in the subject. Because each different learner may expose to different average point values. Therefore, this method overcomes the disadvantage that each learner has a different level of assessment of interest in the course subject. Negative and positive values are obtained to express the meaning of whether the student is good at the subject or not. Technically, the actual result data set, the user-item matrix is often very large. Memory capacity is often not big enough to store all values. In fact, the number of known ratings is usually very small compared to the user-item matrix size, so we often store matrices in the form of a sparse matrix (it only save non-zero values and their positions). Through analyzing the advantages and disadvantages of the methods, we selected the average subtraction method for data normalization.

3.2.2   Forecast the Student's Learning Outcomes

To predict the grades of the courses, we use the Collaborative Filtering (CF) method. There are two different approaches: user-user, item-item. The steps of these two methods are the same, the difference is that the item-item method calculates the correlation between items, the user-user method is to calculate among users. In this article, we present a specific point prediction method based on the item-item method.

In the item based method, to calculate the correlation between items, in this model use the cosine formula.

$$cosine\_similarity(a, b) = cos(a, b) = \frac{a^T \times b}{||a||_2 \times ||b||_2} \tag{1}$$

Applying cosine_similarity function to the calculation for item-item CF, we have the formula to get the similarity of the two vectors as follows:

$$sim(a, b) = \frac{\sum_{u \in U} (r_{u,a} \times r_{u,b})}{\sqrt{\sum_{u \in U} r_{u,a}^2} \times \sqrt{\sum_{u \in U} r_{u,b}^2}} \tag{2}$$

With a and b correspond to two vectors, *item1* (first subject), *item2* (second subject). The value of $r_{u,a}$, $r_{u,b}$ corresponds to the attributes of the item respectively. Applying the formula on calculating the correlation level on all subject data, we obtain the item-item matrix, values in the matrix corresponding to the level of item similarity in the corresponding column and row. After calculating the item-item similar matrix, the next step is to predict the subject's score. The formula is as follows:

$$\hat{y}_{i,u} = \frac{\sum_{u_j \in N(u,i)} \hat{y}_{i,u_j} sim(u, u_j)}{\sum_{u_j \in N(u,i)} |sim(u, u_j)|} \tag{3}$$

Considering the following example, part of the similarity matrix values of a student has id 15021790, as shown in Fig. 1.

	15021790	int1006
int1050	0.97	0.366617
int2202	-0.47	0.386896
int2203	2.86	0.414870
int2204	-4.81	0.343916
int2205	1.62	0.404636
int2209	0.84	0.349245
int3401	2.04	0.306887
mat1093	2.79	0.342839

Fig. 1. A part of the similarity matrix values of a student with id 15021790

The second column (int1006) indicates the similarity of the subject with the ID is 'int1006' for the subjects in the corresponding rows. In this example, the taken threshold of similarity values are greater than or equal to '0.3' proving that the subject has been taken by students (it had grade). There are 8 other subjects that meet the above conditions.

After we have found a list of subjects that meet the criteria, we need to find a grade that has been taken by students (here, students with the ID is '15021790'). The results of students with 8 subjects are listed in the first column (15021790), the score of that cell corresponds to the row is the corresponding subject. Applying formula (3) to predicting student grades with the code '15021790' and subject 'int1006', it has:

$$Predict\ Rate = \frac{0.97 * 0.37 + (-0.47) * 0.39 + \ldots + 2.79 * 0.34}{0.37 + 0.38 + \ldots + 0.34} = 0.77$$

The value of the predicted rating for 'int1006' with students with '15021790' code is '0.77'. This is the predicted value for the matrix (student, subject, point) after normalization. To give a predictable result on a 10-point scale, make the pred point plus with the grade point average. In this example, the average score of 'int1006' is '7.59'. So the predicted result on a scale of '10' resulted form using the formula is 0.77 + 7.59 = 8.36.

3.2.3   Forecast the Student's Grade Point Average

The process of predicting students' GPA is conducted through the following steps: (1) Calculating the similarity between students; (2) Selecting the top-k students with similarities closest to students needing to predict GPA; (3) Making a grade point prediction based on the nearest top-k student.

For example, applying the above formulas for matrix (students, subjects, and grades) obtained the similarity matrix of (students, students) with 580 × 580 - dimensional numbers corresponding to student numbers as described in Fig. 2.

```
MaSv        15022360   15022083   15021317   ...  15021751   15022786   15021988
MaSv                                          ...
15022360    1.000000   0.312940   0.122295   ... -0.747210   0.197527   0.001548
15022083    0.312940   1.000000   0.391410   ... -0.286237   0.000834  -0.012335
15021317    0.122295   0.391410   1.000000   ... -0.280978  -0.098971  -0.076739
...           ...        ...        ...       ...    ...        ...        ...
15021751   -0.747210  -0.286237  -0.280978   ...  1.000000   0.062672   0.018759
15022786    0.197527   0.000834  -0.098971   ...  0.062672   1.000000  -0.336043
15021988    0.001548  -0.012335  -0.076739   ...  0.018759  -0.336043   1.000000
```

Fig. 2. The similarity matrix among the students

To predict a student's GPA, we are interested in the top k students to have simi-larity value is similar to that student. Then, we get the GPA of top-k students. In the example, for top-10 students have similarity values with the student with ID '15021790' is describe in Fig. 3. The predicted GPA for the student with ID '15021790' is in the range of between [6.65, 7.89].

```
                    15021790   GPA
         MaSv
         15021790   1.000000   7.53
         15021779   0.595576   7.49
         15021957   0.562609   5.64
         15021362   0.543263   6.65
         15021804   0.527748   7.08
         15021120   0.523788   7.89
         15021845   0.516756   5.17
         15022829   0.516615   6.49
         15021150   0.489046   6.18
         15021603   0.486534   5.86
```

Fig. 3. The GPA of the top 10 students who have similarity values with the student (15021790).

3.2.4 Suggestions the Subjects to Study in the Next Semester

The recommended subjects to students for the next semester are provided in form of a list of subjects to study in the next semester, after obtaining the list of top-k students with the closest similarity to students need suggestions. The next step is to collect grade data of the subjects that the top-k students have completed. Then, the grade point average of the top-k students will be calculated. Based on the list of top-k subjects with GPA, the subjects with the highest GPA are to be suggested.

For example, considering a student coded 15021790, through the implementation of the model, identify a list of 33 other students with a high similarity threshold (0.3). When calculating the average score of these students and calculating the top-k (for k = 20), in terms of the similarity of these students, it is easy to select the four most successful subjects (int3110, int1006, mat1042, and int2208).

4   Results

We use the Root Mean Square Error (RMSE) method to evaluate the results of the model. The sim value is greater than or equal to 0, the student data set has a minimum score of k subjects in Table 3.

Table 3. RMSE with the sim value ≥ 0

K	0	20	19	18	17	16	15
RMSE (item-item)	0.878	1.162	1.237	1.199	1.189	1.196	1.17
RMSE (user-user)	0.771	1.227	1.173	1.208	1.199	1.226	1.244

The sim value is greater than or equal to 0.3, the student data set has a minimum score of k subjects in Table 4.

Table 4. RMSE with the sim value ≥ 0.3

K	0	20	19	18	17	16	15
RMSE (item-item)	0.656	1.17	1.104	1.108	1.056	1.045	1.003
RMSE (user-user)	0.657	1.167	1.104	1.108	1.056	1.045	1.003

The results show that in both methods, taking the *top-N* value with the condition of sim greater than or equal to 0 and greater than or equal to 0.3 with RMSE is the best. The values, in turn, are 0.878, 0.771 with sim $\geq$ 0 and 0.657 and 0.656 with sim 0.3. It can be seen that with different filter conditions for different results, with sim 0.3, RMSE results are better. And with the CF method, the larger the data set, the better the RMSE result is.

For the test data set, because the number of students is much larger than the number of subjects (in this case, the number of students is about 25 times the number of subjects). The item-item method gives better results because there are more data points of the subjects, so the value of calculating the same level gives better results. However, for the small student data set, the obtained RMSE is not much different. For example, with the case k = 19, item-item method obtains better RMSE (item-item 1,237 and user-user 1.1173).

5   Discussion

Our research results show that the CF method obtains good results for the problem of predicting students' learning outcomes. However, the method proves to be effective with sparse data sets with large numbers of students. For the student data set after filtering with conditions with minimum points of different subjects, the model results are not highly effective.

To improve the performance of the model, sample data should be larger or should involve a large number of students. For the item based method, the more the subject is enrolled, the better the expected outcome. In addition, it is necessary to develop a set of rules, relationships among subjects, such as statistics of data sets, calculating the probability of subjects with the condition that students have learned some subjects of the curriculum.

Another method is to identify some additional attributes of students, such as targeted subject knowledge that the student should acquire, the list of subjects should be taken. Some additional information can be considered such as which year the student is studying that subject; the prerequisite and the relations of the subjects; such as general subjects, specialized subjects, student performance, and the placement test scores. For making a list of suggested subjects, it is necessary to collect data during the four-year study period as well as the student's academic results to suggest appropriate courses for the learners.

The study demonstrates that the recommender system techniques can be applied, in particular, the collaborative filtering method applying to the learning outcome prediction system. In the best case, the results for missing subject prediction in the student's academic program with the test data set is 0.656, a pretty good result for scores in the range [0.10]. For the worst case, the RMSE value for the subject's score prediction is 1,244 which is an acceptable result. From the missing predictions for each student, we can calculate the predicted GPA, the predicted GPA at different times, to give students an overview of the learning process.

6 Conclusion

This research has developed and implemented a model based on the recommender system to forecast the student's learning outcomes. The forecast results are obtained by analyzing the scores of the subjects studied by the student and those of other students who have been participating in the subjects. The forecast results obtained at each specific time are very important. For learners, the course score is one of the factors to calculate GPA. For the forecast the result of the following subject, students can see the results in the future, particularly in the next semester, so that students have more information about possible learning situation in the future. These results also help students prepare and adjust learning methods to get the best results.

Although the model has only been tested on a small scale, the research's results answer questions that students want to learn the subjects when they know: how to predict their learning outcomes; how to choose from a list of subjects in the next semester. This also places a base ground for our further study, as we could apply to larger-scale training institutions in the future. In addition, we will improve the predictive model more effectively by adding information such as the relations among between the courses, the training curriculums, as well as the personal information of the learner model.

References

1. Aud, S., et al.: The condition of education 2013 (2013)
2. Romero, C., Ventura, S.: Educational data mining: a review of the state of the art. IEEE Trans. Syst. Man Cybern. Part C: Appl. Rev. **40**, 601–618 (2010)
3. Elbadrawy, A., Karypis, G.: Domain-aware grade prediction and top-n course recommendation (2016)
4. Saarela, M., Ark Ainen, T.: Analysing student performance using sparse data of core bachelor courses. J. Educ. Data Min. **7**, 3–32 (2015)
5. Sweeney, M., Lester, J., Rangwala, H.: Next-term student grade prediction. In: Proceedings - 2015 IEEE International Conference on Big Data, IEEE Big Data (2015)
6. Thai-Nghe, N., Drumond, L., Horvath, T.: Matrix and tensor factorization for predicting student performance, pp. 69–78, June 2014
7. Iqbal, Z., Qadir, J., Mian, A.N., Kamiran, F.: Machine learning based student grade prediction: a case study 1–22 (2017)
8. Zimmermann, J., Brodersen, K.H., Heinimann, H.R., Buhmann, J.M.: A model-based approach to predicting graduate-level performance using indicators of performance. J. Educ. Data Min. **7**(3), 151–176 (2015)
9. Elbadrawy, A., Polyzou, A., Ren, Z., Sweeney, M., Karypis, G., Rangwala, H.: Predicting student performance using personalized analytics. Computer **49**, 61–69 (2016)

An Overview of Adopting Cloud Computing in E-Learning Systems

Thamer Al-Rousan[1(✉)], Bassam Al-Shargabi[2], and Faisal Alzyoud[1]

[1] Faculty of Information Technology, Isra University, Amman, Jordan
{thamer.rousan, faisal.alzyoud}@iu.edu.jo
[2] Faculty of Information Technology, Middle East University, Amman, Jordan
bshargaabi@meu.edu.jo

Abstract. Education institutions continue to hunt for chances of enhancing the methods of operating and managing their resources. Cloud computing is a modern technology which promises to present opportunities for delivering a wild range of computing services such as software, platform, and infrastructure; in a way that has not been practiced before. Integrating cloud computing into e-learning systems has several advantages and challenges. In this study, we will analyze the effect of cloud computing on e-learning and explore the main factors that made cloud computing to be an attractive method to e-learning systems.

Keywords: Cloud computing · E-Learning systems · Educational institutions

1 Introduction

E-Learning systems are defined as arranged education processes using computer technologies such as networks, Web, video, audio, etc. to accomplish a new way of learning. According to Vaquero [1], the description of e-learning: "E-Learning uses the Internet or other digital content for learning and teaching activities, which takes full advantage of a modern educational technology provided with a new mechanism of communication and resource-rich learning environment to achieve a new way of learning". Based on this description, the learner utilizes the internet for gaining learning experiences and knowledge, improving achievement, increasing connection with other learners, getting support from other educators and learners.

E-learning changed the traditional style of education and its environment which is based on the physical connection. Via E-learning, Learners can be connected to the educator and send and receive the educational contents without the restrictions of time and space. E-learning is an effective and fast way to share knowledge with learners at any time, any place and anything [2].

Despite the benefits of e-learning systems, numerous educational institutions do not have adequate resources and infrastructure to execute the latest e-learning technology. In addition, the new applications releases of modules are cloud computing oriented [3]. As a result, cloud computing is becoming a main part of e-learning systems.

Cloud computing (CC) is a computation model where many IT resources are accessible to the end users via the internet and are offered as services in an adaptive way. CC and e-learning are two varied paradigms. Their combination enables several

L. C. Jain et al. (Eds.): ICICCT 2019, LAIS 9, pp. 98–102, 2020.
https://doi.org/10.1007/978-3-030-38501-9_9

advantages. E-learning can benefit from the unlimited resources and abilities which are offered by Clouds. In the same way, Cloud can profit from e-learning by expanding its range to cope with the new paradigm and to propose new benefits in an extensive variety of real-life scenarios [4].

The increased popularity of employing CC in e-learning is associated with several benefits and challenges. Although several studies have been founded in literature for the usage of CC in e-learning, there are also several issues to be addressed which make their integration efficiently feasible.

The value of cloud computing and the way it benefits e-learning have already been observed by different institutes. This study analysed the deep effects of CC on e-learning systems, as this study will answer its chief question which is: "what are the impacts of using cloud computing on the e-learning system?" The rest of the study is prearranged as follow: Sect. 2, presents the current challenges of e-learning systems. Section 3, gives an overview of CC. While the benefits of using CC are discussed in Sect. 4, followed by the conclusion in Sect. 5.

2 Current Challenges of E-Learning Systems

Compared with traditional classroom-based learning, e-learning offers many benefits such the availability of learning material at anytime and anywhere, the ease of updating the educational material, adding new facilities such as multimedia content for the convenience of the learners, decreasing the learning cost; as there is no need for a physical environment, and offering a friendly and easy framework concept that can be seen as learner-centered style [5].

Despite the benefits offered by e-Learning, there are some weaknesses that must be concentrated on before incorporating e-Learning into the education field. At present, e-Learning models have weaknesses in their infrastructure level. An up-to-date scalable infrastructure is necessary as adding a new content, which frequently includes multimedia, needs a high aspect resource, such as storage and bandwidth. Many institutes suffer from high workloads which make the expense and resource management very costly [6]. In addition, many education institutions are facing difficulties to import ready content into the E-Learning system because this needs to be prearranged earlier by taking into account the needed infrastructure [2].

This main issue is also related to the shortage of maintenance and technical support. The data need to be renewed and the software and hardware need frequent maintenance. Several institutions lack the capabilities to take care of their infrastructure and to react to problems, such as security threats or network outages. In addition, there is a high cost related to technical support, computers, building, and maintenance [7].

3 Cloud Computing

CC is supported by the idea of sharing resources over the internet to achieve consistency and cost-cutting scale. It is a technology revolution that enables hosted services to be delivered to the end user over the Internet based on the end user's demand.

According to [8] CC is described as: "a new style of computing in which dynamically scalable and often virtualized resources are provided as a service over the Internet." NIST described CC as: "a model for enabling ubiquitous, convenient, on-demand network access to a shared pool of configurable computing resources (e.g., networks, servers, storage, applications, and services) that can be rapidly provisioned and released with minimal management effort or service provider interaction" [9].

Based on the nature of the resources delivered to the end user, there are three major service categories of CC [10]:

- "Infrastructure as a service (IaaS)": When a service provider provides computing infrastructure to end users. The infrastructure includes: network connections, bandwidth, virtual server space, load balancers, and IP addresses.
- "Platform as a service (PaaS)": When a service provider supplies platform and environment to end users to allow them to unfold applications and services.
- "Software as a service (SaaS)": When "service supplier hosted computer programs" are made accessible online to the end user.

Depending on the level of resources control by the end users, four models of cloud computing are essentially available: private, public, hybrid, and community [9]. A private model is specially created for one particular company; where the computing resources can be controlled, operated and owned by the same company, and not sharing computing resources with any other companies. When the computing resources are shared by several numbers of collaborated companies which have common concerns, it is referred to as community cloud. In this case, the control, operation and the property are shared amongst the collaborated companies. In the public cloud, computing infrastructures are provisioned for open by end users. Finally, when more than one cloud ("private, community, and public clouds) work collectively; it is referred to as a "hybrid cloud." In hybrid, private, community, and public clouds are separate, but they are linked by interfaces that permit sharing or deploying other clients' data, functions or computational resources.

4 Cloud Computing for E-Learning

According to CC features, like: "on-demand self-service, resource pooling, pay-per-use, or broad network access," CC can be seen as a very attractive paradigm for academic institutions. With CC, the academic institutions can concentrate more on research activities and teaching, rather than wasting their time, effort and money on the complex tasks of creating and hosting IT resources and software systems. Besides, the complexity can be decreased with CC; as cloud features can be used to build a cooperative learning society, and to make collaboration amongst the research groups more attractive and enjoyable [11].

CC can offer education institutions a huge range of modern scalable infrastructures such as bandwidth and storage. Cloud offers easy to use platforms for learners and teachers from anywhere; leading to enhanced readiness [12].

The biggest benefits of cloud computing are cost and resource saving; as the virtualization practices; hardware and software resources can be shared amongst all end users efficiently. CC offers convincing savings in regard to IT resources, which include: less execution and maintenance cost; collaboration amongst users; less software and hardware being required to be bought, reducing power cost, ground space and storage; due to shifting all computing resources to a cloud supplier; in addition to decreasing the cost of operation; and adopting a pay- per -use policy [13]. The advantages of using CC are summarized in Fig. 1, ranked from the top occurrence to the lowest occurrence [14].

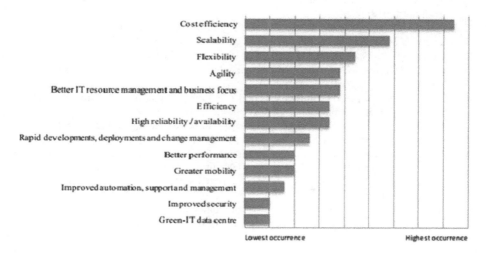

Fig. 1. The main benefits of CC on e-learning systems

5 Conclusion

CC is a modern technology which can provide suitable, accessible, and on-demand access to a wide range of computing resources and infrastructures which can be employed with minimum overhead management and high efficiency. Adopting CC enables the e-learning systems to provide an efficient educational environment to their learners and teachers. In the current study, we analyzed the impact of integrating CC into e-learning systems. Consequently, to answer the question "what are the effects of implementing CC on the e-learning system?" we concluded that CC is an attractive method to provide e-learning services. It offers a lot of benefits to both end users and educational institutions.

Acknowledgments. We would like to convey our thanks to the School of Information Technology, Isra University, Jordan, for providing a conducive environment during the course of our research.

References

1. Vaquero, L.M.: EduCloud: Paas versus Iaas cloud usage for an advanced computer science course. IEEE Trans. Educ. **54**(4), 590–598 (2011)
2. Al-Zoube, M., El-Seoud, S.A., Wyne, M.F.: Cloud computing based e-learning system. Intl. Arab J. e-Technol. **8**(2), 58–71 (2017)
3. Al-Rousan, T.: Cloud computing for global software development: opportunities and challenges. In: Transportation Systems and Engineering: Concepts, Methodologies, Tools, and Applications, pp. 897–908. IGI Global (2015)
4. Buyya, R., Broberg, J., Goscinsky, A.: Cloud Computing: Principles and Paradigms. Wiley, New York (2018)
5. Dong, B., Zheng, Q., Qiao, M., Shu, J., Yang, J.: BlueSky cloud framework: an e-learning framework embracing cloud computing. In: Jaatun, M.G., Zhao, G., Rong, C. (eds.) Cloud Computing. LNCS, vol. 5931, pp. 577–582. Springer, Heidelberg (2018)
6. Ercan, T.: Effective use of cloud computing in educational institutions. Procedia Soc. Behav. Sci. **2**(2), 938–942 (2016)
7. Al-Rousan, T., Al Ese, H.: Impact of cloud computing on educational institutions: a case study. Recent Patents Comput. Sci. **8**(2), 106–111 (2015)
8. Hu, Z., Zhang, S.: Blended/hybrid course design in active learning cloud at South Dakota State University. In: 2nd ICETC, vol. 1, pp. 63–67 (2017)
9. Al-Rousan, T., Al-Zobadi, A., Al-Haj Hassan, O.: The roles of decisions making and empowerment in Jordanian web-based. J. Web Eng. **15**(5&6), 469–482 (2014)
10. Hurwitz, J., Bloor, R., Kaufman, M., Halper, F.: Cloud Computing for Dummies. Wiley, New York (2010)
11. Alhadidi, B., Al-Rousan, T.: Utilization of computer and information technology by physicians. Int. J. Comput. Acad. Res. (IJCAR) **6**(5), 31–37 (2017)
12. Al-Rousan, T., Aljawarneh, S.: Using cloud computing for e-government: a new dawn. In: Advanced Research on Cloud Computing Design and Applications, pp. 15–23. IGI Global (2015)
13. Wheeler, B., Waggener, S.: Above-campus services: shaping the promise of cloud computing for higher education. EDUCAUSE Rev. **44**(6), 52–67 (2009)
14. Sulistio, A., Reich, C., Doelitzscher, F.: Cloud infrastructure & applications – CloudIA. In: Jaatun, M.G., Zhao, G., Rong, C. (eds.) Cloud Computing. LNCS, vol. 5931. Springer, Heidelberg (2009)

Hybrid Feature Selection Algorithm to Support Health Data Warehousing

Md. Badiuzzaman Biplob[1,2(✉)], Shahidul Islam Khan[1,3],
Galib Ahasan Sheraji[1], and Jubayed Ahmed Shuvo[1]

[1] Department of CSE, International Islamic University Chittagong,
Chattogram, Bangladesh
biplob.cse45@gmail.com, nayeemkh@gmail.com,
galibahasan@yahoo.com, jubayedsr@gmail.com
[2] Department of CSE, Chittagong University of Engineering and Technology,
Chattogram, Bangladesh
[3] Department of CSE, Bangladesh University of Engineering and Technology,
Chattogram, Bangladesh

Abstract. Large volumes of data are being generated each day in healthcare. In addition, these huge amounts of data from healthcare datasets cause the issue of proper knowledge discovery. Currently, data integration is an approach, which is increasingly utilized by healthcare data specialists for analyzing the information and data mining. "Which features or attributes should we use to integrate data for data warehouses"-is a difficult question to answer. It requires deep knowledge of the problem domain. Automatic feature selection is the process of selecting a subset of relevant features automatically for later use. In this paper, we proposed a method using four random forest based feature selection algorithm and domain knowledge. Experimental results show that our hybrid method can select a required number of features from a large set of attributes.

Keywords: Hybrid feature selection · Feature selection · Data integration · Data warehouse

1 Introduction

As of late, data becomes bigger and speedier each day which makes it challenging for data researchers to separate and decipher the unpredictability of data in order to discover new information. In addition, a similar issue is also looked by health data system whether its data contain a large amount of health knowledge [1]. This is extremely helpful for the symptomatic motivation to help medicinal services establishments in explaining different social insurance inquire about issues [2].

In most healthcare data mining, the best way to enhance the data quality is by ignoring a sector, which may give better output close to our expectations. Rather, the majority of them just focused on choosing reasonable learning calculations. For example, Social insurance dataset contains many features that can affect the expected execution. In a similar way, the issue of the large data includes if it's being found in high spatial of learning and great process differing nature. Attribute selection, which

© Springer Nature Switzerland AG 2020
L. C. Jain et al. (Eds.): ICICCT 2019, LAIS 9, pp. 103–112, 2020.
https://doi.org/10.1007/978-3-030-38501-9_10

projected by Singh et al., implies the strategy together with the collection of set of features from particular attributes [5, 6]. In any case, the most insightful models are keeping up a strategic way from the reasonable procedures for choosing the best feature. Accordingly, the probability thickness limit of the component vector space is lacking in the midst of the grouping task [7].

Understanding the standing of creating the class arrangements of data, the objective of this examination is to suggest a system based on the hybrid attribute decision methods which can redesign the desired model of patients data for grabbing a superior characterization result.

2   Related Works

Healthcare data Mining (HDM) is a technique for separating useful data and models from a large number of information [2]. This will be engaged around the techniques of delivering quality data. The technique for making efficient data is basically declaring the learning quality that can impact the arrangement procedure [8]. Data quality is frequently delivered in the midst of the preparing ready stage in record mining strategy. Hence, the part-alternative method is regularly used to require care of high dimensional data issue.

Feature selection includes four straightforward steps that set age, set analysis, stopping standard and result approval [4]. Initially, in the set age, the new applicant of feature subsets is conveyed using given procedure. Around then, the competitor subsets assessed and differentiated and past subsets in light of assessment rule. The past grasp sets will be eliminated if the last made subset is ideal. The two methodologies reiterated to the reason once a stopping model is fulfilled. Ultimately, the endorsement procedure on genuine information is foreseen to approve the best-chose grasp subsets. The Plain Steps of feature Assortment process (see Fig. 1).

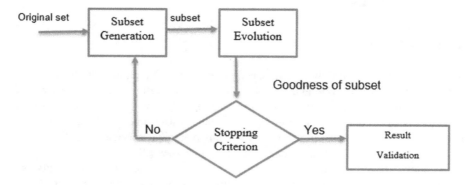

Fig. 1. The simple steps of feature selection procedure

3 Proposed Framework

The agenda of data pre-processing is to change the rough information into gainful data. In the middle of this strategy, the segment determination systems were associated with an upgrade to the idea of data. The four figuring used for Consolidating it will deal with the issue of high spatial data and give imposing accuracy occurs [5, 6]. The methodology of the suggested structure is cleared up in the consequent area. The procedure of feature selection Technique appears in Fig. 2. Our Proposed hybrid feature selection algorithm is presented below:

Input: N features
Output: M features,
Where N > M
Steps:

1. Input N features
2. Using Feature selection with correlation and random forest classification find out the effective feature by domain knowledge
3. Using Univariate feature selection and random forest classification find out the effective feature by each attribute scores.
4. Using Recursive feature elimination (RFE) with random forest finds out the effective feature by running the output at 100 times.
5. Using Recursive feature elimination with cross-validation and random forest classification find out the effective feature by running the output at 100 times.
6. Collect the common features from each Algorithm
7. Get M output, Where N > M

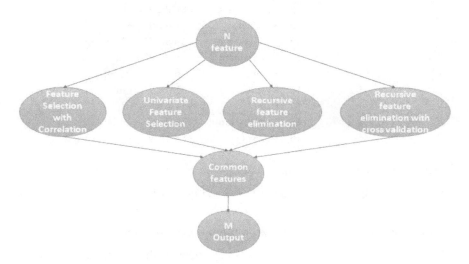

Fig. 2. Feature selection technique

4 Description on Healthcare Dataset and Algorithm

There are several types of healthcare dataset and also many attributes in the dataset. It is needed to find out which features or attributes should we use to integrate data for data warehouses. For this purpose, we are proposing an algorithm where we are showing (see Fig. 2) how to effectively find out the main feature from the dataset. In our dataset 16 types of attributes (see Table 1). Before implementing our algorithm we convert our dataset Categorical Values to numerical Values to easily find out the main feature from the dataset. In the first algorithm, find out the effective feature by domain knowledge. In the second algorithm, Univariate feature selection and random forest classification find out the effective feature by each attribute scores. The first and second algorithm output is not randomly generated. For this reason, we run the code only one time. However, in the third and fourth algorithm because of their random output, we run this code 100 times to find out the effective feature. Finally, showing the most common features with numbering from these four algorithms.

Table 1. Description on healthcare dataset

Category	No	Attributes	Description
Demographic data	1	**Name**	Patients name
	2	**Gender**	Patients sex
	3	**Age**	Patients age
	4	**Contact number**	Patients contact number
	5	**Address**	Patients address
	6	**Date of birth**	Patients date of birth
Medical information	7	**Invoice no**	Medical's invoice no
	8	**Invoice date**	Medical's invoice no
	9	**Test name**	Medical's test name
	10	**Delivery date**	Medical's delivery date
	11	**Department**	Medical's department
	12	**Sample**	Medical's sample
	13	**Result**	Medical's result
	14	**Unit**	Medical's unit
	15	**Test attribute**	Medical's test attribute
	16	**Reference value**	Medical's reference value

1. Feature selection with correlation and random forest classification

   - Select the feature at correlation based and used random forest supervised classification algorithm.

2. Univariate feature selection and random forest classification

   - The single selection of features works by choosing the best features on the basis of uniform statistic testing. It can be considered an estimator preprocessing step. Scikit-learn exposes routines for feature selection as objects implementing the transform method.

- SelectKBest [12] eliminates everything except the K maximum counting features.
- The SelectKBest class just scores the features utilizing a capacity.
- SelectKBest [12] select the best K includes that have the extreme importance with the objective variable. It takes two parameters as data conflicts, "K" (obviously) and the scoring capacity to rate the significance of each segment with the objective variable.

3. Recursive feature elimination (RFE) with random forest

- It is a grasping improvement for discovering the top performing subset of features.
- RFE depends on the plan to more than once manufacture a model, and pick either the finest or most noticeably bad performing feature, setting the features aside and after that reiterating the procedure with whatever is [13] left of the lineaments. This procedure is associated until all aspects in the dataset are depleted. Features are then situated by when they were discarded.
- Essentially discard a particular number or a particular level of the minimum situating features in the model and retrains. This makes the supposition that takes features are not critical, regardless, so they can be discarded. It, by then continues taking out features in that limit, to the point that your stop premise is come to.

4. Recursive feature elimination with cross-validation and random forest classification

- Cross-validation is a framework to survey perceptive models by dividing the first precedent into a planning set to set up the model, and a test set to evaluate it. In k-fold cross-validation, the principal model is subjectively separated into k measure up to estimate subsamples.

5   Results and Discussion

Firstly, we have collected the data from the Dataset and showing the first five rows from the datasets (see Table 2). Then showing which types of data are available in the Dataset (see Fig. 3). Next step we are showing only the object columns (see Table 3). Then we convert the dataset from Categorical Values to numerical Values for easily find out the main feature from the dataset (see Fig. 4). In the first algorithm, we find out the effective feature and showing the output (see Table 4) by domain knowledge. In the second algorithm, we find out the effective feature and showing (see Fig. 5) by each attribute scores. In the third algorithm (see Fig. 6) and fourth algorithm (see Fig. 7) output are randomly generated. Therefore, we run this code 100 times to find out the effective feature and showing the descending order result in Figs. 6 and 7. In final step is to find out the common feature from all of the output of the algorithm and showing with numbering (see Fig. 8).

In this table (see Table 2) shows only the first 5 rows from the dataset. In the dataset, there are 16 attributes.

Table 2. Show only first 5 rows from the dataset

Invoice date	Date of birth	Invoice no	Gender	Test name	Age	Delivery date	Department	Sample	Contact number	Patient name	Unit	Reference value
1/1/2018	22/1/1990	900	M	555	28	8/1/2018	909	776	112233	akkas	123	111
2/1/2018	20/9/1989	901	F	501	29	9/1/2018	991	667	990077	nusrat	321	123
3/1/2018	1/11/1993	902	M	502	19	10/1/2018	992	555	123456	kalam	132	124
3/1/2018	2/11/1990	903	M	503	23	10/1/2018	993	777	880965	robiul	120	246
4/1/2018	22/9/1988	904	M	504	24	11/1/2018	994	445	235765	jamal	123	467

After showing only the first 5 rows now we showing how many types of attributes (see Fig. 3) in our dataset. Our dataset mixed with categorical and numerical values. Before using our dataset, all the data move into numerical.

```
id                int64
diagnosis         object
Invoice Date      object
Date of birth     object
Invoice No        int64
Gender            object
Test Name         int64
Age               int64
Delivery Date     object
Department        int64
Sample            int64
Contact number    int64
patient name      object
Unit              int64
Reference Value   int64
Address           object
Test Attribute    int64
Result            int64
dtype: object
```

Fig. 3. Data types in the dataset

Pandas incorporate a helpful select_dtypes perform that we will use to create a replacement data frame containing exclusively the item columns. In our dataset, there are several types of data. In Table 3 only showing the object columns attribute.

In our dataset, there are several types of that. But for implementation, we need to change all the data into arithmetical format. For this purpose, we Encoding all the Categorical Values and convert all the data into a numerical format (see Fig. 4).

Table 3. Shows only object columns

	Diagnosis	Invoice date	Date of birth	Gender	Delivery date	Patient name	Address
0	M	1/1/2018	22/1/1990	M	8/1/2018	akkas	ctg
1	B	2/1/2018	20/9/1989	F	9/1/2018	nusrat	dhaka
2	M	3/1/2018	1/11/1993	M	10/1/2018	kalam	Korimpur
3	M	3/1/2018	2/11/1990	M	10/1/2018	robiul	birampur
4	M	4/1/2018	22/9/1988	M	11/1/2018	jamal	syria

Before making numerical data:

```
X before making numerical:
[[111 'M' '1/1/2018' '22/1/1990' 900 'M' 555 28 '8/1/2018' 909 776 112233
  'akkas' 123 111 'ctg' 7765 1122]
 [112 'B' '2/1/2018' '20/9/1989' 901 'F' 501 29 '9/1/2018' 991 667 990077
  'nusrat' 321 123 'dhaka' 4788 1456]
 [113 'M' '3/1/2018' '1/11/1993' 902 'M' 502 19 '10/1/2018' 992 555 123456
  'kalam' 132 124 'Korimpur' 6754 5532]
 [114 'M' '3/1/2018' '2/11/1990' 903 'M' 503 23 '10/1/2018' 993 777 880965
  'robiul' 120 246 'birampur' 7754 5322]
 [115 'M' '4/1/2018' '22/9/1988' 904 'M' 504 24 '11/1/2018' 994 445 235765
  'jamal' 123 467 'syria' 6754 6543]
 [116 'B' '4/1/2018' '31/12/1982' 905 'F' 505 26 '11/1/2018' 995 336 345865
  'kamal' 564 756 'agrabad' 8534 6433]
 [117 'B' '22/2/2018' '15/9/1993' 906 'M' 506 40 '28/2/2018' 996 365 349065
  'nizam' 546 456 'gec' 7643 7898]
 [118 'B' '25/2/2018' '26/5/1997' 907 'M' 507 41 '28/2/2018' 997 334 357852
  'amin' 789 797 'halisohor' 7543 7946]
 [119 'M' '3/3/2018' '11/2/1978' 908 'M' 508 43 '13/3/2018' 998 975 120956
  'onik' 987 567 'dinajpur' 7643 4567]
```

After making numerical data:

```
X after making numerical:
[[111 'M' 0 8 900 1 555 8 7 909 776 112233 1 123 111 4 7765 1122]
 [112 'B' 4 7 901 0 501 9 8 991 667 990077 10 321 123 5 4788 1456]
 [113 'M' 7 0 902 1 502 4 0 992 555 123456 6 132 124 0 6754 5532]
 [114 'M' 7 6 903 1 503 5 0 993 777 880965 13 120 246 3 7754 5322]
 [115 'M' 9 9 904 1 504 6 1 994 445 235765 5 123 467 13 6754 6543]
 [116 'B' 9 13 905 0 505 7 1 995 336 345865 7 564 756 1 8534 6433]
 [117 'B' 5 3 906 1 506 10 4 996 365 349065 9 546 456 7 7643 7898]
 [118 'B' 6 12 907 1 507 11 4 997 334 357852 2 789 797 8 7543 7946]
 [119 'M' 8 1 908 1 508 12 2 998 975 120956 11 987 567 6 7643 4567]
 [120 'M' 10 4 909 0 509 13 2 999 365 437865 8 897 466 12 7426 5476]
 [121 'M' 11 11 910 1 510 0 3 1001 309 192830 12 879 467 11 3678 45768]
 [122 'B' 1 10 911 1 511 1 3 1101 936 706123 0 907 432 2 3468 4578]
 [123 'M' 2 2 912 1 512 2 5 1189 854 490123 3 107 678 10 8653 353]
 [124 'B' 3 5 913 0 513 3 6 1230 468 358524 4 309 665 9 8643 2356]]
```

Fig. 4. Convert all the data into a numerical format

Now, our dataset only has numerical format data and ready to execute. Using this dataset now Implement this 4 random algorithm to find out the efficient features:

1. The first algorithm used for find out the effective feature by domain knowledge and showing the output result in Table 4.

Table 4. Shows the selected feature

	Date of birth	Gender	Age	Sample	Contact number	Patient name	Address	Result
0	8	1	8	776	112233	1	4	1122
1	7	0	9	667	990077	10	5	1456
2	0	1	4	555	123456	6	0	5532
3	6	1	5	777	880965	13	3	5322
4	9	1	6	445	235765	5	13	6543

2. The second algorithm used for find out the effective feature by each attribute scores (see Fig. 5).

```
Score list: [  6.75000000e-01   7.14285714e-01   2.70750000e+00   9.65217160e+01
     2.28528451e+05   2.75625000e+00   1.36290323e-01   1.70545910e+04]
Feature list: Index(['Date of birth', 'Gender', 'Age', 'Sample', 'Contact number',
     'patient name', 'Address', 'Result'],
     dtype='object')
```

Fig. 5. Univariate feature selection output

3. In the third algorithm, the output is randomly generated. So, run this code 100 times to find out the effective feature and showing the descending order result in Fig. 6.

```
[('Contact number', 89), ('patient name', 79), ('Result', 73), ('Age', 72), ('Sample', 58), ('Addres
s', 58), ('Date of birth', 41), ('Gender', 30), ('Invoice Date', 0), ('Invoice No', 0), ('Test Nam
e', 0), ('Delivery Date', 0), ('Department', 0), ('Unit', 0), ('Reference Value', 0), ('Test Attribu
te', 0)]
```

Fig. 6. Recursive feature elimination (RFE)

4. In the fourth algorithm, this algorithm also generates the result randomly. So, run this code 100 times to find out the effective feature and showing the descending order result in Fig. 7.

```
[('Contact number', 90), ('patient name', 86), ('Result', 80), ('Address', 78), ('Date of birth', 7
6), ('Sample', 76), ('Age', 72), ('Gender', 63), ('Invoice Date', 0), ('Invoice No', 0), ('Test Nam
e', 0), ('Delivery Date', 0), ('Department', 0), ('Unit', 0), ('Reference Value', 0), ('Test Attribu
te', 0)]
```

Fig. 7. Proper feature selection process.

In final step is to find out the common feature from all of the output of the algorithm and showing with numbering (see Fig. 8).

```
[('Contact number', 228528.45051245467),
 ('Result', 17054.590964434476),
 ('patient name', 2.75625),
 ('Age', 2.7075000000000022),
 ('Address', 0.13629032258064525)]

1 :Contact number
2 :Result
3 :patient name
4 :Age
5 :Address
```

Fig. 8. Numbering of proper feature selection process

6 Conclusions

Selection of a proper subset of features is very important in many types of research such as data integration and machine learning. It is always difficult to choose which features or attributes should we use for data integration for data warehousing. It requires deep knowledge of the problem domain. Automatic feature assortment is the procedure of choosing a subset of applicable features automatically for later use. In the future, we will add more security issue to increase efficiency. We will use more healthcare dataset. When the dataset will increase that time security and efficiency issue is more important. We will build and update our algorithm to ensure security and efficiency. We will also enhance the noise reduction technique.

References

1. Shen, F., et al.: Bearing fault diagnosis based on SVD feature extraction and transfer learning classification. In: 2015 Prognostics and System Health Management Conference (PHM). IEEE (2015)
2. Khan, S., Latiful Haque, A.: Towards development of national health data warehouse for knowledge discovery (2016). https://doi.org/10.1007/978-3-319-23258-4_36
3. Acharya, A., Sinha, D.: Application of feature selection methods in educational data mining. Int. J. Comput. Appl. 103(2) (2014)
4. Bidgoli, A.-M., Parsa, M.N.: A hybrid feature selection by resampling, chi squared and consistency evaluation techniques. World Acad. Sci. Eng. Technol. 68, 276–285 (2012)
5. Singh, B., Kushwaha, N., Vyas, O.P.: A feature subset selection technique for high dimensional data using symmetric uncertainty. J. Data Anal. Inf. Process. 2(04), 95 (2014)
6. Ramaswami, M., Bhaskaran, R.: A study on feature selection techniques in educational data mining. J. Comput. 1, 7–11 (2009)

7. Jishan, S.T., et al.: Improving accuracy of students' final grade prediction model using optimal equal width binning and synthetic minority over-sampling technique. Decis. Anal. **2** (1), 1 (2015)
8. Tang, J., et al.: Feature selection for classification: a review. In: Data Classification: Algorithms and Applications, vol. 37 (2014)
9. Kandel, S., Heer, J., Plaisant, C., Kennedy, J., Van Ham, F., Riche, N.H., Buono, P.: Research directions in data wrangling: visualizations and transformations for usable and credible data. Inf. Vis. **10**(4), 271–288 (2011)
10. Lin, T.Y., Cercone, N. (eds.): Rough Sets and Data Mining: Analysis of Imprecise Data. Springer, New York (2012)
11. Chuang, L.Y., Ke, C. H., Yang, C.H.: A hybrid both filter and wrapper feature selection method for microarray classification. arXiv preprint arXiv:1612.08669 (2016)
12. Ghaemidizaji, M., Derakhshi, F., Reza, M.: Classifying different feature selection algorithms based on the search strategies (2014)
13. Ong, T.C., et al.: Dynamic-ETL: a hybrid approach for health data extraction, transformation and loading. BMC Med. Inform. Decis. Mak. **17**(1), 134 (2017)
14. Khan, S.I., Hoque, A.S.M.L.: Development of national health data warehouse Bangladesh: privacy issues and a practical solution. In: 2015 18th International Conference on Computer and Information Technology (ICCIT), pp. 373–378. IEEE (2015)
15. Biplob, M.B., Sheraji, G.A., Khan, S.I.: Comparison of different extraction transformation and loading tools for data warehousing. In: 2018 International Conference on Innovations in Science, Engineering and Technology (ICISET), pp. 262–267. IEEE (2018)

A Survey on Methods of Ontology Learning from Text

Ravi Lourdusamy and Stanislaus Abraham[(✉)]

Department of Computer Science, Sacred Heart College, Tirupattur 635 601,
Tamil Nadu, India
astanislaus@gmail.com

Abstract. Ontologies continue to emerge in performing a greater function in
various business processes and knowledge sharing in the field of modern
information systems. They help to exchange and extend, from syntax to
semantic, data and knowledge. They are viewed as a remedy for interoperable
semantics which power semantic web. Since the turn of the millennium,
ontology learning (OL) from text has received a considerable attention with a
sudden surge of textual information for promising research. Remarkably, with
intermingling of various disciplines like text mining (TM), statistical analysis
techniques (SAT), machine learning (ML), knowledge representation (KR),
natural language processing (NLP), etc., more research is on in the area of
Ontological Engineering and OL from text. This survey brings out several
outcomes of foregone research, discusses challenges and limitations of research
in the past decade and points to new challenges in future.

Keywords: Ontology learning · Ontology learning methods · Ontology
learning from text

1 Introduction and Motivation

The term, "ontology" has been studied by various philosophers in the past twenty three
centuries. Since mid-1980's it has become a buzzword in Artificial Intelligence (AI).
Initially, AI community used it to refer to a theoretical representation of an adapted or
modeled world. In computer science, ontology was perceived broadly as a collection of
concepts, axioms, properties, relations that illustrates a specific domain.

In 1960s search on the net began with syntax format, in 1990s it moved to a
structured format and since 2010 it has changed into semantic based, which is an
extension of WWW. After the emergence of Semantic Web, active research in ontology
received a great support, popularized by Tim Berners-Lee. Ontologies form an essential
component and are the back bone to Semantic Web. The Semantic Web uses domain
ontologies to conceptualize domain through terms, concepts, relationships, axioms and
rules. Considering the vastness of data available in the web, it becomes a laborious task
to process manually. Mostly it is cost-intensive with respect to time and efficiency and
sometimes it is impossible to process. Therefore, a large volume of data can be handled
in an automatic and semi-automatic process.

© Springer Nature Switzerland AG 2020
L. C. Jain et al. (Eds.): ICICCT 2019, LAIS 9, pp. 113–123, 2020.
https://doi.org/10.1007/978-3-030-38501-9_11

Syntactic-based knowledge is declarative. Language with human characteristics is implicit, inferential or intuitive. Therefore, it is very difficult to bridge semantic gap between anthropomorphic language and highly formalized knowledge. The data source is more embedded and implied. Hence, to achieve good results, techniques needed to extract, buried and unspoken meanings, demand an array of techniques which draw out the nuances in the text.

The objectives of the article are to survey and present, discuss and summarize recent research and the latest trends in several prominent methods of OL from text since the year 2001 until 2015. The focus is to bring out the current challenges and potential areas for future research and in conclusion to propose a four-layer framework to semi-automate OL from textual resources.

The survey undertaken brings out the end result of past research, deliberates on challenges of recent research and leads to challenges of the future. In what follows, Sect. 2 deals with preliminary concepts in ontology and its learning from texts; Sect. 3 explains works related to the study. In Sect. 4 review on OL methods are deliberated, a discussion is carried out on the outcome of the survey and research gaps are brought forth. Finally, Sect. 5 concludes the survey focusing on areas of potential research for future.

2 Preliminaries

2.1 Ontology

The term ontology belongs to the field of western philosophy and is most commonly defined as 'Theory of Existence.' In contrast, this term is borrowed and adopted into the field of AI and Knowledge Engineering. "Ontologies in computer science are specifications of shared conceptualizations of a domain of interest that are shared by a group of people" [1]. The popular definition often quoted is by Gruber which is defined as "An ontology is a formal, explicit specification of a shared conceptualization" [2]. In the words of Studer et al. [3], 'conceptualization' means identifying relevant concepts of an abstract and theoretical model or adaptation of some phenomena in the world. "'Explicit' refers to the type of concepts and constraints derived and used are explicitly defined. 'Formal' states the fact that the ontology should be machine-readable. 'Shared' reflects the notion what an ontology captures as consensual knowledge" [3] and is broadly accepted by various communities. Ontologies contain modelling primitives such as concepts, relations between concepts, and axioms.

2.2 Ontology Learning

There exists a robust bond between ontologies and written form of text in all natural languages. Ontologies capture knowledge of a domain in a formal language. Various approaches to extract ontologies from such texts have been developed. OL adapts Data Mining (DM), ML, NLP, SAT to extract terms, concepts, relations and axioms, from large-volume of data.

The input for text based OL techniques can be distinguished by completely unstructured text or semi-structured text such as Web documents. The output of the OL methods are grouped together as "terms, synonyms, concepts, taxonomy, (concept hierarchies), relations, axioms and rules" [4]. Buitelaar [5] presents the OL Layer Cake in the Fig. 1.

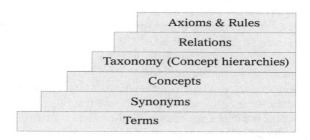

Fig. 1. Ontology leaning layer cake

OL is a process of developing a conceptualization of a domain with less human intervention and more automatic or semi-automatic approach, i.e., generating ontology from the beginning, enrichment, or populating the current ontology. It also means automating a process by which discovering, creating, extracting of ontological knowledge takes place using ML techniques.

2.3   Ontology Learning from Text

OL from text extracts information by means of TM focusing on automatic/semi-automatic generation of lightweight taxonomies. "Ontology learning from text is a process of identifying terms, concepts, relations, and axioms from textual information and using them to construct and maintain an ontology" [6]. Techniques used in ML, information retrieval, knowledge acquisition, TM, AI, DM, NLP, reasoning and database management etc., are essential in the progress and expansion of OL systems.

Different types of OL approaches are generally classified as: OL from Text, Linked Data Mining, Concept Learning in Description Logics, Web Ontology Language (OWL), Crowdsourcing ontologies, Dictionary, Knowledge Base, Semi-structured schemata and Relational schemata. The field of OL from text deals mainly with ontological elements, i.e., learning concepts and their relationships. In order to establish this, related elements have to be obtained from the input source.

As OL is a complex process, manual building and maintenance of ontologies are highly tedious, time-consuming, error-prone, and labor-intensive. In traditional perspective after the completion of building ontologies, the ontology engineers tended to think that the task has come to an end. But in reality it is significantly different. As contexts change meanings change. Diverse and rapid changes affect text in use for OL. Ontology evolution will never cease to exist. It evolves dynamically. Hence, the approaches to learn ontologies from texts continue to undergo change.

With limited resources manual approaches are unrealistic to be realized. Never-theless, more and more research is carried out in areas, "including the life sciences, is moving to an industrialized e-science paradigm, with the rapid generation of large volumes of new data requiring machine-readable annotations. In such a context, we explore the use of techniques developed to learn ontologies from text to reduce effort in the early stages of ontology development" [7].

Textual web sources differ from representative textual corpus and adaptation of old learning techniques, or development of new ones is necessary. The web resources are multilingual, and development of multilingual OL techniques is needed. To overcome this knowledge acquisition bottleneck Data-driven Knowledge Acquisition and OL from text are proposed as a possible solution.

As text is massively available on the Web, OL from text is an attractive option. It is one of the approaches for partial automation of knowledge acquisition process. One of the main aims in OL is complete automation of all processes involved, but as ontology is "shared conceptualization", complete automatic learning probably is impossible at the moment.

3   Previous Surveys – A Glance

The related surveys carried out by various authors is classified under Methods and Approaches & Models and Frameworks, are depicted in Fig. 2.

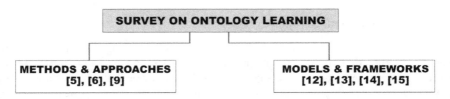

Fig. 2. Classification of ontology learning methods and models between 2001–2015

3.1   Methods and Approaches

The report survey titled "OntoWeb Consortium" [8], presents eighteen approaches for OL from text. These OL methods are compared based on the following criteria: techniques used, reuse other ontologies, sources used for learning, tool associated, and evaluation.

Buitelaar et al. [5] while combining research articles on OL and knowledge acquisition published at two workshops, popularized the *expression "ontology learning layer cake."* It best described several subtasks to be carried out in OL.

Gómez-Pérez and Manzano-Macho [9] claimed to have reviewed thirteen methods and fourteen tools for constructing ontologies from texts semi-automatically. They studied each method and its relationships with techniques used. Three groups were categorized based on techniques followed viz., linguistics, statistics, and ML.

An overview of the most promising OL methodologies as well as used approaches, algorithms for OL, evaluation and evolution are deliberated [1, 10]. It is learnt that techniques such as concept identification and hierarchy extraction are adapted by many approaches [11]. An exhaustive review of methods, the wide variety of techniques is presented by Wong et al. [6]. The authors also bring out dearth of shared evaluation platforms, research in progress to date, and unresolved challenges.

3.2 Models and Frameworks

Maedche and Staab [12] have developed an OL framework. The framework is designed consisting of various phases viz., import, reuse, extract, prune, refine and apply. Based on the framework an architecture for learning ontologies is constructed. Further, as a proof of concept, a tool named "Text-onto" is developed for OL from text.

The survey by Shamsfard and Barforoush [13] states to have researched seven important systems. The survey's key objective was to present a new framework which is used to classify and compare seven OL systems. "Hasti", a tool, is an outcome of this project, developed as OL system to test ontology building approach in an automated manner. "The ontology in it is a small kernel at the beginning. It extracts lexical and ontological knowledge from Persian texts. It learns concepts, taxonomic and non-taxonomic conceptual relations, and axioms, to build ontologies upon the existing kernel. The learning approach in HASTI is a hybrid symbolic approach, a combination of linguistic, logical, template driven and semantic analysis methods. It performs online and offline clustering to organize its ontology" [11].

Zhou "proposed a learning-oriented model for the development of ontology" [14]. The author highlights various relevant issues: Human understandable vs. machine-understandable; Learning specific relations; Learning higher-degree relations; Learning definitions; Term filtering; Mapping to high-level ontology; Evaluation benchmark; Incremental ontology learning; Levels of ontology learning; Multi-agent learning; Learning beyond text. The author opined that "expert-curated domain knowledge is no longer adequate and highlighted the fact that researchers are turning to other sources on the Web for the content of ontologies" [14].

A study conducted by Browarnik et al. [15] "shows shortcomings, and proposes alternative models for ontology learning based on linguistic knowledge and existing, wide coverage syntactical, lexical and semantic resources, using constructs such as clauses" [15]. The authors determine that the OL Layer Cake has a low maximum F measure which paves way for further alternatives to be explored.

4   Review on Ontology Learning Methods

An analysis of important OL methods and approaches are summarized and presented in Table 1. OL methods are tabulated based on the criteria such as: input type, tools associated and techniques used. The input types used by methods to learn ontologies vary from method to method depending on their domain. Input types used are structured, semi-structured, unstructured and existing ontologies. Input for OL methods is only text and no video, image or audio. For improved ontology output, quality of text corpora used towards learning ontologies ought to be considered.

The tools associated to process the input types are: NLP, SAT, text pre-processing tools, clustering, Welkin, LEXTER, GEDITERM, TERMINAE, Differential Ontology Editor, SubWordNet Engineering tool, Promethee, Cameleon, ML, Text-To-Onto, Search, Filtering & Extraction tools, and OntoLearn.

Most of the methods focus on concept and taxonomy learning. It is learnt from survey that NLP is mostly common among all techniques. Methods exploit NLP, statistical and computational inference based techniques to extract concepts, relations and axioms from voluminous corpora of text. Some methods are purely statistical approach based and most others have a combination of statistical and linguistic and/or NLP based. Linguistic or NLP methods mostly rely on text corpora for exploring and extracting of ontology entities.

4.1   Research Gaps

The key challenging task in the construction of ontology from texts is discovering base units, viz., a set of terms used as labels in class hierarchy of ontology. OL approach is to overcome knowledge acquisition bottleneck. Through literature survey on various OL methods a number of issues are noted which are unsatisfactorily resolved. The following are a few prominent areas: (a) text understanding (b) knowledge extraction, (c) ontological structures labelling and filtering, (d) ontology evaluation.

Knowledge acquisition, ML, NLP etc. techniques become the very base for OL. Even though this field of research has seen growth in the past decades, it is yet to reach the goal of full automation of OL process. Another field of vast growing area is web resources, which are multilingual in nature. Evaluation of OL methods is yet another area of study.

The significant findings of the literature survey point to "the issue of noise, authority, and validity in Web data for OL; the integration of social data into the learning process to incorporate consensus into ontology building; the design of new techniques for exploiting the structural richness of collaboratively maintained Web data, ontology learning process from text does not have a well-defined methodology" [6]. Although discovering taxonomic relations is dealt exhaustively on one hand, non-taxonomic relations receive less consideration [13]. Fully automated OL is impossible whereas semi-automated OL from text is researched and implemented satisfactorily [6, 10, 13, 16].

Other key issues for future research should necessarily adapt current techniques applicable for "the design of new techniques for exploiting the structural richness of collaboratively maintained Web data; the representation of ontological entities as language-independent constructs; the applicability of existing techniques for learning ontologies for different writing systems (e.g., alphabetic, logographic); the efficiency and robustness of existing techniques for Web-scale ontology learning; the increasing role of ontology mapping as more ontologies become available" [6]. Other research directions are to move from existing lightweight ontologies to formal ones, from small static text collections to large volume of text viz., Web resources, and invent new techniques to suit for huge volumes of web sources [5, 6, 10].

Table 1. Summary of ontology learning methods from text

Sl. No.	Methods	Input type	Tools associated	Techniques used
1.	Aguirre and Colleagues' [17]	Text; Domain text, WordNet	Clustering tools	Statistical approach, topic signatures, semantic distance metrics, clustering methods
2.	Alfonseca and Manandhar's [18]	Existing ontology, Domain text, WordNet	Welkin	Topic signatures, semantic distance metrics
3.	Aussenac-Gilles and Colleagues' approach [19]	Texts, existing ontologies, human expert knowledge for validation; Domain text, domain ontologies	LEXTER, GEDITERM, TERMINAE	NLP tools, linguistic analysis, clustering, term extraction based on distributional analysis, relation extraction based on linguistic patterns, and knowledge extraction with syntactic patterns
4.	Bachimont's [20]	Text, terms proposed by the expert; Domain text	Differential ontology editor	NLP techniques, semantic relativeness, linguistic techniques
5.	Faatz and Steinmetz approach [21]	Corpus from WWW; domain corpus, domain ontology	Any ontology workbench	Statistical approach, semantic distance, semantic relativeness, clustering
6.	Gupta and Colleagues' approach [8]	Textual documents; Domain text, WordNet	SubWordNet Engineering tool	NLP techniques, linguistic techniques, term-extraction techniques
7.	Hahn and Colleagues' [22]	Domain text, Domain Knowledge base; Domain text	NLP tools	NLP, semantic relativeness, concept hypothesis based on linguistic and conceptual quality labels, statistical approach
8.	Hearst's approach [23]	The original ontology and a corpus of texts; Domain text WordNet	Welkin, Prométhée and Caméléon	NLP, linguistic patterns; statistical approach, topic signature, semantic distance metrics, semantic relativeness, ML techniques, extraction, use of word

(continued)

Table 1. (*continued*)

Sl. No.	Methods	Input type	Tools associated	Techniques used
				patterns to learn new relationships between the concepts
9.	Hwang's [24]	Text; Domain text	NLP & ML	NLP techniques, ML techniques, statistical approach
10.	Khan and Luo's [25]	Text documents; Domain text, WordNet	SOTA outperforms HAC	Clustering techniques, statistical approach
11.	Kietz and Colleagues' [26]	Text, existing ontologies; Domain and nonspecific domain, domain ontologies, WordNet	Text-To-Onto	NLP techniques, statistical approach, semantic relativeness, clustering, ML techniques, linguistic techniques, extraction
12.	Lonsdale and Colleagues' [27]	Terminological databases, domain ontologies, WordNet, text documents; Domain text	Search, filtering & extraction tools	NLP mappings, linguistic technique mappings, several linguistic heuristic, graph theory, extraction
13.	Missikoff and Colleagues' [28]	Text; Domain text, WordNet	OntoLearn	NLP, statistical approach, ML techniques, linguistic techniques, extraction
14.	Moldovan and Girju's [29]	Non-specific domain corpus, lexical resources, and dictionaries; Domain text, WordNet	Knowledge intensive & NLP tools	NLP techniques, ML techniques, linguistic techniques, extraction
15.	Nobécourt approach [30]	Text; Domain text	TERMINAE	NLP, linguistic analysis, extraction
16.	Roux and Colleagues' approach [31]	Text; Domain text, domain ontology	Tools developed in Xerox labs	NLP, linguistic techniques, extraction, verb-patterns
17.	Wagner approach [32]	Domain corpus; WordNet	NLP tool	Statistical approach, verb patterns
18.	Xu and Colleagues' approach [33]	Linguistically annotated and pre-classified text; annotated text corpus, WordNet	TFIDF	NLP, statistical analysis, linguistic techniques and text-mining approach

Most systems focus on static background knowledge, building domain ontologies and heavy use of domain-specific patterns; scant regard is given to the portability of the systems across various domains [13]. Axiom extraction technique needs to be strengthened for yielding of better results [5, 13, 16]. Multilingual OL techniques need to be developed in order to process multilingual text resources [10].

As of now, there doesn't seem to exist a gold standard evaluation method for OL and there are no systems in place to compare the results produced by the emerging OL system and it is still an open problem [5, 6, 13, 16].

5   Conclusions and Future Works

OL is a complex process. Research in OL has progressed in recent years. In spite of advances made, key challenges still remain unaddressed. Some of the areas include full automation of OL development process, ontology interoperability, multi-lingual diversity, semantic web standard and practices, engineering methodologies are just to name a few. This article is a survey and summary of current issues in the area of OL from text and open questions that continue to intrigue the field of ontology engineering.

Some of the areas to work on in future are designing OL system by building suitable deep network to discover relations and rules. Improved Parts-of-Speech tagging and parsing can increase the efficiency to form better ontologies. Research in text classification can bring new algorithms in choosing highly discriminative features among the classes. Term selection algorithms is another area to explore for OL systems to improve in unearthing terms and concepts.

Since OL borrows many techniques from ML such as clustering and association rule mining, advances in integrating the domain of DL into these algorithms would naturally enhance OL from text. Besides, the phenomenal rate of growth of textual data on the web is powerfully shaping several methods adapted at various stages of OL. The future of OL depends on the enormous quantity of unstructured web data. The following are future directions in OL process: dependence on social media for data authentication, language independent OL, scalability of existing OL techniques to accommodate huge data sets, use of crowdsourcing and human-based computation games to perform ontology post-processing, and development of more formal or heavyweight ontologies.

Ontology creation needs broad expertise. Need for full automation of OL development process is evident but far-fetched approach. Hence, as an edge of our research study, we propose a four-layer framework consisting of the following components to semi-automate the development process of OL: Text Extraction, Text Pre-processing, Ontology Construction, Ontology Enrichment and Evaluation. As a proof of concept a semi-automated tool will be developed, evaluated and compared with the existing OL systems to generate ontology from textual resources.

References

1. Biemann, C.: Ontology learning from text: a survey of methods. In: LDV Forum, vol. 20, no. 2 (2005)
2. Gruber, T.: Towards principles for the design of ontologies used for knowledge sharing. Int. J. Hum. Comput. Stud. **43**(4–5), 907–928 (1995)
3. Studer, R., Benjamins, V.R., Fensel, D.: Knowledge engineering, principles and methods. Data Knowl. Eng. **25**(1–2), 161–197 (1998)
4. Cimiano, P.: Ontology Learning and Population from Text: Algorithms, Evaluation and Applications. Springer, Heidelberg (2006)
5. Buitelaar, P., Cimiano, P., Magnini, B.: Ontology learning from text: an overview. In: Buitelaar, P., et al. (eds.) Ontology Learning from Text: Methods, Evaluation and Applications. IOS Press, Amsterdam (2005)
6. Wong, W., Liu, W., Bennamoun, M.: Ontology learning from text: a look back and into the future. ACM Comput. Surv. (CSUR) **44**(4) (2012). Article no. 20.3
7. Brewster, C., Jupp, S., Luciano, J., Shotton, D., Stevens, R.D., Zhang, Z.: Issues in learning an ontology from text. BMC Bioinform. **10**(5), S1 (2009)
8. Gomez-Perez, A., Manzano-Macho, D.: OntoWeb deliverable 1.5: a survey of ontology learning methods and techniques. Universidad Politecnica de Madrid (2003)
9. Gómez-Pérez, A., Manzano-Macho, D.: An overview of methods and tools for ontology learning from texts. Knowl. Eng. Rev. **19**(3), 187–212 (2004)
10. Ivanova, T.: Ontology learning technologies-brief survey, trends and problems. In: Proceedings of the International Conference on Information Technologies (2012)
11. Drumond, L., Girardi, R.: A survey of ontology learning procedures. In: WONTO, vol. 427, pp. 1–13 (2008)
12. Maedche, A., Staab, S.: Ontology learning for the semantic web. IEEE Intell. Syst. Special Issue: Semantic Web **16**(2), 72–79 (2001)
13. Shamsfard, M., Barforoush, A.A.: Learning ontologies from natural language texts. Int. J. Hum. Comput. Stud. **60**(1), 17–63 (2004)
14. Zhou, L.: Ontology learning: state of the art and open issues. Inf. Technol. Manage. **8**(3), 241–252 (2007)
15. Browarnik, A., Maimon, O.: Departing the ontology layer cake, pp. 167–203 (2015). https://doi.org/10.4018/978-1-4666-8690-8.ch007
16. Shamsfard, M., Barforoush, A.A.: The state of the art in ontology learning: a framework for comparison. Knowl. Eng. Rev. **18**(4), 293–316 (2003)
17. Agirre, E., Ansa, O., Hovy, E., Martinez, D.: Enriching very large ontologies using the WWW. In: Proceedings of ECAI (2000)
18. Alfonseca, E., Manandhar, S.: An unsupervised method for general named entity recognition and automated concept discovery. In: Proceedings of the 1st International Conference on General WordNet, Mysore, India (2002)
19. Aussenac-Gilles, N., Biébow, B., Szulman, S.: Corpus analysis for conceptual modelling workshop on ontologies and text, knowledge engineering and knowledge management: methods, models and tools. In: 12th International Conference EKAW, Juan-les-pins, France. Springer (2000)
20. Bachimont, B., Isaac, A., Troncy, R.: Semantic commitment for designing ontologies: a proposal. In: Gomez-Perez, A., Benjamins, V.R. (eds.) EKAW 2002. LNAI, vol. 2473, pp. 114–121. Springer, Heidelberg (2002)
21. Faatz, A., Steinmetz, R.: Ontology enrichment with texts from the WWW. In: Semantic Web Mining 2nd Workshop at ECML/PKDD, Helsinki, Finland (2002)

22. Hahn, U., Schnattinger, K.: Towards text knowledge engineering. In: AAAI 1998/IAAI 1998, Madison, Wisconsin, pp. 524–531. AAAI Press/MIT Press, Menlo Park, Cambridge (1998)

23. Hearst, M.A.: Automated discovery of WordNet relations. In: Fellbaum, C. (ed.) WordNet: An Electronic Lexical Database, pp. 132–152. MIT Press, Cambridge (1998)

24. Hwang, C.H.: Incompletely and imprecisely speaking: using dynamic ontologies for representing and retrieving information. In: Proceedings of KRDB 1999, Linköping, Sweden (1999)

25. Khan, L., Luo, F.: Ontology construction for information selection. In: Proceedings of 14th IEEE International Conference on Tools with Artificial Intelligence, Washington, D.C., pp. 122–127 (2002)

26. Maedche, A., Volz, R.: The Text-To-Onto ontology extraction and maintenance environment. In: Proceedings of the ICDM, San Jose, California, USA (2001)

27. Lonsdale, D., Ding, Y., Embley, D.W., Melby, A.: Peppering knowledge sources with SALT; boosting conceptual content for ontology generation. In: Proceedings of the AAAI, Edmonton, Alberta, Canada, July 2002

28. Navigli, R., Velardi, P., Gangemi, A.: Ontology learning and its application to automated terminology translation. IEEE Intell. Syst. **18**(1), 22–31 (2003)

29. Harabagiu, S.M., Moldovan, D.I.: Enriching the WordNet taxonomy with contextual knowledge acquired from text. In: Shapiro, S., Iwanska, L. (eds.) Natural Language Processing and Knowledge Representation: Language for Knowledge and Knowledge for Language, pp. 301–334. AAAI/MIT Press, Cambridge (2000)

30. Biébow, B., Szulman, S., Clément, A.J.B.: TERMINAE: a linguistics-based tool for the building of a domain ontology. In: Fensel, D., Studer, R. (eds.) EKAW 1999, pp. 49–66. Springer, Heidelberg (1999)

31. Roux, C., Proux, D., Rechermann, F., Julliard, L.: An ontology enrichment method for a pragmatic information extraction system gathering data on genetic interactions. In: Proceedings of the ECAI 2000, OL 2000, Berlin (2000)

32. Wagner, A.: Enriching a lexical semantic net with selectional preferences by means of statistical corpus analysis. In: Proceedings of the ECAI 2000, Berlin, pp. 37–42 (2000)

33. Xu, F., Kurz, D., Piskorski, J., Schmeier, S.: A domain adaptive approach to automatic acquisition of domain relevant terms and their relations with bootstrapping. In: Proceedings of LREC 2002, Canary Island, Spain (2002)

Improving Energy Consumption by Using DVFS

M. Iyapparaja[(⊠)], L. Abirami, and M. Sathish Kumar

SITE School, Vellore Institute of Technology, Vellore, Tamil Nadu, India
iyapparaja85@gmail.com,
abirami.2018@vitstudent.ac.in,
sathishkumar.mani@vit.ac.in

Abstract. Power management has turned into a basic plan parameter as more transistors are coordinated on a solitary chip. Bringing down the supply voltage is one of the alluring ways to deal with spare intensity of the variable outstanding task at hand framework, and to accomplish long battery life. DVFS is one of the effective methods to lessen the vitality utilization. The principle thought behind DVFS plan is to powerfully scale the supply voltage of CPU, to give enough circuit speed to process the framework remaining burden so as to meet the time and execution, along these lines lessening power. The problem faced by users in cloud computing is energy consumption. Some resources in cloud stay idle for some time. This idleness of resources consume energy thereby causing more consumption of energy. Both the hardware and software reduces the energy consumptions. The results shows that proposed model reduces the consumption of energy by using DVFS.

Keywords: Power management · DVFS · Energy consumption · Cloud computing

1 Introduction

Dynamic voltage frequency scaling, is used for decreasing the normal power utilization. This is cultivated by decreasing the exchanging misfortunes of the framework by specifically lessening the recurrence and voltage of the framework. To reduce the complexity in scheduling and to save the energy, energy efficient based scheduling algorithm were proposed. The DVFS technique helps to reduce the energy consumption and scheduling in clusters.

2 Motivation

The frequency of cpu gets higher when there is an enormous usage of power. So to reduce the cpu frequency DVFS technique is used. By using this technique the cpu frequencies can be lessened. The challenging task is to maintain the relationship between the server and frequency of server for given workload is difficult.

© Springer Nature Switzerland AG 2020
L. C. Jain et al. (Eds.): ICICCT 2019, LAIS 9, pp. 124–132, 2020.
https://doi.org/10.1007/978-3-030-38501-9_12

3   Background*

DVFS is a technique mainly reduces the power consumption by dynamically adjusting the frequency. Cloud datacenters is used by many users. But the problem faced by the users is scheduling efficient energy conducted a real cluster bed experiment and shows improved DVFS performance oriented and power consumption model depends on the frequency and usage.

4   Problem Statement

The problem faced by users in cloud computing is energy consumption. Some resources in cloud stay idle for some time. This idleness of resources consume energy thereby causing more consumption of energy. This energy consumption reduces both the hardware and software devices. But also by software. The results shows the proposed model reduce the consumption of energy by using DVFS.

5   Objective

A fundamental challenge exists pertaining to how the probability of power consumption done during the execution. A decision problem is created when one realizes that we cannot simply minimize the power consumption during the execution of the applications. Likewise, we cannot change the running time of an application without considering the amount of workload to be done in tat Applications. We must look into the data which can be used in the application and time taken during the application execution.

6   Innovation

The problem faced by users in cloud computing is energy consumption. Some resources in cloud stay idle for some time. This idleness of resources consume energy thereby causing more consumption of energy. This energy consumption reduces both the hardware and software devices. But also by software. The results shows the proposed model reduce the consumption of energy by using DVFS. So that no resource can stay idle.

7   Proposed System Overview

In this paper, improves the DVFS algorithm by Analyzing the task size & complexity and choose the right processor to complete the task, So that the power consumption and the performance of the system will be improved. At the same time we use caching mechanism to improve the performance for repetitive task so that we could give the result without invoking the processor. This algorithm is suitable for multi-core processor. It will also consider the task priority to give preference to task completion. In idle time the processor will go to sleep mode and only active if Job is assigned to it.

8 Challenges

Determining the power consumption done by the application during the execution will make the decision for the energy efficiency. The challenges faced is that the allocating the power according to the size of the application.

9 Assumptions

The work assumes the power consumption can be changed during the execution of application based on the size of the application considered.

Architecture Specification
See Fig. 1.

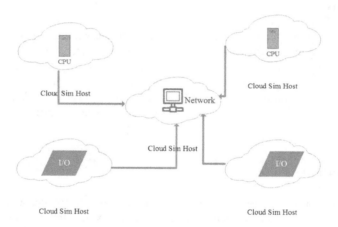

Fig. 1. Architecture specification

System Design
See Fig. 2.

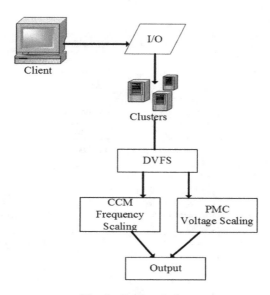

Fig. 2. System design

10 Literature Survey and Review

See Table 1.

Table 1. Literature survey and review from different years

Paper name	Year	Advantages	Disadvantages	Limitations
Towards energy efficient scheduling for online tasks in cloud data centers based on DVFS [1]	2015	The advantage of this paper is that the consumption of power will be reduced when compared to other models because of the new proposed DVFS technique and PTU Algorithm	Scheduling will be difficult because only one Task can be done at a time	Only if the PTU algorithm is based on the Rack-Sized unit the scheduling will be easy
An energy-efficient scheduling algorithm using dynamic voltage scaling for parallel applications on clusters [2]	2007	The energy consumption in clusters is reduced in this paper by reducing the idle resources that consume energy by using Energy efficient scheduling algorithm (TDVAS)	More time will be required to perform parallel tasks since the processor voltages of idle resources is reduced	The time is reduced by using the task precedence constraints which is one of the feature of TDVAS. The TDVAS along with the DVS technique the time consumed for parallel task can be reduced

(continued)

Table 1. (*continued*)

Paper name	Year	Advantages	Disadvantages	Limitations
Latency-aware dynamic voltage and frequency scaling on many-core architectures for data-intensive applications [3]	2013	The 31% of the energy is saved in large amount by considering the latency features of the DVFS. The latency features plays a major role in energy consumption. The performance of execution is also increased up to 22.4%	The proposed technique will consider two steps. So more time is required to identify the latency features and to implement the DVFS algorithm	The proposed algorithm is implemented in intel SCC platform by using data intensive benchmark
Towards energy efficient data centers: a DVFS-based request scheduling perspective [4]	2013	The dynamic allocation problem of DVFS in reducing power consumption in Np-Hard is reduced	Only an average of 12-14% of power consumption is reduced in this paper	First Fit Decreasing algorithm is used to compare the proposed algorithm in order to derive an conclusion
Energy efficient scheduling of parallel tasks on multiprocessor computers [6]	2012	An heuristic algorithm for good results is developed in this paper to solve scheduling problems. The algorithm will produce better results and the scheduling lengths is also reduced	During the scheduling problems like power supply, system partitioning, task scheduling will arises	Based on dynamic voltage and speed in parallel task on microprocessors computers is investigated in this paper
Energy efficient scheduling of mapreduce workloads on heterogeneous clusters	2011	To increase the energy efficiency in Mapreduce workloads an investigation is done on energy level scheduling heuristics. 27% better energy efficiency with our heuristics compared with the default Hadoop scheduler is achieved	High performance Intel Atom and Sandy Bridge processors were used for heterogeneity. They propose an heuristics algorithm. Considering low and high performance systems in map reduce workloads is difficult	The systems considered are more energy efficient for cpu and I/O bound workloads
Data center energy efficient resource scheduling	2014	Power consumption is reduced by powering off the redundant elements that are not used	Minimum Cost multi commodity flow problem arises during servicing the current workload	To solve the MCMCF problem Benders algorithm were used

(*continued*)

Table 1. (*continued*)

Paper name	Year	Advantages	Disadvantages	Limitations
		presently in data centers. The DCEERS algorithm is used in this paper		
Energy-aware task scheduling in heterogeneous computing environments [9]	2014	Minimizing duplication technique were used to reduce the task duplication problem during the scheduling	To copy, search and delete the redundant data it consumes more time	To test the algorithm real world and randomly generated graphs application were used. The performance in makespan is reduced
Improving energy efficiency of IO-intensive MapReduce jobs	2015	HPc hubs occupied with IO devices and the CPUs were not utilized, bringing about a terrible showing. This issue can be overcome by changing the power setting and it increases the IO devices performance. MapReduce occupations by playing out an intensive observational investigation	If the cpu frequency changes then the performance will be low there by the system consumes more energy	A lot of relapse models to anticipate the vitality utilization of IO based jobs for CPU recurrence to given information volume
An energy-aware bi-level optimization model for multi-job scheduling problems under cloud computing [11]	2014	Cloud computing energy consumption will be reduced by improving the energy efficiency of the servers. The usage of energy is reduced by considering the energy consumption in terms of performance, bandwidth and scheduling into account	Since it is in cloud scheduling will be more as ten to more number of tasks to be rescheduled. It leads to a large-scale optimization problem	To prove the effectiveness of the problem integer bi level algorithm were used
Energy efficient utilization of resources in cloud computing systems [12]	2012	Reducing the power consumption in task consolidation will increase the resource utilization. The task	An heuristics should be assigned to each task on a resource	To increase the resource utilization with reduced power consumption both

(*continued*)

Table 1. (*continued*)

Paper name	Year	Advantages	Disadvantages	Limitations
		consolidation is not only used in increasing the resource utilization but it also used to free up the resources that sits idle in a network		idle and active method were used
Energy efficiency for large-scale mapreduce workloads with significant interactive analysis [13]	2012	Berkeley Energy efficient mapreduce (BEEMR) algorithm were used to decrease the energy consumed for Map reduce interactive analysis workloads. The BEEMR algorithm will saves up to 40–50% of energy in large data constraints	Even though mapreduce interactive analysis holds large volume of data the operation of interactive jobs takes place in a small fraction of data	Expanding equipment use improves proficiency, yet is trying to accomplish for MIA remaining burdens
Energy-aware hierarchical scheduling of applications in large scale data centres [14]	2011	An hierarchical scheduling algorithm (HSA) is proposed. The advantage of using this algorithm is that it reduces the energy. For both network and Servers	If the number of servers of datacentre increased, then the time complexity of HAS also increases	Hierarchical crossing-switch adjustment reduces the data transfer and running servers
Energy efficiency for MapReduce workloads: an in-depth study [15]	2012	Four factors that decreases the energy of map reduce is identified and based on the factors the energy consumption for various clusters is measured and defined	The implementation is done in Hadoop. The response time of Hadoop in map reduce will be low	Various clusters used to measure the energy consumption will work better when compared with the previous cluster algorithms proposed
Exploiting spatio-temporal tradeoffs for energy-aware mapreduce in the cloud	2012	An unique spatio temporal tradeoff is sets to reduce the tradeoffs obtained by Map reduce clusters by using VM placement algorithm	Placing or fitting Virtual Machines in servers in machine resources and balanced fitting is complex	The Vms are placed for map reduce clusters in private cloud environment. The clusters are distributed based on the repetition and batch execution jobs

11   Results Analysis

The simulation were dine using Cloud Sim with the initial parameter were set with the file size. Then simulation uses the improved DVFS algorithm by Analyzing the task size & complexity and choose the right processor to complete the task, So that the power consumption and the performance of the system will be improved (Fig. 3).

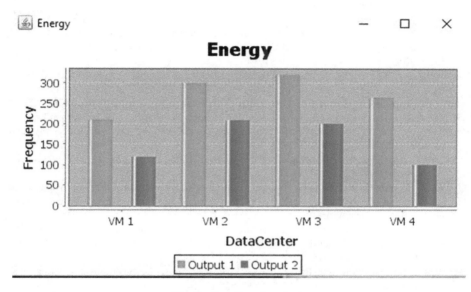

Fig. 3.   Result and analysis

12   Conclusion and Future Work

In this work, DVFS is a prominent procedure for lessening force and vitality utilization under unique working conditions by focusing on various power and execution modes in a solitary plan. In this paper, we showed that DVFS will permit the examination of existing methodologies and their effect on vitality, power, and execution. Further the framework can be created is to utilize both vitality mindful strategies and nature of administration approaches to create green booking calculations for various classes of utilizations.

Another future work of this work is to utilize both vitality mindful systems and nature of administration strategies to create green booking calculations for various classes of uses.

References

1. Huai, W., Huang, W., Jin, S., Qian, Z.: Towards energy efficient scheduling for online tasks in cloud data centers based on DVFS. In: 2015 9th International Conference on Innovative Mobile and Internet Services in Ubiquitous Computing (IMIS), pp. 225–232. IEEE (2015)
2. Ruan, X., Qin, X., Zong, Z., Bellam, K., Nijim, M.: An energy-efficient scheduling algorithm using dynamic voltage scaling for parallel applications on clusters. In: Proceedings of 16th International Conference on Computer Communications and Networks, ICCCN 2007, pp. 735–740. IEEE (2007)
3. Lai, Z., Lam, K.T., Wang, C.-L., Su, J., Yan, Y., Zhu, W.: Latency-aware dynamic voltage and frequency scaling on many-core architectures for data-intensive applications. In: 2013 International Conference on Cloud Computing and Big Data (CloudCom-Asia), pp. 78–83. IEEE (2013)
4. Huai, W., Qian, Z., Li, X., Lu, S.: Towards energy efficient data centers: a DVFS-based request scheduling perspective. In: 2013 Seventh International Conference on Innovative Mobile and Internet Services in Ubiquitous Computing (IMIS), pp. 299–306. IEEE (2013)
5. Iyapparaja, M., et al.: Coupling and cohesion metrics in Java for adaptive reusability risk reduction. In: IET Chennai 3rd International Conference on Sustainable Energy and Intelligent Systems (SEISCON 2012), pp. 52–57 (2012)
6. Li, K.: Energy efficient scheduling of parallel tasks on multiprocessor computers. J. Supercomput. **60**(2), 223–247 (2012)
7. Nezih, Y., Datta, K., Jain, N., Willke, T.: Energy efficient scheduling of MapReduce workloads on heterogeneous clusters. In: Green Computing Middleware on Proceedings of the 2nd International Workshop, p. 1. ACM (2011)
8. Iyapparaja, M., Sharma, B.: Augmenting SCA project management and automation Framework. IOP Conf. Ser. Mater. Sci. Eng. **263**, 1–8 (2017). https://doi.org/10.1088/1757-899x/263/4/042018. 042018
9. Mei, J., Li, K., Li, K.: Energy-aware task scheduling in heterogeneous computing environments. Cluster Comput. **17**(2), 537–550 (2014)
10. Iyapparaja, M., Tiwari, M.: Security policy speculation of user uploaded images on content sharing sites. IOP Conf. Ser. Mater. Sci. Eng. **263**, 1–8 (2017). https://doi.org/10.1088/1757-899x/263/4/042019. 042018
11. Wang, X., Wang, Y., Cui, Y.: An energy-aware bi-level optimization model for multi-job scheduling problems under cloud computing. Soft Comput. (2014)
12. Lee, Y.C., Zomaya, A.Y.: Energy efficient utilization of resources in cloud computing systems. J. Supercomput. **60**(2), 268–280 (2012)
13. Chen, Y., Alspaugh, S., Borthakur, D., Katz, R.: Energy efficiency for large-scale MapReduce workloads with significant interactive analysis. In: Proceedings of the 7th ACM European Conference on Computer Systems, pp. 43–56. ACM (2012)
14. Wen, G., Hong, J., Xu, C., Balaji, P., Feng, S., Jiang, P.: Energy-aware hierarchical scheduling of applications in large scale data centers. In: 2011 International Conference on Cloud and Service Computing (CSC), pp. 158–165. IEEE (2011)
15. Feng, B., Lu, J., Zhou, Y., Yang, N.: Energy efficiency for MapReduce workloads: an in-depth study. In: Proceedings of the Twenty-Third Australasian Database Conference-Volume 124, pp. 61–70. Australian Computer Society, Inc. (2012)

Educational Video Classification Using Fuzzy Logic Classifier Based on Arabic Closed Captions

Shadi R. Masadeh[1(✉)] and Saba A. Soub[2]

[1] Isra University, Amman, Jordan
shadi.almasadeh@iu.edu.jo
[2] MoE, Amman, Jordan
saba_soub@yahoo.com

Abstract. The process of adding closed-captions to the E-Learning courses offers several benefits to the hard of hearing or deaf people. Closed captioning allows many different populations of viewers to access the contents of videos. Traditionally deaf/hard of hearing users are the population of this service, but captioning also helps non-native language speakers and also viewers with learning disabilities. Closed captions can be useful to the viewers with no hard of hearing such as: improving the comprehension dialogues for non-native speakers, providing the ability to observe videos in sound-aware environments such as libraries and offices, and recognizing the technical terminologies and brand names behind noise environments. In this paper, a video classification technique for Arabic closed captions to create semantic categories using Fuzzy logic is introduced. This classification is important to apply indexing and searching as meta-data which makes the involved videos text-searchable. This contribution will be focused to help the layer of society which has a defect hearing problem but not deaf, by integrating them with the normal students in the regular schools and keep them in a good level of education without the need to join them to special needs schools.

Keywords: Closed captions · Open captions · Video classification · Video contents analysis

1 Introduction

The process of adding closed captions (CCs) to the E-Learning courses offers several benefits to the hard of hearing or deaf people. If the captions are not embedded into the design then it is called closed captions and usually they can be switched on or off by the learner as shown in [1] (Fig. 1).

Recently, it has been noticed that the viewers who are not hard of hearing use the CCs [1] as well. For instance, CCs are useful to the viewers who are not native speakers; so it is easier to come up with the speech, also they can improve the comprehension.

© Springer Nature Switzerland AG 2020
L. C. Jain et al. (Eds.): ICICCT 2019, LAIS 9, pp. 133–138, 2020.
https://doi.org/10.1007/978-3-030-38501-9_13

Fig. 1. An example of closed captions.

The challenge of this research originates from the ability to apply video contents analysis to build efficient browsing, searching and summarization system using Arabic CCs. Furthermore, video classification into semantic category for Arabic CC presents a great problem of interest [2]. A video classifier for news videos was proposed by [3] as shown in Fig. 2.

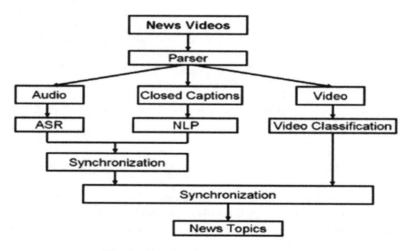

Fig. 2. The classification of news videos

The main aspects of this research are classified into the below issues:

1. Closed captions need many videos to covering the curriculums.
2. Need to find a way to management the curriculums and classify them.

3. Need to find a way to make the searching process during the closed caption not the videos.
4. Need to work in classification and retrieval in my search about the Closed Caption.
5. Make different categories for different level such as Elementary, middle and high school.
6. Apply the idea in math, which contains numbers and equations
7. Apply the idea in the language of art, which contains letters, words, different accents, and pronunciation.
8. Consider the Arabic language as the prime stone of this research.

2   Literature Review

The most challenge was retrieving and classifying the needed data from big collections of digital video. Thus, the CC text has been used as an option to improve the video classification and retrieval. In the research, the results have revealed that Arabic language is unlucky compared with English language and other languages for the reason that a number of systems are measured based on English closed caption.

The paper showed that the presented layout was able for classifying Arabic videos and also for retrieving Arabic video scenes into a set of predefined semantic groups including, economics, politics, tourism, religion, sports, social, weather, and health [4].

The KNN classifier method categories the input query and retrieves all the video scenes with the same category of the input query. Results were collected and prepared dataset about 2500 Arabic words are entered in the dictionary [4]. The closed-caption text facilitates the acquisition of the video transcript. The research was proposed a novel scheme for retrieving video scenes by exploring the Arabic closed-caption text. The proposed system was implemented and tested using a self-collected and prepared dataset.

The video transcript of closed caption represents the spoken dialogue and it includes also sound effects, speaker identification, music, and other non-speech information. The new proposed method was also designed to retrieve video scenes not the whole video. The evaluation of the developed system has indicated that the proposed approach was efficient in terms of both precision and recall [5].

There are three main methods in indexing and retrieval of videos in multimedia databases. Firstly, is the content-based indexing in which videos are indexed and retrieved using spatial and temporal characteristics of the video [6].

Secondly, is the keyword-based indexing which uses descriptive text or keywords to index video documents? Retrieval is performed by matching the query, given in the form of keywords, with the stored keywords [5].

Thirdly, is the multi-modal-based indexing which is a combination of the content – based indexing and keyword – based indexing however, the use of different components to represent the video still has its problems [5].

The research challenge was classified video scenes into semantic category. However, the main interest of the research video scene classification that is the categorization of video sequence into one of a few predetermined scene class and consternating on its semantic features. The approach was used the Arabic closed caption text which consists the speech transcript of the video [2].

Another work can be useful in providing the service of closed captions by providing the facial expressions for the talker. This can be applied by integrating the one of the researches presented in [7] or [8]. The main works were to create a method for recognizing the facial expressions. Also another information can be extracted from the used camera of recording by determining the location of recordings and describing the environments for the blind audiences as the research that was proposed in [9].

Closed captions can be involved in E-learning courses by integrating the use of the Internet as a communications medium between the instructors and the recorded video by considering the work presented in [10]. Furthermore, e-learning systems for the educational organisations have grown exponentially. The competition involves providing a secure connection, a novel authentication and authorization model (NAAM) was presented in [11] to securely transmit the data in e-learning environment without affecting the system resources.

A research to process the programming information to render the caption data at the specified location and to set up the programming on a display was introduced in [12] by providing a high level of configurability and the 3D experience while displaying captions.

A system to rapidly create high-quality CCs and subtitles for live TV shows using automated components namely automatic speech recognition was described in [13].

The presented work was on how the system feeds the human edits and corrections back into the different components for improvement.

3   The Research Methodology

The main aim of this paper is to develop CC system to help the layer of society that has a defect hearing problem but not deaf, by integrating them with the normal students in the regular schools and keep them in good level of education without the need to join them to special needs schools, and in this way it is expected to achieve two important issues as follows:

1. Avoiding these people from any psychological problems or introverted.
2. Involve them with the community as normal people, and after their studies; they are expected to participate in all life activities with normal people.

On the other side, CC will be very important in the government schools especially in middle east area, which as it known contain a high number of the students in each class, that's make the teacher need to use a high voice all the time without any consideration of the differences between them and the health cases that might be effect in the explanation process, also CC in these classes will give a higher chance of understanding for the students.

The main significant of the closed caption as follows:

1. **Enhanced Indexing and Searching:** Making video to be text-searchable enabled as meta-data.
2. **Viewer Flexibility:** The spread growing of mobile devices; accessing to the audio in sound-sensitive environment is required. CC allows watching videos in libraries, concerts, on a noisy bus, or hospitals.
3. **Improved Scalability:** The total number of viewers with a second English language is more than the English native speakers. CCs enforce the messages to reach more people.
4. **Enriched Comprehension:** CCs enrich comprehension and remembering of video contents for all viewers.

4   Conclusion and Future Works

The main issue of this paper is to present the proposed approach of building an efficient classification tool to browse, search, and categorizes videos with Arabic CCs. Also in this approach will be concentrated to help the layer of society which has a defect hearing problem but not deaf, by participating them with the normal students in the regular schools.

The Future work can be by using more efficient technique of classifiers in the concept of Arabic closed captions to help deaf students in all levels of education.

Acknowledgments. The author would like to thank you so much for the Isra University to fully cover for this research.

References

1. 3PlayMedia.   https://www.3playmedia.com/2015/08/28/who-uses-closed-captions-not-just-the-deaf-or-hard-of-hearing/. Accessed May 2018
2. Nassar, H., et al.: Classification of video scenes using Arabic closed-caption. In: Proceedings of the Third International Conference on Intelligent Computing and Information Systems, Cairo, Egypt (2007)
3. Luo, H., et al.: Personalized news video recommendation. In: International Conference on Multimedia Modeling. Springer (2009)
4. Anwar, A., Salama, G.I., Abdelhalim, M.: Video classification and retrieval using Arabic closed caption. In: ICIT 2013 the 6th International Conference on Information Technology VIDEO (2013)
5. Nassar, H., et al.: Retrieving of video scenes using Arabic closed-caption
6. Geetha, P., Narayanan, V.: A survey of content-based video retrieval (2008)
7. Zraqou, J., Alkhadour, W., Al-Nu'aimi, A.: An efficient approach for recognizing and tracking spontaneous facial expressions. In: 2013 Second International Conference on e-Learning and e-Technologies in Education (ICEEE). IEEE (2013)
8. Zraqou, J., Alkhadour, W., Al-Nu'aimi, A.: Robust real-time facial expressions tracking and recognition. Int. J. Technol. Educ. Mark. (IJTEM) **4**(1), 95–105 (2014)

9. Zraqou, J.S., Alkhadour, W.M., Siam, M.Z.: Real-time objects recognition approach for assisting blind people (2017)
10. Masadeh, S.R., Turab, N., Obisat, F.: A secure model for building e-learning systems. Netw. Secur. **2012**(1), 17–20 (2012)
11. Masadeh, S.R., Zraqou, J.S., Alazab, M.: A novel authentication and authorization model based on multiple encryption techniques for adopting secure e-learning system. J. Theor. Appl. Inf. Technol. **96**(6) (2018)
12. John, A.C.I.: Systems and methods for providing closed captioning in three-dimensional imagery. Google Patents (2012)
13. Sawaf, H.: Automatic speech recognition and hybrid machine translation for high-quality closed-captioning and subtitling for video broadcast. In: Proceedings of Association for Machine Translation in the Americas–AMTA (2012)

A Two-Level Authentication Protocol for Vehicular Adhoc Networks

M. Bhuvaneswari[1], B. Paramasivan[2(✉)], and Hamza Aldabbas[3]

[1] Department of Computer Science and Engineering,
National Engineering College, Kovilpatti, Tamilnadu, India
itsbhuvana@gmail.com
[2] Department of Information Technology, National Engineering College,
Kovilpatti, Tamilnadu, India
bparamasivan@yahoo.co.in
[3] Department of Software Engineering, Al-Balqa Applied University,
Al-Salt, Jordan
aldabbas@bau.edu.jo

Abstract. The technology growth nowadays has made its way into the world of transportation by establishing smart vehicles and smarter vehicular transportation named Vehicular Ad-hoc networks (VANET). VANET gains its fame day by day by the way how it facilitates the efficient and safe transportation to the voyagers. However, the prime requirement of VANET is authentication. The main aim of the proposed work is to develop a practical and efficient pseudonymous authentication protocol which augments the conditional privacy preservation. Two-Way authentication is provided using local alias and global alias name. The alias name has offered to all the vehicles who register with the certification authority. The simulation result shows that the Two-Way protocol provides authentication with optimal packet delivery ratio (PDR), end-to-end delay and propagation delay.

Keywords: Vehicular Ad-hoc Networks · Security · Pseudonymous authentication

1 Secure Communication in VANET: Overview

Vehicular Ad-hoc Network (VANET) is a special category of mobile ad-hoc standard. It uses the proliferation of wireless technology to perform vehicle to vehicle vehicular communication (V2VC) and Vehicle to Infrastructure communication (V2IC). Road Side Units (WRSU) also involve in providing infrastructure to vehicle communication. They are installed at a fixed distance in roads and they are interconnected with each other. In VANET, vehicles (nodes) have configured with couple of sensors, transceiver, processors and limited size of memory which helps to transfer packets among the vehicles/nodes. As per the standard "Dedicated Short Range Communication (DSRC)", a vehicle/node periodically forwards beacon messages which holds traffic and safety information. In addition, the beacon messages contain information like velocity of the

© Springer Nature Switzerland AG 2020
L. C. Jain et al. (Eds.): ICICCT 2019, LAIS 9, pp. 139–147, 2020.
https://doi.org/10.1007/978-3-030-38501-9_14

vehicles, location/position of the vehicle, direction of node mobility, congestion and accident. This information helps the drivers to avoid congested routes or accident by providing contextual view of traffic conditions.

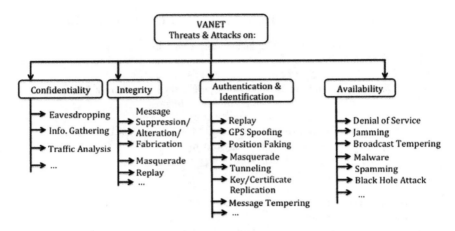

Fig. 1. Types of threats and attacks in VANET.

Since the aforementioned information reveals the context of the legal user (which may be irrelevant to the intruders), it is very critical to ensure the privacy of such information [3, 4]. Hence VANETs have high chance to expose with many attacks and threads. Figure 1 illustrates the different types of threats and attacks in VANET. Moreover, the OBU need not bother limited battery life like other mobile devices because the vehicle itself has enough source of electricity. The unique characteristics of VANET also bring specific vulnerabilities and various kinds of attacks [1, 2, 10]. Hence it is required to have a detailed classification about the attacks. The categorization classifies the attack based on the security requirement, confidentiality, integrity, and availability. It may also extend to further classification based on authentication and accountability [5] criteria. In this work, it is aimed to introduce a two-level authentication algorithm for authenticating a vehicle with other vehicles in the network during the communication. If there is any occurrence of malicious activity, immediately corresponding actions has been taken to track the perpetrator. Once the perpetrator node was tracked, it is averted from the network topology.

The remaining sections of the paper have organized as five sections. Section 2 discusses the bench mark survey in providing secure communication in VANET. Section 3 explains the proposed two-level authentication system. The experimental setup, simulation parameter configuration and result discussion of the proposed security model has been discussed in fourth section. Finally, Sect. 5 discusses the contribution of the proposed work.

2 System Initiation

2.1 Global Alias Acquisition

Each vehicle has provided with an identification number denoted by VID. The vehicle first gets its global alias from the Certification Authority. To reach its global alias, it registers itself with Certification Authority. Each vehicle is identified with its unique vehicle ID say its number plate. The global alias expires after some time to enhance its security. The random number has generated by cellular automaton algorithm. The random number along with the Vehicle ID identifies the vehicle uniquely. The public key of Certification Authority is known to all the vehicles which enter into the communication zone.

Step 1. Vehicle gets global alias from certification authority

$$\text{Vehicle} \rightarrow \text{CA} : (\text{VID}|\text{rand}|)\text{CAPK}$$

Step 2. Certification Authority issues global alias to vehicle

$$\text{CA} \rightarrow \text{Vehicle} : ((\text{rand})\text{Pkey}|\text{Tslot}|)\text{CASK}$$

2.2 Local Alias Acquisition

The RSU periodically broadcasts a message to notify the vehicles about its presence within its communication range. The transmitted message includes RSU's public key. Once a vehicle/node enters in to the RSU communication range, it makes a request message to get a local alias.

i. Vehicle/node request a local alias from RSU

$$\text{vehicle} \rightarrow \text{RSU} : ((T_{exp}|T_{slot}|(\text{rand})P_{Key})\text{CA}_{SK}|V_{PK^1}|-\text{rand})\text{RSU}_{PK}$$

2.3 RSU Processing

- RSU verifies CA's signature
- Since Certification authority opts for a homomorphic cryptosystem, RSU takes the homomorphic sum calculation of the random numbers. When the homomorphic sum nullifies then the identity of vehicle is confirmed.

$$R = (\text{rand})P_{key} + (-\text{rand})P_{key}$$

- RSU transmits $(R)P_{key}$ to Certification Authority (CA) for verification.
- Certification authority decrypts "R" and transmits a verification message to road side unit. Otherwise it sends a not verified message.

The RSU prepares a local alias, once it is getting verification message came from VID. Subsequently it generates the expiration timer (T_{R-exp}). The created timer embed with the newly generated PK^1. The PK^1 encrypts and sends it to vehicle/node.

Step 3. RSU sends local alias to Vehicle

$$RSU \rightarrow Vehicle : ((T_{R-exp}||V_{PK^1})RSU_{SK})V_{PK^1}$$

Once the vehicle receives its local alias it broadcasts a beacon message to

$$VID \ Broadcasts : (beacon)V_{SK^1}||(T_{R-exp}||V_{PK^1})SK_{RSU}$$

The message receiver/node verifies the alias by checking with the signature of RSU. It then validates the beacon message by matching Vi's sign with V_{PK^1}. The V_{PK^1} generally enclose with the secondary pseudonym.

3   Vehicle/Node Revocations

The message was checked with the associated key provided in the local alias by the received node/vehicle. Suppose the received node found that message is bogus, it reports the local alias to the LEA by sending the beacon message. The received messages will be relinquished by the received node, if any one of the signatures has not properly verified. If any node is trying to broadcast false packets, it is planned to perform the following steps to track the perpetrator and revoked him out from the network. Figure 2 illustrates the revocation process.

- The receiver sends the malicious activity logs to the LEA.
- Road side unit signed each secondary pseudonym. The signed copy is attached with the bogus message. LEA pings RSU to get that details.
- The road side unit provides the corresponding primary pseudonym to the LEA.
- CA has been instructed by the LEA in providing the decryption key of the vehicle identification number (VID) when it has malicious local alias.
- Then the Vehicle ID is placed in the CRL list, and its certification is revoked. Once a vehicle/node is averted, it cannot be a part in the network.

The beacon packets have been used to find out the traffic view. Hence a VANET has coverage limit of each vehicle/node is 300 m; the bogus traffic analysis report verification can be done in a short period. The beacon messages are logged for a short time period of 30–60 min (Table 1).

Table 1. System elements and its notations

S. No.	Notation	System element
1	VID	Vehicle identification number
2	RSU	Road side unit
3	CA	Certification authority
4	LEA	Law enforcement agency
5	CRL	Certification revocation list
6	V_{Pk}	Vehicle/node private key
7	V_{Sk}	Vehicle/node public key
8	RSU_{Pk}	RSU public key
9	RSU_{Sk}	RSU private key
10	P_{key}	Pailler key

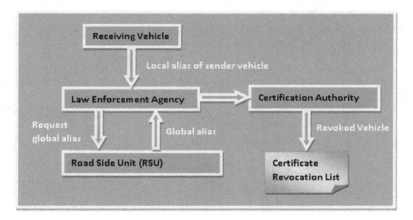

Fig. 2. Vehicle revocation

4 Attack Model

The road side unit (RSU) does not aware about the real identity of the vehicle/node inherent. The broadcasted beacon messages are encrypted by the private key (V_Pk) of the node whose public key (V_sk) is inbuilt within the local alias. RSU does not have the information regarding the real identities of the user. In case of a side channel attack on RSU, no valuable information about the used is revealed to the malicious node/attackers. It is also noted that privacy guarantees are provided to all legitimate user. The protocol offers privacy-preserving authentication by maintaining the original context of the vehicle with the certification authority itself. Only the certification authority has the rights to reveal the identity of the vehicle to the Law Enforcement Agency. Even the RSU does not have any clue on the original identity/status of the vehicle. The protocol also handles the integrity of the broadcasted message by encrypting it by the vehicle private key.

The local alias is set a shorter expiry time when compared to the global alias. Therefore, it is enough to implement the revocation policy only at the global end of encryption that is only the global alias is revoked. It is found that there was no problem, if the environment is a pervasive RSU environment. But in case the network environment has sparse deployment of RSU, then averted node of the network can broadcast beacon messages until TRSU expires. To handle such sparse RSU distribution trust values is appended with the local alias when a message is found to be malicious the receiver vehicle broadcasts the local alias of malicious vehicle by adding a negative trust value indicating that the holder of this local alias would be a malicious vehicle.

5   Results and Discussion

The proposed protocol has been compared with Vehicular Authentication Security Approach (VASA) and PDR Ad-hoc On-Demand Distance Vector (PDR-AODV) protocol by conducting simulation using NS2 and SUMO. The simulation setup is as follows (Table 2).

Table 2. Simulation parameter setup

Sl. No.	Simulation parameters	Values
1	Simulation area (m^2)	1500 × 1500
2	Simulation time (s)	6000
3	Antenna type	Sectorized
4	No. of vehicles	200
5	No. of RSUs	10
6	No. of attackers	5, 10, 15, 20
7	Transmission range (m)	250 and 100
8	Network interface type	IEEE 802.11p wirelessPhyExt
9	MAC type	IEEE 802.11p MacExt
10	Packet size (Bytes)	1024
11	Network traffic generator	UDP or CBR (1024 bytes, rate: 2)
12	Vehicle speed (Kmph)	NH - 80 to 120 and City - 30 to 60

The Fig. 3 proves that proposed work has less propagation delay than existing works. It has also noted that the propagation delay is increasing gradually when the number of attackers increased. The Fig. 4 exhibits that the proposed work outperforms well than VASA and PDR-AODV in terms of PDR. Here also the presence of attackers has made an impact in packet routing. From the Fig. 5, it has been concluded that the routing overhead is higher for the proposed work than its related work.

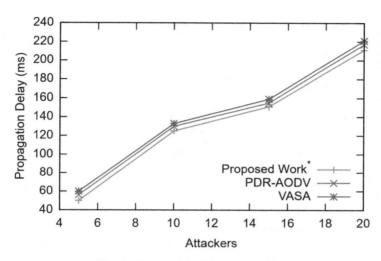

Fig. 3. Propagation delay vs. attackers

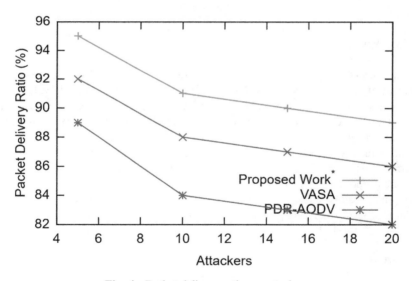

Fig. 4. Packet delivery ratio vs. attackers

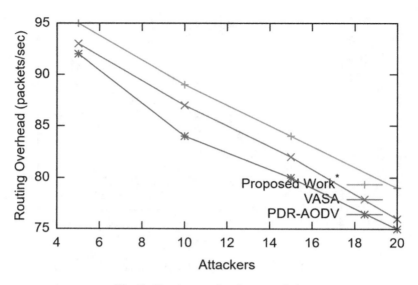

Fig. 5. Routing overhead vs. attackers

6 Conclusion

The main aim of the proposed work is to introduce a conditional privacy preservation augmented two-level pseudonymous authentication technique. A novel two-level authentication algorithm has been introduced with a different lifetime. The proposed two-level authentication protocol outperforms over the existing secure communication algorithms in terms of trust. It also ensures that there is less chance in disclosing of valuable data. Moreover, the protocol provides privacy to each network users. The malicious nodes or vehicles are revealed by the proposed work. The security analysis shows the proposed protocol has optimal resistance against all the security threats. The experimental results and discussion portray the optimal performance of the proposed algorithm in terms of PDR, propagation delay and routing overhead.

References

1. Sun, Y., Yu, W., Han, Z., Liu, K.J.R.: Information theoretic framework of trust modeling and evaluation for ad hoc networks. IEEE J. Sel. Areas Commun. **24**(2), 305–317 (2006)
2. Sen, J.: A robust and efficient node authentication protocol for mobile ad hoc networks. In: International Conference on Computational Intelligence Modelling and Simulation, Bali, pp. 476–481. IEEE (2010)
3. Ellcy Priana, M.: Trust based clustering and secure authentication for multicast in ad-hoc. Int. J. Comput. Appl. (0975-8887) **108**(19) (2014)
4. Muqtadir, A.: Secure routing protocol for VANET using hashing with session key. Int. J. Res. **3**(1), 687–696 (2019)

5. Mohammad, A.A.K., Mirza, A., Vemuru, S.: Cluster based mutual authenticated key agreement based on chaotic maps for mobile ad hoc networks. Indian J. Sci. Technol. **9**(26) (2016)
6. Muqtadir, A., Sattar, S.A.: Network security based on chaotic maps authentication in VANET. J. Appl. Sci. Comput. **5**(12), 280–292 (2018)
7. Verma, U.K., Kumar, S., Sinha, D.: A secure and efficient certificate based authentication protocol for MANET. In: International Conference on Circuit, Power and Computing Technologies (ICCPCT), Nagercoil, pp. 1–7 (2016)
8. Roy, D., Das, P.: Trusted and secured routing protocol for vehicular ad-hoc networks. Indian J. Sci. Technol. **10**(17), 1–12 (2017)
9. Jeyaprakash, T., Mukesh, R.: A new trusted routing protocol for vehicular ad hoc networks using trusted metrics. Int. J. Netw. Secur. **19**(4), 537–545 (2017)
10. Baqar, M.A., Aldabbas, H., Alwadan, T., Alfawair, M., Janicke, H.: Review of security in VANETs and MANETs. In: Network Security Technologies: Design and Applications, pp. 1–27. IGI Global (2014)

Bengali Poem Generation Using Deep Learning Approach

Md. Kalim Amzad Chy[1], Md. Abdur Rahman[2],
Abdul Kadar Muhammad Masum[1(✉)], Shayhan Ameen Chowdhury[1],
Md. Golam Robiul Alam[1], and Md. Shahidul Islam Khan[1]

[1] Department of Computer Science and Engineering, International Islamic
University Chittagong, Chittagong, Bangladesh
kalim.amzad.chy@gmail.com, akmmasum@yahoo.com,
shayhan.ameen@gmail.com, rabiul.alam@bracu.ac.bd,
nayeemkh@gmail.com
[2] Department of Computer Science and Engineering, Dhaka Commerce College,
Dhaka, Bangladesh
mardcc@gmail.com

Abstract. This research proposes a deep learning based effective Bangla poem generation model that can generate any famous Bengali poet style poem. Unigram, bigram, and trigram word level approach has been followed to make the poem generation model, which is a very early study for this language. From various online website and social media about 35K Bangla poem containing more than 90 million words has been scrapped using python programming language. After preprocessing and tokenizing the data, text data is converted to the n-dimensional numerical vector space using word embedding tool. To build the language model various extension of Recurrent Neural Network (RNN) like LSTM, BiLSTM has experimented. Also, another merge mode for BiLSTM such as sum, avg, mul, concat has been tested. The generated poems are evaluated via various modern evaluation methods like BLEU, GLEU, and a Turing test. Among diverse neural network, BiLSTM performs better for unigram, bigram, trigram where for bigram it gives the best outcomes against BLEU and GLEU more than 87% and Turing test 89% satisfactory level. This study is an excellent contribution to the Bengali text generation research.

Keywords: Web scrapping · Word embedding · Text generation · Natural language processing · LSTM · BiLSTM · GLEU

1 Introduction

Natural Language Generation (NLG), a subset of Natural Language Processing (NLP) that leverages knowledge in Artificial Intelligence (AI) and Computational Linguistics (CL) to generate human understandable text [1]. Text generation techniques can be applied to improve language models, summarization, machine translation, and captioning. About 245 million population speak in the Bengali language that ranked seventh out of the top 30 languages of the world [2]. Bengali text generation, more specifically Bengali poem generation adds an extra dimension to the Bangla NLP research community.

© Springer Nature Switzerland AG 2020
L. C. Jain et al. (Eds.): ICICCT 2019, LAIS 9, pp. 148–157, 2020.
https://doi.org/10.1007/978-3-030-38501-9_15

The primary purpose of this research is to build a Bengali poem generation model that creates the great Bengali poet style poem. Hidden Markov Models (HMM) can be used to generate text but difficult to create a whole sentence that's why RNN, a type of neural network, deals with sequence data that fit for our text generation model as a poem is a sequence of sentences or words. Due to the vanishing gradient problem, despite RNN, Long Short-Term Memory (LSTM), Bidirectional LSTM (BiLSTM), etc. architectures are chosen. Because of the scarcity of the Bangla data, a Bangla Poem Corpus (BPC), collected from various website and social media platform, is built via Web Scrapping techniques. After preprocessing the BPC, the text data is converted into numerical data via Keras word embedding. Among character level, word level, sentence level model word-level model performs better. After compiling and training the various type of neural network model, the model successfully generated the famous Bengali poet style poem.

The rest of the paper constructed with five sections, including the introduction. Section 2 discusses previous related research. Section 3 contains a methodology where the working procedure, data pre-processing, sentence representation, various language model, and model evaluation methods are described. Section 4 carries out an experimental result, and Sect. 5 concludes the paper.

2   Related Works

First text generation work appeared in the 1970s, and in 1980s McDonald used text generation in decision-making problem [1]. Nowadays, deep learning-based text generation is a great attraction of the NLP researcher. In past decades most of the poet generation work was limited to only templated based, but deep learning based current research model performs much better. Ghazvininejad et al. designed an interactive poem generation system at word level based upon RNN decoder [3] while Yan uses RNN encoder and decoder at character level [4]. Zhang and Lapata worked with Chinese poem to generate poem using Recurrent Context and Generation model [5], and Yi et al. followed mutual reinforcement learning that not only learns from the rewarder but also from each other [6]. Kutlugun et al. deal with Turkish poem to generate meaningful text based upon trigram model [7] while Dogan et al. tried to create original Turkish text about a specific subject employing text mining and deep learning methods [8]. Ismail et al. deal with Arabic text that generates a summarized text of an input document via Rich Semantic Graph (RSG) using Ontology [9]. A corpus-based poetry generation system using templates is proposed to generate poems under the given constraints on rhyme, stress, meter, sentiment, word similarity, and frequency [10]. In neural machine transliteration, Jadidinejad et al. proposed character level sequence to sequence model wherein encoder is based on BiLSTM and decoder RNN with attention [11]. Islam et al. proposed a sequence to sequence Bangla sentence generation using only LSTM that considers only 12.5K words [12] ignoring various better neural network architecture. Most of the system just limited to a few specific neural network architectures in spite of testing various type of extension of RNN. Evaluating the performance of the model through various method name BLEU, GLEU, etc. is missed in a great number of models.

Though several text and poem generation research has been found on English, Arabic, Turkish, etc. languages, in perspective of Bangla language, there doesn't exist any decent deep learning-based poem generation research. To the best of our study, this proposed system is the first effort to build a Bangla poem generation model using various type of neural network named RNN, LSTM, and BiLSTM. Also, a different mode of BiLSTM, sum, mul, ave, concat, is experimented to make an outstanding model. Different types of text generation model evaluation procedure like BLEU, GLEU, Turing test has been adopted to understand the performance of the trained model.

3   Methodology

The text generation model is a chain of the consecutive following six phases:

(1) Data Collection
(2) Data Pre-processing
(3) Sentence Representation
(4) Preparing data for machine
(5) Language Model
(6) Model Evaluation

Figure 1 demonstrates the whole picture of the proposed deep learning-based poem generation model that is described in following discussions.

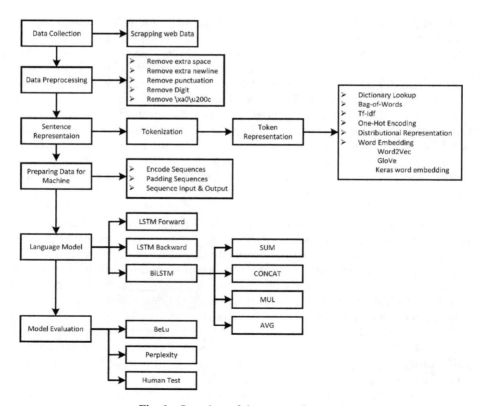

Fig. 1. Overview of the proposed system

3.1   Data Collection

A well-structured dataset is a driving force for any machine learning predictive model. In Bangla natural language processing perspective, still there have a great lacking of Bangla dataset. By scrapping using python programming language form various Bangla poem web site and social groups, we have collected about 30K poems containing more than 9 million words. For simplicity, we have just filtered out famous poems of the national poet of Bangladesh, Kazi Nazrul Islam, to generate his style poem. To generate another poet style poem, the same procedure is applicable. Name of the poem, poet name, category of the poem, poem, comments against each poem is considered to build the first Bangla poem corpus that is still growing by adding daily basis poem data. Figure 2 is an overview of the Bangla Poem Corpus (BPC).

```
No of Poem:   400
No of Lines:  21029
No of Words:  101204
No of Character:  708308
No of Unique Word:  20622
```

Fig. 2. Statistics of Bangla Poem Corpus (BPC) Corpus

Figure 3 is a pie visualization based upon the found nine poem categories wise, collected poem of Kazi Nazrul Islam, where "Not Found" refers unknown category.

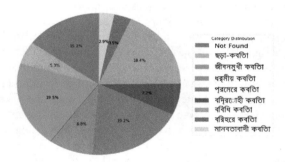

Fig. 3. Category wise poem visualization

3.2   Data Preprocessing

The raw dataset contains lots of garbage like extra space, newline, non-breaking space, etc. need to be cleaned. From Fig. 1, data preprocessing phase shows up the undertaken steps to clean the data. Non-breaking space (\xa0), half space character (\u200c), digits are deleted. Extra space is replaced by a single space while extra newlines replace by a single newline. As the punctuation doesn't have any contribution to make a good text generation model, Bangla dari, comma, hyphen, etc. type of punctuation is removed from the corpus.

3.3   Sentence Representation

Text data need to be defined numerically, as a machine can't handle text data directly. Tokenization is one kind of splitting the sentence or poem into small elements that can be in character level, word level, or sentence level. Table 1 demonstrates different level tokenization of a part of a poem. Word level tokenization better denote the meaning of a text and have benefited over others like poorly generated sentences can be processed, fewer strings can be generated.

Table 1. Tokenization in different level

Sentence	'এই শিকল পরা ছল মোদের\nএ শিকল পরা ছল'
Char Level	['এ', 'ই', ' ', 'শ', 'ি', 'ক', 'ল', ' ', 'প', 'র', 'া', ' ', 'ছ', 'ল', ' ', 'ম', 'ো', 'দ', 'ে', 'র', '\n', 'এ', ' ', 'শ', 'ি', 'ক', 'ল', ' ', 'প', 'র', 'া', ' ', 'ছ', 'ল']
Word Level	['এই', 'শিকল', 'পরা', 'ছল', 'মোদের\n', 'এ', 'শিকল', 'পরা', 'ছল']
Sentence Level	['এই শিকল পরা ছল মোদের', ' এ শিকল পরা ছল']

After tokenizing, the tokens need to be changed into numbers so that the machine can compute with. Among various techniques, dictionary lookup is the simplest where each unique word in the vocabulary is assigned an ID. Because of the random numbering, the greater value may be considered as more important to the deep learning model while the assigned ID doesn't refer to that sense. One hot encoding, another approach, generate a vocabulary size vector filled with zeros but one. For a word, only respective column is 1 and the rest 0 valued. Though it free from bias problem but one-hot encoded vector ignores the context of the word used in the corpus and doesn't store any meaningful information. In Bag-of-Words representation, word frequency is the primary key to convert into numerical form, while one-hot encoding just focuses on the existence of a word in a sentence. Extending this representation, TF-IDF also considers the inverse frequency i.e. the frequency of the word in other sentences that gives higher value to the important words ignoring common words. One of the main drawbacks of this type of representation is that they don't consider the word consequences, thereby discarding the context of the sentence is not suitable for an NLG model.

Word Embedding, best fit for NLG model, defines words in a coordinate system where correlated words, based on a corpus of relationships, are placed closer together. Word2Vec takes a large text corpus and generate n-dimensional vector space for each unique word. It is famous for capturing the context of the word. GloVe, an extension of Word2Vec, constructs an explicit word co-occurrence matrix using statistics across the whole text corpus. Keras embedding layer is also a great choice to the NLG researcher that is set with random weights at the beginning and will learn an embedding for all of the words in the corpus. Keras embedding layer must specify three arguments named input dimension, output dimension, input length, and in output, an embedded 2-dimensional vector with one embedding for each word in the input document.

3.4 Preparing Data for Machine

The Keras embedding layer anticipates input sequences as integers and can be mapped each unique word of the vocabulary as a unique integer. From the entire training data, the Keras Tokenizer API finds out all unique words and assigns a unique integer. After training, the fit tokenizer converts each sequence of the list of words to a list of integers. To define the embedding layer, need to know the vocabulary size that can be achieved by the tokenizer word index attribute. As the length of each sentence is not equal padding is essential to make each length equal. Keras preprocessing has a pad sequence method by which each sentence can be padded by the max length of the sentences by assigning extra zero to the extended list at pre or post position. Through array slicing, the sequence is separated as input (X) and output (y). Output (y) pass through the one-hot encode process, the number of classes so that the model learns to forecast the probability distribution for the next word.

3.5 Language Model

A language model can foresee the likelihood of the following word in the sequence based upon the already observed words. Neural networks are a favored strategy because of their distributed representation where similar meaning words closely represented and maintaining a large context of already experimental words at the time of predictions.

RNN, a short-term memory, can't carry out the information for a long sequence. Because of the vanishing gradient problem, RNN can't significantly update the initial weight. LSTM, and BiLSTM are the solution of the RNN because of their internal mechanisms to regulate the flow of information.

Long Short-Term Memory (LSTM), Fig. 4, is very much similar to the RNN except the internal gate concepts. The cell state can transmit related information all over the processing of the sequence that's why former time steps information can make its way to next time steps. Here, the previous hidden state and the current input get concatenated, say merge, fed into the forget layer to remove non-related data. By using merge, a candidate layer is formed that holds possible values to add to the cell state. Merge also fed into the input layer to decide what data from the candidate should be added to the new cell state. By the outcomes of the candidate, forget, and output layer, the cell state is calculated. Point-wise multiplying of the computed output and the new cell state gives the new hidden state. In the case of backward LSTM, it processes the input sequence backward and returns the reversed sequence.

LSTM only preserves information of the past while BiLSTM runs the inputs in two directions, one from past to future and another from future to past. This bi-direction approach helps the model to understand the context better. There has mainly four types of merge mode name sum, mul, concat, ave by which outputs of the forward and

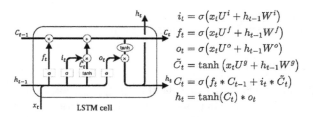

$$i_t = \sigma\big(x_t U^i + h_{t-1} W^i\big)$$
$$f_t = \sigma\big(x_t U^f + h_{t-1} W^f\big)$$
$$o_t = \sigma\big(x_t U^o + h_{t-1} W^o\big)$$
$$\tilde{C}_t = \tanh\big(x_t U^g + h_{t-1} W^g\big)$$
$$C_t = \sigma\big(f_t * C_{t-1} + i_t * \tilde{C}_t\big)$$
$$h_t = \tanh(C_t) * o_t$$

Fig. 4. LSTM network

backward RNNs will be merged. From the experimental view, concat and sum work much better for our poem generation model where concat mode works better than the best.

3.6 Model Evaluation

BiLingual Evaluation Understudy (BLEU), most popular to evaluate machine translation also for text generation, determines the number of words matched in a given sentence with a reference sentence, giving higher scores to consecutive words (recall) [13]. But it doesn't consider different types of errors like substitutions, paraphrase, insertions. The General Language Evaluation Understanding (GLEU) used to evaluate and analyze the accuracy of the models across various range of current Natural Language Understanding (NLU) tasks, a simple variant of BLEU, much more closely to human judgments [14]. Correlates well with BLEU metric on a corpus metric but does not have its drawbacks for per sentence reward objective. Turing Test, testing the outcomes by a group of human methods also adopted to evaluate the model. A group of ten people, including five technical and five non-technical people, score the outcomes out of five.

4 Experimental Results and Discussion

Among the three major types of text generation approach like character level, word level, sentence level text generation, word level text generation method gives the most promising output. In word level text generation, poem sequences have experimented on unigram, bi-gram, and tri-gram levels. To maintain a good vocabulary size, unique words are down from about 20K to 15K as the infrequent words have fewer examples, and model can't learn the proper use of them. In every LSTM, and BiLSTM, at the top, a vector space of 20 dimensions embedding layer, two LSTM/BiLSTM layer with 0.5 dropout layer and two dense layers are placed sequentially to configure the model. Two hundred fifty-six neurons are set in hidden layer while batch size (at a time the amount of data to be processed in parallel) is 50 and epochs are 250. At the output layer, a softmax activation function is utilized to confirm the outputs have the features of normalized likelihoods. The efficient Adam optimizer to mini-batch gradient descent is chosen. After training the network model architecture and weights are saved in HDF5 file format that will help later to generate the text just by loading the pre-trained model.

The evaluation of the performance of the trained model is demonstrated in Table 2, where the BLEU, GLEU, and human testing is observed for every kind of used neural network. The table illustrates that at unigram level LSTM forward and backward works similarly while BiLSTM works better. Similarly, BiLSTM performs better in other word level text generation model. Overall bigram word level approach with BiLSTM network architecture gives us outstanding results against BLEU and GLEU more than 87% and Turing test 89% satisfactory level.

Table 2. Model evaluation

Word level	Evaluation metric	LSTM forward	LSTM backward	BiLSTM
Unigram	BLEU	76.57	76.64	81.48
	GLEU	76.61	76.62	81.35
	Turing test	80.50	80.50	83.00
Bigram	BLEU	84.65	84.69	87.43
	GLEU	84.67	84.72	87.45
	Turing test	86.00	86.50	89.00
Trigram	BLEU	82.62	82.71	86.33
	GLEU	82.68	82.74	86.54
	Turing test	82.00	85.00	88.00

Tables 3 and 4 are two sample observation of bigram word level approach where the top row is the original poem, and the below column are generated poem via respective architecture.

Table 3. A sample observation of Bigram model

ওরে ভয় নাই আর দুলিয়া উঠিছে হিমালয় চাপা প্রাচী গৌরশিখরে তুহিন ভেদিয়া আগিছে সব্যসাচী ছাপর যুগের মৃত্যু ঠেলিয়া আগে মহাযোগী নয়ন মেলিয়া মহাভারতের মহাবীর আগে বলে আমি আসিয়াছি				
LSTM Forward	**LSTM Backward**	**GRU**	**BiLSTM**
ওরে ভয় নাই আর দুলিয়া উঠিছে হিমালয় চাপা প্রাচী গৌরশিখরে তুহিন ভেদিয়া আগিছে সব্যসাচী ছাপর যুগের মৃত্যু ঠেলিয়া আগে মহাযোগী নয়ন মেলিয়া মহাভারতের মহাবীর আগে না অমৃতর	ওরে ভয় নাই আর দুলিয়া উঠিছে হিমালয় চাপা প্রাচী গৌর শিখরে তুহিন ভেদিয়া আগিছে সব্যসাচী ছাপর যুগের মৃত্যু ঠেলিয়া আগে মহাযোগী নয়ন মেলিয়া মহাভারতের মহাবীর আগেনা	ওরে ভয় নাই আর দুলিয়া উঠিছে হিমালয় আজ আমি সেই সে জাতি পাপবিদগ্ধ তৃষিত ধরার লাগিয়া আনিল যারা সে এ কোন	ওরে ভয় নাই আর দুলিয়া উঠিছে হিমালয় চাপা প্রাচী গৌরশিখরে তুহিন ভেদিয়া আগিছে সব্যসাচী ছাপর যুগের মৃত্যু ঠেলিয়া আগে মহাযোগী নয়ন মেলিয়া মহাভারতের মহাবীর আগে বলে

Table 4. Another sample observation of Bigram model

বন্দি তোমায় ফন্দি কারার গতিমুক্ত বন্দিবীর লম্খিলে আজি ভয়দানবের ছয় বছরের জয় প্রাচীর বন্দি তোমায় বন্দিবীর জয় জয়ন্ত বন্দিবীর							
LSTM Forward		**LSTM Backward**		**GRU**		**BiLSTM**	
বন্দি	তোমায়	বন্দি	তোমায়	বন্দি	তোমায়	বন্দি	তোমায়
ফন্দি	কারার	ফন্দি	কারার	ফন্দি	কারার	ফন্দি	কারার
গতিমুক্ত	বন্দিবীর	গতিমুক্ত	বন্দিবীর	গতিমুক্ত	বন্দিবীর	গতিমুক্ত	বন্দিবীর
লম্খিলে	আজি	লম্খিলে	আজি	লম্খিলে	আজি	লম্খিলে	আজি
ভয়দানবের	ছয়	ভয়দানবের	ছয়	ভয়দানবের	ছয়	ভয়দানবের	ছয়
বছরের জয়প্রাচীর		বছরের জয়প্রাচীর		বছরের জয়প্রাচীর		বছরের জয়প্রাচীর	
বন্দি	তোমায়	বন্দি	তোমায়	বন্দি	তোমায়	বন্দি	তোমায়
বন্দিবীর	জয়	ফন্দি	কারার	বন্দিবীর	জয়	বন্দিবীর	জয়
জয়ন্ত	বন্দিবীর	গতিমুক্ত বন্দিবীর		জয়ন্ত বন্দিবীর		জয়ন্ত বন্দিবীর	

A few words are used as seed text to generate the poem via the trained model. From these outcomes, we have seen that, the developed text generation model using various kind of deep learning approach more closely generate poem that is similar to the original poem in context of the literal and metaphorical meaning of the poem.

5   Conclusion

Natural language processing techniques are very efficient way to establish communication between humans and a computer. Deep learning-based NLP research able to attract to the NLP researcher for its outstanding performances. Dealing with Bengali text generation is a challenging topic due to its diversity in language and corpus. Our proposed deep learning-based poem generation model can generate famous poet style poem. To handle the sequential data various deep learning approach like LSTM, BiLSTM, has been experimented where BiLSTM performs better for its bidirectional architecture. A large number of poem corpus will increase the accuracy of the poem generation model. In future attention based RNN is under our consideration to build better model. Increasing the number of epoch and tuning more better hyper parameter are our future scope. Developing a multilingual text generation model is our future scope. We hope the proposed model will pave the way to more advance research to the field of Natural Language Generation (NLG) for Bengali language.

References

1. Zhang, L., Sun, J.-T.: Text generation. In: Liu, L., Özsu, M.T. (eds.) Encyclopedia of Database Systems, pp. 3048–3051. Springer, New York (2009)
2. Mandal, A.K., Sen, R.: Supervised learning methods for Bangla web document categorization. arXiv preprint arXiv:1410.2045 (2014)
3. Ghazvininejad, M., Shi, X., Priyadarshi, J., Knight, K.: Hafez: an interactive poetry generation system, pp. 43–48 (2017)

4. Yan, R.: i, Poet: automatic poetry composition through recurrent neural networks with iterative polishing schema, pp. 2238–2244 (2016)
5. Zhang, X., Lapata, M.: Chinese poetry generation with recurrent neural networks, pp. 670–680 (2014)
6. Yi, X., Sun, M., Li, R., Li, W.: Automatic poetry generation with mutual reinforcement learning, pp. 3143–3153 (2018)
7. Kutlugün, M.A., Şirin, Y.: Turkish meaningful text generation with class based n-gram model, pp. 1–4. IEEE (2018)
8. Doğan, E., Buket, K., Müngen, A.: Generation of original text with text mining and deep learning methods for Turkish and other languages, pp. 1–9. IEEE (2019)
9. Ismail, S.S., Aref, M., Moawad, I.F.: A model for generating Arabic text from semantic representation, pp. 117–122. IEEE (2015)
10. Colton, S., Goodwin, J., Veale, T.: Full-FACE poetry generation, pp. 95–102 (2012)
11. Jadidinejad, A.H.: Neural machine transliteration: preliminary results. arXiv preprint arXiv: 1609.04253 (2016)
12. Islam, M.S., Mousumi, S.S.S., Abujar, S., Hossain, S.A.: Sequence-to-sequence Bangla sentence generation with LSTM Recurrent Neural Networks. Procedia Comput. Sci. **152**, 51–58 (2019)
13. Papineni, K., Roukos, S., Ward, T., Zhu, W.-J.: BLEU: a method for automatic evaluation of machine translation, pp. 311–318. Association for Computational Linguistics (2002)
14. Wu, Y., Schuster, M., Chen, Z., Le, Q.V., Norouzi, M., Macherey, W., Krikun, M., Cao, Y., Gao, Q., Macherey, K.: Google's neural machine translation system: bridging the gap between human and machine translation. arXiv preprint arXiv:1609.08144 (2016)

Incorporating Digital "Teach-nology" and Mobile Learning Application in Teacher Education in VNU, Hanoi

Ton Quang Cuong[1(✉)], Pham Kim Chung[1], Nguyen Thi Linh Yen[2], and Le Thi Phuong[1,2]

[1] VNU University of Education, Vietnam National University, Hanoi, 144, Xuan Thuy, Cau Giay, Hanoi, Vietnam
{cuongtq, chungpk, phuopnglt}@vnu.edu.vn
[2] VNU University of Languages and International Studies, Vietnam National University, Hanoi, Pham Van Dong, Cau Giay, Hanoi, Vietnam
linhyen.nguyen@vnu.edu.vn

Abstract. Today's students are digital natives who learn best when they have in hand digital tools for various interactive experiences. Digital transformation in education, mobile/wearable devices, and applications, BYOD/BYOTs do this, but it can be a complex process, especially in pedagogical aspects. There is also change the conception of ICT use in education by well-known educational technology preferably digital trend in advance. In Vietnam, mobile device ownership is increasing day by day. Relatively, discussions on the benefits of mobile device applications in education.

Keywords: Digital education · Mobile learning application · Teacher education

1 Introduction

Vietnam has a competitive and hard-working, young workforce of nearly 56 million people of total up to 97 million population with large number of mobile subscribers, internet and active social media users of 143,3, 64 and 62 million respectively (by statistics on January 2019). Importantly, the ability to lifelong learning is a required skill today in changing world. Moreover, it should focus on soft skills training as core human strengths together with digital know-how, re-skilling or up-skilling at work.

Efficiency vs effectiveness, usage vs usability, equity vs disparity access, pedagogy vs technology etc. make serious educational dilemma today. Mobile applications and devices in the classroom and digital literacy are critical aspect of young student's learning, experience and skill development. Playing not only a role of ubiquitous tool supported such learning, these devices can incorporate students creative thinking, expand learning experience and environment with so they can effectively participate in the classroom.

© Springer Nature Switzerland AG 2020
L. C. Jain et al. (Eds.): ICICCT 2019, LAIS 9, pp. 158–166, 2020.
https://doi.org/10.1007/978-3-030-38501-9_16

Thus, it may rise the needs to research and implement the followings:

- What the scientific arguments to formulate a theoretical framework of E-pedagogy or digital pedagogy?
- Which policies for mobile devices use can be developed in schools?
- How specific learning apps and digital skills can develop for teachers and students in whole education? and
- How digital technology can effort implementation whole-education curriculum for the VNU-University of Education?

Furthermore, teachers, learners, as well as educational institutions, are beginning to discover the many benefits of utilizing mobile technology in classrooms and toward the balance of teaching integrated by digital technology ("teach-nology").

2 E-Pedagogy and Mobile Device Use in Education

In Vietnam currently there is still no official permission of mobile devices use inside the classroom and K-12 education overall. However, MOET, provincial authorities and schools encourage teachers and students to use properly ICT infrastructure, digital devices in particular in some specific lessons or class activities.

The University of Education, Vietnam National University, Hanoi (VNU-UED) is leading teacher education institution, which providers teacher training in both mode of pre-service and in-service training, research on cutting edge educational technology, digital implementation into teaching and learning process for all levels. Since 2018 the Faculty of Educational Technology has established towards research, training interactive learning experience, image technology, mobile learning as well as new conception of digital pedagogy. Among VNU-UED teacher education programs there has been widespread experimentation with active learning practices in lectures, such as the mini-lectures by small group discussions, problem-solving tasks, project based learning and the "flipped", blended classes, where the instruction is done between classes and the "lecture" session is devoted to interactive problem solving. In these prospectus mobile devices usage on makes priority for all teaching practices. Getting beyond the instructional model, especially at the VNU-UED, has been the focus of several efforts, such as the digital education content development, mobile learning apps for teachers and principals towards new professional standards (2018).

2.1 E-Pedagogy

Since 2018 Vietnam MOET announced New National K-12 Education Program proposing 6 multidimensional characters and 10 learning competences for future learners [5] (Table 1).

Table 1. The competencies and characteristics of learners.

Multidimensional characters	Common learning competencies	Subject area competencies
Patriotism	Self-control and self-study	Natural and social sciences
Compassionateness	Communication and cooperation	Technology
Honesty	Problem-solving and creativity	Aesthetics
Industriousness		Physical education
Responsibility		Informative technology
		Calculation
		Language

These findings require a big change for educators and teacher education on all levels. To allow this to happen, VNU-UED take advantage of leading-edge pedagogies, teaching/learning technological tools, and optimal utilization of spaces (real and virtual) for teacher training programs exclusively. We change and/or renew existing curricula emerging components of new instructional models such as effective group work, metacognition and reflection, competence-based backward course design, comprehensive teaching training on evidence-based practices, and deliberate development of interpersonal and intrapersonal skills, authentic assessment etc. In this way, the students can learn, grow and reach their best potential regarding their way of learning style.

The latest educational technology trends considered to replace passive teaching-learning methodologies by more active, collaborative and personalized ones including more student-focused learning, the co-creation of knowledge, and peer review assessment strategies. For instance, social software, mobile apps, virtual/augmented/mixed/extended or cinematic reality (VR/AR/MR/XR/CR) in the classroom now significantly change the way the learners access knowledge and acquisition, and interaction with their teachers and peers [2, 3]. Furthermore, the process of digitization, digitalization and digital transformation in education in the last decade enables full realization of the advantages that technology offers in classrooms for both teachers and learners. The classroom set of device, gamification, smart learning spaces, AI, IoT, Blockchain and Big data application etc. consider the new paradigm for 21st century education.

TPACK framework proposes a new productive approach of implementing educational technologies in classrooms [4]. By separating the three types of knowledge (technological knowledge - TK, pedagogical knowledge - PK, and content knowledge - CK), the TPACK framework enforces combination and recombination in various ways within the components inside (TPK, PCK, TCK). Finally, based on the relationships and intersections among technologies, content and pedagogy, the triangulated dimensions TPACK considers complex space and may be new concept of E-pedagogy as a effective edtech integration.

In term of various learning theories and tandem with digital transformation in education it may be referred as digital pedagogy framework or *"Teach-nology"* (the new way of teaching with technology embedded), which combined the Art of teaching, the Art of using technology and the Art of knowledge creation.

E-/Digital Pedagogy = "Teach-nology"

[Art of Teaching + Art of using Technology + Art of Knowledge creation]

Furthermore, in the digital transformation context traditional learning theories (cognitive, constructivism, interactive etc.) may be discussed in an aspect in which knowledge is acquired in non-linear manner, technology support many cognitive operations and activities previously performed by learners (information search, storage, sharing and creation). The connectivism makes learning and cognition available not only among people, but also artifacts and digital devices that enables highest efficiency in teacher education performance.

To verify the relationship and intersections among technologies, content and pedagogy following triangulated dimensions TPACK, a survey was conducted in collaboration between two research teams of VNU-UED and VNU-ULIS (2018).

The results are presented as follows (Table 2):

Table 2. Reliability of the scores

Doman of TPACK and "Teach-nology"	Internal consistency (alpha)
Art of Teaching	
Pedagogy Knowledge (PK)	.85
Pedagogical Content Knowledge (PCK)	.89
Art of using Technology	
Technology Knowledge (TK)	.82
Technological Content Knowledge (TCK)	.91
Technological Pedagogical Knowledge (TPK)	.89
Art of Knowledge creation	
Content Knowledge (CK)	
Teaching Physics Methodology	.86
Teaching Biology Methodology	.79
English Translation Practice	.72
Learning Methodology and Instructional Technology	.95
"Teach-nology"	
Technological Pedagogical Content Knowledge (TPACK)	.89

In order to find out participants' perceptions of incorporating Digital "Teach-nology" a survey was administered to the 129 fourth-year pre-service teachers of VNU-UED, who took the computer and mobile to practice course in 2018 (http://moodle.ued.vnu.edu.vn). The survey is according to Technology Acceptance Model (TAM), which can be stated in the perspective of learning as follows: easy to use; useful to use (in learning), want to use. The items of the questionnaire have five levels: (1) Strongly disagree; (2) Disagree; (3) Neither Agree or Disagree; (4) Agree; and (5) Strongly agree. The survey outcomes are shown in Table 3 (Tables 4 and 5).

Table 3. Pre-service teachers' perceptions incorporating digital "Teach-nology"

	Mean	Std. deviation
Easy to use Platform	3.36	0.82
Platform is useful to learn	3.56	0.87
Want to use Platform to learn	2.96	0.86
N = 129	**3.37**	**0.89**

Table 4. Online course design issues

Online course design requirements	Limitations
Acceptance and use	*Content issues*
Friendly interface and interaction	Attractive
High order thinking tasks	Format
Content delivery format	Value
Flexibility	Efficiency
Structure content presented	*Technical issues*
Gained personalized and adaptive learning	Utility
Instant feedback	Web 2.0 tools embedded
Differentiation, collaboration	Various activities
	Interaction, sharing tools
	Cloud technology; Multimedia tools
Authentic assessment tasks	*Assignment issues*
Outcome/competence based assessment	Time
	Group/Personal tasks
	Support
	Sharing
	Grade and feedback

Table 5. Challenges for online course designing

	Mostly diff.	Somewhat diff.	Somewhat ease	Mostly ease
Outcome based design	71%	21%	4%	4%
Multimedia content presented	87%	8%	3%	2%
High order thinking learning activities design	93%	7%	0%	0%
Interactive activities design	87%	13%	0%	0%
Structure of content design	68%	24%	7%	1%
Learning resources design	67%	26%	7%	0%
Various assessment tasks	75%	12%	11%	2%
Course delivery format	45%	28%	18%	9%
Digital learning format integrated	75%	15%	10%	0%
Prior technology skills	62%	29%	8%	1%

2.2   Mobile Learning Applications in Teacher Education

Researchers proposed to define "mobile learning" as a learning type where the teachers and learners do not need a specific, predetermined location while learning and taking advantage of mobile technologies or any kinds of learning that carries the employment of wireless and/or mobile devices. The application of mobile learning enables accessibility at anytime from anywhere using the wireless and connected network with ease. The mobile learning system runs on the basis of mobile devices and their ICT-used capabilities. Whether online or offline, the accessibility of the system empowers teachers-and-students communication, exchangeable information, which consists of learning and administrative resources, activities, and E-learning standards [5].

Mobile learning should not be considered as a conjunction of "mobile" and "learning" or as a stand apart element from other learning types. It makes learning process more ubiquitous, localized, authentic, situated and personalized in digital transformation aspects. Furthermore, the mobile application in teacher education imposes challenges the adaptation of traditional educational culture, society's perception, and other factors that influence both learners' and teachers' adoption of mobile learning approaches and essential skills relating to the applications of mobile devices. In general, mobile applications in initial teacher preparation proposed newly introduced learning opportunity in such ways as: (1) contingent learning (where students respond to changes in environments and experiences); (2) situated learning (where students are put in their comfortable environment regarding context or culture); (3) authentic learning (where students' learning connects directly to their learning goals); (4) context-aware learning (where students interact with the environment with the help of mobile tools); and (5) personalized learning (where learning is customized to the preferences and needs of each student). Mobile applications and devices support student-centric approach by encouraging students to access the Internet with ease, seek for information (where the process of knowledge creation happen), enhance synchronous or asynchronous activities (communicate/interact/share) as pedagogic style based on mobile technology [6, 7].

By improving the learning experiences, we encourage student-teachers to use mobile applications and devices at every stage and for every experience in cooperative, personalized and adaptive learning, STEAM, immersive (physic-cyber) environment that continuously pushes them to confront their limits. In the class and practicum by using mobile devices, VNU-UED students deeply acquire and adaptively apply knowledge in their subject, develop whole-person skills including cognitive skills (e.g., creative thinking, critical thinking, comprehensive synthesizing), interpersonal skills (e.g. leadership, listening, decision making, problem solving), and intrapersonal skills (e.g. open-mindedness, self-motivation, self-awareness, self-evaluation of one's ability to learn).

Nowadays, using mobile applications and devices in various courses or lessons in teacher education program at VNU-UED can be deployed in the following application orientations:

– Group of tools to support learning process, create online courses, specific learning management system (LMS) aiming to increase interactivity, learning opportunities anytime, anywhere, with anyone and learn anything that students care;
– Group of illustrative tools, support for visual presentation, interactive video and audio creation, 3D image creation, virtual reality, augmented reality, mixed reality aiming to increasing visual, creating expressions literary and artistic objects, concepts and emotions for learners;
– Group of tools to support communication, social sharing, personal interaction, use of personal handsets towards increasing opportunities for practical communication, experience in everyday communication (texts, speech activities, speaking, etc.), manipulating multimodal texts etc.;
– Group of text/speech creation, illustration tasks and multi-function integration tools (add-ons) in the system towards increasing opportunities for creating documents, authentic assessment presentation;
– Group of tools to support storage, distribution and sharing of data and learning materials, increase opportunities to store information and multimedia documents;
– Creative tools and personal design tools (authoring tools) towards increasing opportunities for developing aesthetic skills and language activities (reading, writing, speaking and listening) (Table 6).

Table 6. Current capacities and implementation mobile apps examples

Mobile apps	Functions	Easy to use	Useful to use	Frequency of use
Edmodo	Personalized mini LMS	3, 7/5	3/5	4, 6/5
Kahoot	Interactive presentation; sharing	3, 5/5	4, 2/5	3, 5/5
Prezi	Presentation	4, 3/5	4, 5/5	4, 5/5
ThingLink	Interactive presentation; Simple AR resource building			
DrawChat	Interactive presentation; livestream	4, 5/5	3, 7/5	4, 1/5
Google docs	Interactive presentation; resources sharing	4, 7/5	5/5	4, 8/5
Loom	Interactive presentation; livestream; resources sharing	4, 9/5	5/5	4, 8/5
Other subject specific Apps	Multimedia creation; interactive presentation	3, 8/5	4, 1/5	3, 5/5

3   Limitation and Discussion

The current research reveals some limitations that need to be put into consideration. Firstly, the already ongoing applications of mobile learning (including blended learning) course design and delivery activities into the newly adapted learning environment (i.e. fully online, flipped, etc.) might enforce the process of *rethinking* of pedagogy as "radically and comprehensively", which contrasts the traditional approach of how we imagine the implementation of mobile devices in learning. Understanding factors of this approach and master prerequisite related skills enable teacher-students to maximize their mobile devices to the highest capacity in the future.

Secondly, full comprehension of the complex or periodicity of mobile learning course aspects may not be fully captured in this study. Hence, in the aspect of digital pedagogical teacher education program at VNU-UED, the study results should particularly be considered as an external experimental evidence to examining the relationship of digital transformation activities. Currently, there is no tool available for the identification of essential components of electronic pedagogy and the processes of applications and mobile devices.

Further in-depth discussions might be held to study the integration of how students value, respond, accept and adopt the technologies of digital learning that includes the examination of *cognitivism, behaviorism, activism*, and *constructivism*. This process generates the clarification of digital technology acceptance and impact on mobile applications and devices in learning on multiple levels.

4 Conclusion

The results of this study may raise hypotheses of the ultimate question for research on education: How to optimize mobile applications and other technological devices in learning with the purpose to maximize students' opportunities and successes? The "perfect" conditions to best apply these approaches have yet to be explicitly stated.

To support future success of teacher-students and their career with digital "Teach-nology" as well as perceptions of educational quality assurance, educators would prepare students with new technological capacity on mobile demands. To create the highest efficiency in the teaching/learning process for students and staff, the leaders should and must create an encouraging environment towards mobile learning values and usage in practice. Relatively, the government needs to formulate policies that encourage the practical application of mobile devices in schools and higher education. The MOET currently manages and guides the school system to develop a digital technology plan aligned with the National Education Program by addressing pedagogy, technology infrastructure and community engagement in teacher education organizations as the vitals to a sustainable support for students and educators.

References

1. Dron, J., Anderson, T.: Teaching Crowds: Learning and Social Media. Athabasca University Press, Edmonton (2014)
2. Fraillon, J., Ainley, J., Schulz, W., Friedman, T., Gebhardt, E.: Preparing for Life in a Digital Age. Springer, Heidelberg (2013)
3. Kolb, L., Tonner, S.: Mobile phones and mobile learning. In: McLeod, S., Lehmann, C. (eds.) What School Leaders Need to Know About Digital Technologies and Social Media, pp. 159–172. Jose-Bass, San Francisco (2012)

4. Mishra, P., Koehler, M.J.: Technological pedagogical content knowledge: a framework for integrating technology in teachers' knowledge. Teach. Coll. Rec. **108**(6), 1017–1054 (2006)
5. National Education Program, MOET, Vietnam (2018)
6. Rekkedal, T., Dye, A.: Mobile distance learning with PDAs: development and testing of pedagogical and system solutions supporting mobile distance learners. Int. Rev. Res. Open Distance Learn. **8**(2), 1 (2007)
7. Traxler, J.: Research essay: mobile learning. Int. J. Mob. Blended Learn. (IJMBL) **3**(2), 57–67 (2011). https://doi.org/10.4018/jmbl.2011040105

Whale Optimization Algorithm for Traffic Signal Scheduling Problem

Thaer Thaher[1(✉)], Baker Abdalhaq[2(✉)], Ahmed Awad[2(✉)],
and Amjad Hawash[2(✉)]

[1] Department of Information Technology, At-Tadamun Society,
Nablus, Palestine
thaer.thaher@gmail.com
[2] College of Engineering and Information Technology, An-Najah National
University, Nablus, Palestine
{baker,ahmedawad,amjad}@najah.edu

Abstract. Traffic congestion is one of the most important problems with respect to people daily lives. As a consequence, a lot of environmental and economic problems were emerged. Several works were proposed to participate in problem solving. Traffic signal management introduced a promising solution by minimizing vehicles average travel times and hence decreasing traffic congestion. Studying vehicles' activities on roads is non-deterministic by nature and contains several continuously changing parameters which makes it hard to find an optimal solution for the mentioned problem. Therefore, optimization techniques were intensively exploited with respect to Traffic Signal Scheduling (TSS) systems. In this work, we propose TSS control methodology based on Whale Optimization Algorithm (WOA) in order to minimize Average Travel Time (ATT). Experimental results show the superiority of WOA over other related algorithms specially with the case of large-scale benchmarks.

Keywords: Whale Optimization Algorithm · WOA · Metaheuristics · Traffic Signal Scheduling · Simulation

1 Introduction

With the rapid increase in population and the lack of infrastructure availability for transportation systems in the urban areas, traffic congestion has become a major challenge for both urban planners and researchers [2]. Heavy traffic congestion usually causes various serious problems that impact different aspects of people personal lives, such as high pollution levels, increased fuel consumption, safety problems, time wasting, frustration, disaster management, and consequent economic loss [3]. Therefore, planners seek to utilize the existing infrastructure in an optimal way [4]. In this manner, efficient traffic signals scheduling approaches became highly required to contribute in solutions for traffic congestion problem [2].

Traffic Signal Scheduling (TSS) is defined as the mechanism of finding the optimal time schedule of traffic lights to enhance the overall traffic conditions (e.g. minimizing travel times for vehicles). However, TSS is considered a difficult optimization problem

© Springer Nature Switzerland AG 2020
L. C. Jain et al. (Eds.): ICICCT 2019, LAIS 9, pp. 167–176, 2020.
https://doi.org/10.1007/978-3-030-38501-9_17

because traffic system consists of various interconnected subsystems with dynamic and stochastic behavior that leads to non-deterministic outcomes which results in huge solutions space that makes exhaustive search infeasible for such a problem [8].

Exact optimization methods are impractical to handle complex problems with multi-dimensional search spaces. Therefore, heuristic methods are employed in tackling TSS problem instead. The framework of TSS optimization consists of two main stages: (1) the search process that considers the different time schedules, and (2) the evaluation approach to asses these schedules. In this stage, microscopic models for traffic simulation are exploited [3].

Heuristic methods have become popular for TSS optimization problems such as Genetic Algorithms (GAs), Swarm intelligence, and others [6]. However, TSS optimization requires exploiting recent optimization algorithms to handle largescale cases. In this paper, a simulation-based approach is proposed to solve TSS optimization problem based on a recent swarm optimization algorithm, namely, Whale Optimization Algorithm (WOA). Our contributions are summarized as follows:

1. A WOA-based Traffic Signal Scheduling (TSS) optimization methodology is proposed to find the best duration for different phases for TSS systems in an input network topology.
2. The proposed algorithm has been evaluated in terms of Average Travel Time (ATT) on small, medium, and large-scale benchmarks.
3. The proposed algorithm has been compared with other state-of-art algorithms and have outperformed them with significant improvement in large scale benchmarks.

2   Related Work

Several techniques and algorithms have been proposed to tackle the Traffic Signal Scheduling (TSS) problem including mathematical optimization models and simulation-based approaches. However, due to the complexity and dynamism of the transport network, most researchers turned to propose simulation-based approaches.

Recently, Meta-heuristic (MH) algorithms have been widely employed in the field of traffic signal timing like evolution-based (e.g. GA), swarm-based (e.g. PSO and ACO), physics-based (e.g. SA) and human-based (e.g. HS) algorithms. The work of Kachroudi [5] is related to PSO area. Two versions of multi-objective PSO algorithm were applied on a virtual urban network to optimize the cycle programs for private and public vehicles. A complex model was built to determine the optimal green splits and the offsets of the traffic lights.

Signal timing optimization methodology of phase combination was proposed in [14]. The objective function of the work is to minimize the number of stops at intersection by combining the least different flow ratio of two streams into one stream using PSO where the number of stops is decreased by 19.04%.

A robust heuristic was proposed in [1] to use the history information in predicting the future traffic load on each street leading to an intersection controlled by a traffic light. Simulation results show that the proposed algorithm optimized the traffic flow up to 18% more than standard traffic systems.

Recent efficient swarm optimization algorithms have not been exploited in TSS problem. Moreover, most methods are investigated on a special traffic network with limited elements, and thus the scalability of such algorithms have not been tested. However, based on the No-Free-Lunch theorem which states that there is no universal algorithm which outperforms the other methods for all problems [12], the door is still open to investigate new algorithms in the field of traffic optimization.

Mirjalili and Lewis have recently developed an innovative meta-heuristic algorithm which is called the Whale Optimization Algorithm (WOA) [9]. Our work is related to applying WOA on a set of predefined parameters to introduce a solution for the traffic congestion problem. Promising results were achieved that add a value to the set of proposed solutions to the mentioned problem.

3   Traffic Signal Scheduling (TSS) Problem Formulation

The design of a Traffic Signal Scheduling System (TSS) should ensure minimizing the stop delay for different vehicles without violating the core requirement of avoiding conflicts between opposing through traffic [2].

3.1   TSS Basic Terminology

Universally, each conjunction is controlled by a set of traffic lights to control the flow of vehicles in which a round robin technique is used to give traffic light a slice of time to allow the flow of its related vehicles. A state (or phase) for a given conjunction is defined by the set of current lights for all of the light signals installed for the conjunction. Each light state (Red, Yellow and Green) has a predefined interval of time [3].

The key component in the design of a TSS system for an intersection is to separate the conflicting movements of the traffic into different phases. Typically, a phase includes the state of allowed movements, followed by the yellow and red states for the opposing movement. For example, consider the four-legged intersection shown in Fig. 1, the traffic movement in this intersection can be controlled by a 4-phase signal system as shown in Fig. 2 left side.

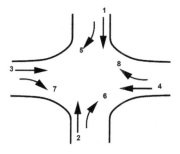

Fig. 1. An intersection with 8 different traffic approaches.

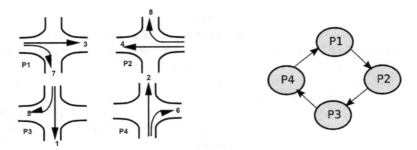

Fig. 2. The right figure represents the transitions between phases for the TSS shown in the left figure that has the 4-phase signaling system for an intersection with 8 different traffic approaches.

3.2   TSS as an Optimization Problem

The functionality of a TSS for an intersection can be described as a sequence of phases represented as a state diagram. Right part of Fig. 2 shows the state diagram for the 4-phase TSS shown in left part of the figure.

Optimization of the timing of traffic signals aims to find the best interval for each phase in the sequence of functionalities in the TSS, such that the overall flow of the vehicles in the intersection is enhanced. Thus, our key objective is to find the best interval allocated for each phase in each intersection in a network topology such that the average travel arrival time for all vehicles in the topology is minimized.

Solution Representation: Given a network topology that consists of n intersections. Each intersection is controlled by a number of phases. Let P_i represent the number of phases for intersection i. Each solution for TSS optimization problem is represented as a one-dimensional vector of integers X such that each element in X represents the duration of a phase in an intersection, as illustrated in Fig. 3.

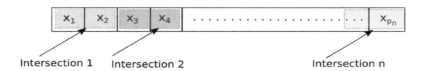

Fig. 3.   Solution representation.

Fitness Function: A candidate solution is evaluated in terms of Average Travel Time (ATT). ATT is defined as the total trip time divided by the total number of vehicles that reached their predefined destinations. Thus, TSS optimization problem aims to find the duration for all phases in all intersections in a network such that the ATT is minimized. This optimization problem is formulated in Eq. 1 where $f(X) = ATT(X)$. Notice that the X should be bounded by predefined lower bound L and upper bound U.

$$\text{Minimize} \quad f(X)$$
$$\text{subject to } L \leq X \leq U \tag{1}$$

4 Proposed TSS Optimization Methodology

Figure 4 illustrates our proposed optimization methodology. The input for our algorithm is a network topology that consists of number of intersections. An initial random population including vector of durations for different phases is generated. The WOA algorithm is applied following fitness function evaluation for each candidate solution by simulation. The Average Travel Time (ATT) guides the next iteration in the WOA recipe until a satisfactory solution in terms of the durations for different phases in the TSS systems is outputted.

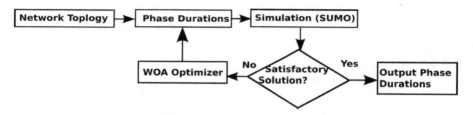

Fig. 4. Optimization methodology.

5 Results and Discussion

In this work, our focus is to study the efficiency of WOA in tackling TSS problem. For this purpose, we compare WOA with other modern meta-heuristic algorithms from the literature. Thereafter, the performance of WOA is validated and compared with other traditional optimization models. In the following subsections detailed description of the experiments is explained.

5.1 Experimental Setup

Three test cases (sites) were developed. These sites range in their scale and topology. The first test site is relatively small-sized which simulates the real signalized segment at the center of Nablus city, Palestine, while the other two are software created. Second and third sites can be classified as medium and large size networks, respectively. The number of parameters that need to be optimized in the created sites are 13, 34, and 141 respectively.

The experimental work involves two main phases; sensitivity analysis of WOA, and comparison phase. In the first phase, the real network was used while investigating the sensitivity of WOA to common parameters. In the second phase, four well-known algorithms were implemented to confirm the performance of WOA. Finally, we validated the proposed WOA-based model with common optimization models (e.g. Highway Capacity Manual-HCM, Webster and Synchro) [10].

The networks were modeled using SUMO v1.1.0 [7], and all optimizing algorithms were implemented using python 3.7.3. We tested the experiments on the RMACC Summit supercomputer[1]. To have a fair comparison, all optimizing algorithms were evaluated using the same common parameters; 100 iterations, and population size of 30. Table 1 presents the setting of specific parameters of optimizing algorithms. These parameters were selected based on the settings recommended in previous researches (PSO [11], BAT [13]). Due to the stochastic behavior of MH algorithms, we presented the average results of 20 independent runs for each conducted experiment.

Table 1. The used settings of algorithms.

Algorithm	Parameter	Value
WOA	Exploration/exploitation switch factor a	From 2 to 0
GA	Crossover prob.	0.9
	Mutation prob.	0.01
	Selection method	Roulette wheel
	Elite	2
PSO	Inertia weight	From 0.9 to 0.4
	Acceleration coefficients	2
BAT	Qmin, Qmax	0, 2
	A loudness	0.5
	r pulse rate	0.5

5.2 Sensitivity Analysis of WOA

In this section, the real test site is considered to evaluate the sensitivity of WOA. We tuned both (1) the common parameters (i.e. population size and the number of iterations), and (2) the internal parameter a (in Table 1) which is the main control parameter of WOA. This is useful for getting the best performance of WOA in handling the problem.

Convergence Behavior: To examine the convergence behavior of WOA, different number of iterations (from 20 to 200) with fixed search agents (population size of 30) are used. The reported results in Table 2 show that increasing the iterations from 20 to 100 significantly improves the average travel time from 58.5 to 55.91. However, the further increase in the number of iterations (more than 100) does not significantly improve the results. Therefore, 100 iterations (i.e. 3000 evaluations) are enough to get a good solution within an acceptable time. We need to mention that the evaluation of the objective function includes running a simulation which means it is computationally

[1] The RMACC Summit supercomputer, which is supported by the National Science Foundation (awards ACI-1532235 and ACI-1532236), the University of Colorado Boulder, and Colorado State University. The Summit supercomputer is a joint effort of the University of Colorado Boulder and Colorado State University.

Table 2. The average travel time results of WOA for case 1 using population size of 30 and different numbers of iterations.

# Iterations	20	50	70	100	150	200
AVG	58.50	56.72	56.38	55.91	55.71	55.65
STD	3.63	2.41	2.24	2.31	2.35	2.37

heavy problem. Therefore, the number of iterations is an important factor to be considered.

Impact of Population Size: In this part, we are interested to assess the sensitivity of WOA to the initial population. For this purpose, the number of evaluations was fixed to 3000 while the population size was used with 6 different values (5, 15, 30, 50, and 150). The extensive experiments revealed that WOA with the population size of 30 is able to give the best average result.

Using a small number of different solutions will result in the loss of diversity. Therefore, during the limited evaluations, a trade-off between the population size and the number of iterations is highly demanded. Consequently, the next experiments and comparative studies are conducted using 100 iterations and population size of 30 (i.e. 300 evaluations).

Impact of Main Controlling Parameter of WOA: The WOA algorithm has the ability to change the searching pattern from exploration to exploitation based on the parameter (a). This parameter is decreased linearly inside the interval $[a_0 \ 0]$ where the initial value a_0 was set to 2 in the original paper [9]. In this experiment, WOA is tested with five different values of a_0 including 0.5, 1, 2, 3, and 4 to investigate how the behavior of WOA can be influenced. The obtained results are visualized in Fig. 5.

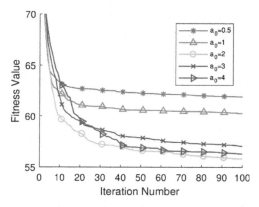

Fig. 5. Convergence behaviour of WOA with different initial values of parameter (a).

According to the convergence patterns in Fig. 5, we can see that when $a_0 < 2$ the algorithm is stuck early into local optima because all iterations are utilized for exploitation. While, when $a_0 > 2$ the algorithm shows better performance to avoid local optima since most iterations are used for exploration. However, WOA with a_0 equals to 2 can make a balanced shifting between the exploratory and exploitative potentials.

5.3   Performance of WOA Versus Other Optimization Algorithms

To validate the performance of WOA we conducted a comparison study with four well-known algorithms from literature. These algorithms are Particle Swarm Optimization (PSO), Genetic Algorithm (GA), Bat algorithm (BAT), and Slap Swarm Algorithm (SSA). The obtained numerical results in terms of average and standard deviation of fitness value are presented in Table 3.

Table 3. Comparison between the travel time results obtained by WOA versus other algorithms in terms of average and standard deviation.

Cases	WOA		PSO		GA		BAT		SSA	
	AVG	STD	AVG	STD	AVG	STD	AVG	STD	AVG	STD
Case 1	55.91	2.31	54.88	2.73	65.79	4.16	76.58	5.98	55.38	1.96
Case 2	131.80	11.61	132.81	20.35	241.59	47.20	333.61	55.64	155.98	37.88
Case 3	175.36	6.21	210.33	11.74	270.80	11.13	318.08	12.35	255.79	17.70
Mean rank (F-test)	1.67		1.67		4.00		5.00		2.67	
Overall rank	1		1		3		4		2	

The cases above are sorted according to the number of variables that need to be optimized, or search space size. Table 3 shows that PSO and SSA outperform and are very close to WOA in the first case while GA and BAT do not perform well. From Fig. 6a BAT does not converge at all. It is stuck and does not reach good values. It seems that BAT stays around local minima. GA starts to converge at the beginning but it stops in better minima compared with BAT but still is not as good as other algorithms. PSO and WOA show very similar convergence pattern. SSA is interesting, it shows unusual conversions pattern. It starts slow almost like GA but it continues to converge to reach PSO and WOA at the end.

In test case 2, BAT and GA show the same poor behaviour. By observing the convergence trends in Fig. 6b, we see that SSA has the same interesting behaviour as in case 1. The curve of SSA seems to have two phases; the first phase is slow convergence that may be exploration phase, while in the second phase it accelerates its convergence most probably towards some minima. The difference between PSO and WOA from one side and SSA from another side starts to be clear in this case. PSO and WOA outperform SSA slightly.

Regarding test case 3, which is the most complex one we experimented in this work, the reported travel time values of compared algorithms in Table 3 show that WOA achieves the best result, followed by PSO, SSA, and BAT respectively. As per convergence curves in Fig. 6c, BAT starts and ends in the same neighborhood. It does not seem to make any improvement. GA suffers from slow convergence. SSA despite its interesting behaviour but still can not perform as PSO and WOA. It starts slower than GA but it switches its behaviour and outperforms GA slightly. From Fig. 6c it is obvious that WOA has superior performance if compared with other algorithms in this case.

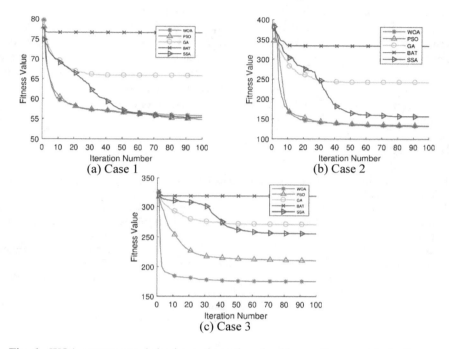

(a) Case 1

(b) Case 2

(c) Case 3

Fig. 6. WOA convergence behavior against other algorithms with respect to the 3 cases.

As per F-test results, it is clear that both WOA and PSO are ranked first, followed by SSA, GA, and BAT respectively. This comparative study proves that WOA can perform better and very competitive results compared to other algorithms in dealing with TSS problem. Both the implementation of PSO used in this work and WOA use an adaptive control parameter that starts the iterations with high amount of exploration then it reduces exploration to give more emphasis on exploitation. WOA differs from PSO that it uses 50% of times a spiral movement. This movement searches for more places around the believed optimal solution so far found. This may overcome the deceptive behavior of some objective functions.

Finally, we compared our work with traditional TSS methods: HCM, Webster and Synchro. The resulting average travel time was 149 s, 202 s and 92 s respectively. These results are far from the result we obtained using WOA which was 55.9 s. The superiority of WOA-based method is due to its effective exploration and exploitation behavior in the entire solution space if compared with traditional isolated section-based optimization algorithms.

6 Conclusion

In this work, a WOA-based simulation methodology was investigated to tackle traffic congestion problem. We compared WOA with widely used algorithms in this domain namely GA, PSO, SSA and BAT as well as traditional methods. Three test cases were developed: small, medium and big. This work emphasizes that WOA, PSO and SSA performed almost the same in small and medium size cases, while WOA shows preponderance with bigger size problems. This research realizes that WOA works fine with big traffic light scheduling problems and can be relied on with big cities. Our future work will be related to studying the effect of spiral motion on WOA. In addition to the reformulation of objective function in the case of emergencies.

References

1. Yousef, K., Shatnawi, K., Latayfeh, M.: Intelligent traffic light scheduling technique using calendar-based history information. Future Gener. Comput. Syst. **91**, 124–135 (2018)
2. Gao, K., Zhang, Y., Sadollah, A., Su, R.: Optimizing urban traffic light scheduling problem using harmony search with ensemble of local search. Appl. Soft Comput. **48**, 359–372 (2016)
3. Garcia-Nieto, J., Olivera, A.C., Alba, E.: Optimal cycle program of traffic lights with particle swarm optimization. IEEE Trans. Evol. Comput. **17**(6), 823–839 (2013)
4. Hu, W., Wang, H., Yan, L., Du, B.: A swarm intelligent method for traffic light scheduling: application to real urban traffic networks. Appl. Intell. **44**(1), 208–231 (2015)
5. Kachroudi, S., Bhouri, N.: A multimodal traffic responsive strategy using particle swarm optimization. In: 12th IFAC Symposium on Control in Transportation Systems, pp. 531–537 (2009)
6. Kennedy, J., Eberhart, R.: Particle Swarm Optimization, pp. 760–766. Springer, Boston (2010)
7. Lopez, P.A., Behrisch, M., Bieker-Walz, L., Erdmann, J., Flötteröd, Y.P., Hilbrich, R., Lücken, L., Rummel, J., Wagner, P., Wießner, E.: Microscopic traffic simulation using SUMO. In: The 21st IEEE International Conference on Intelligent Transportation Systems, pp. 2575–2582. IEEE (2018)
8. Lpez-Neri, E., Ramrez-Trevio, A., Lpez-Mellado, E.: A modeling framework for urban traffic systems microscopic simulation. Simul. Model. Pract. Theory **18**(8), 1145–1161 (2010)
9. Mirjalili, S., Lewis, A.: The whale optimization algorithm. Adv. Eng. Softw. **95**(5167), 51–67 (2016)
10. Shehab, R.A.: Benchmark for tuning metaheuristic optimization technique to optimize traffic light signals timing. Master's thesis, Al-Quds University, Jerusalem (2015)
11. Shi, Y., Eberhart, R.C.: Empirical study of particle swarm optimization. In: Proceedings of the 1999 Congress on Evolutionary Computation, CEC 1999, vol. 3, pp. 1945–1950 (1999)
12. Wolpert, D.H., Macready, W.G.: No free lunch theorems for optimization. IEEE Trans. Evol. Comput. **1**(1), 67–82 (1997)
13. Yang, X.S.: A New Metaheuristic Bat-Inspired Algorithm, pp. 65–74. Springer, Heidelberg (2010)
14. Zhao, Y., Fang, W., Qing, C.: A PSO based signal timing optimization approach of phase combination. In: Proceedings of the 2016 International Conference on Civil, Transportation and Environment (2016)

Peer-to-Peer Lookup Process Based on Data Popularity

Sarra Cherbal$^{(\boxtimes)}$ and Bilal Lamraoui

University of Ferhat Abbas Setif 1, 19000 Setif, Algeria
sarra_cherbal@univ-setif.dz,
lamraouibilall995@gmail.com

Abstract. The term "Peer-to-Peer" (P2P) refers to a computer network model whose elements (nodes or "peers") are both clients and servers during exchanges. With the evolution of P2P, users express the need of exchanging resources in the shortest possible time, which encouraged the development and optimization of many of resource (data) search mechanisms in order to offer a good quality of research. In this field, we find lookup mechanisms based on data popularity. However, the routing of P2P requests and the methods used for estimating popularity of shared files can generate communication redundancy and unnecessary network traffic. Therefore, the objective of this work is to develop a P2P lookup mechanism that estimates popularity of shared data and uses it in order to increase the search success rate and reduce the lookup path without generating excessive traffic or increasing the storage spaces. The implementation results show the effect of the proposed approach in reducing the lookup path and the nodes storage space.

Keywords: Peer-to-Peer · Popularity · Lookup · Replication

1 Introduction

Peer-To-Peer (or P2P) makes it possible to establish direct communications between the different nodes of the network, which can then exchange different types of data. Decentralized P2P can be divided into two architectures, the first occurred is the unstructured one with random distribution of nodes on the network and a flooding based routing, e.g. Gnutella [1]. The next occurred is the structured architecture that is based on a restricted network topology (ring, tree ...) and a well-defined routing protocol, e.g. Chord [2].

P2P systems have enjoyed great prominence and widespread use by internet users through their file-sharing feature. In this type of system, the optimization of research (lookup) services always presents a major challenge in order to offer a good quality of service to the users [3–6], in terms of much criteria as: lookup latency, lookup hops number, overhead (traffic network) ...etc. In this context, researchers are interested in optimizing the lookup services in P2P systems using different methods, among which we find lookup mechanisms based on data popularity [7–10]. However, besides the advantages presented by this mechanism, the methods used for estimating popularity of shared files can generate communication redundancy and unnecessary network traffic. In addition of increasing storage spaces when replicating data.

© Springer Nature Switzerland AG 2020
L. C. Jain et al. (Eds.): ICICCT 2019, LAIS 9, pp. 177–186, 2020.
https://doi.org/10.1007/978-3-030-38501-9_18

Therefore, the objective of this work is to develop a P2P lookup mechanism in which we propose a method to estimate popularity of shared data and we present a replication approach, in order to minimize the lookup path without generating excessive traffic or excessive storage spaces.

2 Background and Related Work

P2P Research in Unstructured Systems. The random search in the unstructured architecture does not take into account the network characteristics and it is usually done by flooding or random walk. As traditional protocols that existed firstly in this approach, there are: Gnutella [11, 12], GIA [13] and Swaplinks [14]. Thereafter, researchers have proposed alternative solutions to control flooding and to reduce the large amount of unnecessary generated traffic, such as: Expanding Ring Search (ERS) [15], random walk search [16] and Intelligent Walks [17].

Popularity in P2P Research Mechanisms. In p2p, there are a numerous search mechanisms proposed to improve research considering data popularity. In this section, we review some papers that use data popularity and mention the methods of measuring the data popularity.

- In [7] the distribution of requests is load balanced among the nodes based on the popularity of searched ressources. In other words, the resource popularity is defined according to the number of requests seeking this resource.
- In 2P lookup [8] each time the network needs to calculate a resource popularity, peers communicate with each other to exchange their local popularities of this resource and calculate the ressource global popularity. Thus, the calculation in this work is based only on the parameter of received search query packets.
- In [9] the idea of popularity is based on history of searched videos, thus, the authors use the past to predict the future demand. While, in [10] the popularity is measured through a decentralized approach, where, for a set of videos, each node broadcasts a popularity query to other nodes with a predefined TTL counter.

The authors of [18] mention different categories to measure the popularity, here, we mention the disadvantages (deficiencies) of this works.

When the number of requests is the only considered parameter, there are two deficiencies to discuss: the dataset creation time and when it started to be functioning are not considered in the calculation methods, i.e. the mean lifetime of dataset is ignored, which can lead to a wrong estimation of dataset popularity.

As shown in the second scenario of Table 1, the creation of dataset is done in the fifth period. In the last three periods, we notice that dataset of the first scenario which is cretated in the first period is searched less than that of the second scenario. Hence, the results show the contrary, i.e. the first scenario is appeared as more popular because the only factor considered here is the dataset lifetime.

Table 1. Example of the deficiency: when neglecting the dataset lifetime.

Scenario	P1	P2	P3	P4	P5	P6	P7	#Requests
First scenario	5	5	5	5	5	5	5	35
Second scenario					8	10	12	30

In Table 2, we can see how the dataset popularity results can be affected incorrectly by the old requests. In this example, the dataset of the second scenario is more requestes in the last periods, thus, it should be more popular. Consequently, recent requests should be considered to be more important than old ones.

Table 2. Example of the deficiency: when neglecting the requests distribution over time.

Scenario	P1	P2	P3	P4	P5	P6	P7	P8	#Requests
First scenario	25	25	25	20	15	10	5	0	125
Second scenario	0	0	0	10	15	20	25	30	100

From these observations, we propose an approach that aims to avoid these deficiencies when measuring the popularity and also to avoid some other deficiencies that are mentioned all along the rest of this paper. Wherein, the proposed approach, its processes and its functionalities are explained in details.

3 Description of the Proposed Approach

Our solution applied in an unstructured p2p architecture, aims to improve lookup mechanism quality: reduce lookup latency and lookup hops number, without generating more network messages to avoid increasing overhead. Therefore, we propose a replication method based on a proposed popularity mechanism.

As an overview of the proposed approach, we adopt 3 processes to ensure the efficiency of the approach, which are: popularity table construction, popularity factor calculation and replication method.

The functioning of these processes, the linking between them and the lookup process are explained in the following of this chapter.

3.1 Popularity Table Construction

In order to calculate the popularity factor, we need firstly to construct what we call a "Popularity table" which will be found in each node of this overlay network. This table presents the number of received requests of each resource in each defined period of time (the last recent periods). As shown in Table 3, the first cologne of this table presents the names of searched resources, the other colognes present the periods (e.g. the last three periods) and the values in the table present the number of received requests of each resource in each period.

Table 3. Example of popularity table of node N.

Resources	Periods		
	P1	P2	P3
A	1	4	7
B	3	5	2

A period presents a time interval, during which we measure the number of received requests for each node. Here, we choose the number of three periods as an example, but it can be extended to a larger number. When the time of the fourth period begins, the information of the first period will be crashed, so we always keep the information of the three periods only.

The time interval in each period can be determined according to the network size and the number of lookup requests circulating in this network. I.e. when these two parameters are high, the interval time will be bigger and vice versa.

3.2 Popularity Factor Calculation

To calculate the popularity factor of a resource, there are several methods proposed in the literature. We have mentioned the most known ones in Sect. 2 and we have given a comparative discussion in the second section of this Chapter, in order to avoid some of their existing drawbacks in our proposed method.

There are some approaches that proposed to calculate the local popularity factor of a resource (in each node) and also the global one by exchanging the local factor of each resource between all the nodes (or some nodes) of the network. This global factor provides the popularity of a resource in the network. However, its calculation requires exchanging a large number of messages between nodes, so increasing the overhead. Therefore, in our approach we propose to measure the local popularity from the information found in the local popularity table, as we need to know the popularity regarding the node only and not regarding the whole network and in the same time to avoid the overhead generated by global factor calculation. Thus, this approach doesn't require sending new messages; it just uses the information of the received lookup requests.

Moreover, as we see in the second section of this chapter, the methods of popularity factor calculation that are based only on the number of requests cannot be sufficient to give a right indication of dataset popularity. That is because they did not take into consideration the mean lifetime of datasets (as shown in Table 1). However, these two parameters are not yet sufficient, another parameter is missing, it is the requests distribution over time, hence, we have to give recent requests more importance regarding the former ones (as shown in Table 2).

Therefore, our proposed calculation method aims to consider all of these parameters to get an indication close to reality. Thus, we propose the following formula:

$$P(X) = \frac{\sum_1^m Nb\_Req(X).Coeff}{m \times NbT\_Req} \tag{1}$$

Such that:

- P(X): Popularity of resource X.
- m: Number of used periods (e.g. in Table 3: m = 3)
- Nb_Req(X): Number of received requests that look for X in each period.
- Coeff: Coefficient that takes values ("1" in P1, "2" in P2 and "3" in P3 ...etc). This coefficient is used in order to give importance to resources searched in the recent periods.
- NbT_Req: total number of requests received by this node (e.g. in Table 3: NbT_Req = 22).

Remark 1. Periods are used in order to take into consideration the dataset lifetime and the distribution of requests over time.

3.3   Replication Method

Replication is one of the most known mechanisms used in distributed systems, especially in works that aim to improve the lookup process quality [6]. It serves to create more than one copy of the resource and distribute them through the network, thus, increasing the availability of resources. However, replicating the resource itself can increase the storage space of the replica node, especially for heavy data and when its content is useless for this replica node. Also, methods that apply a replication mechanism need firstly to determine what data to replicate and on which replica nodes. Therefore, in our approach we:

- Apply a replication method that is based on the resource popularity in order that the node replicates the most searched (popular) resources using the calculated popularity factor.
- Don't generate new messages to replicate data. It is just applied using the existing lookup and reply requests.
- Don't replicate the resource itself, but the index of this resource (its identifier and its node owner).

Each time a node receives lookup requests of a resource, it calculates its popularity factor according to our proposed formula. When this factor exceeds a defined threshold, then, this resource is considered as popular for this node, as it's so searched compared to other resources. In this case, when a resource is popular, the node replicates the index of this resource, i.e. the resource identifier and the node owner where this resource is found.

According to the proposed formula, the value of the calculated factor is limited with the interval (0, 1) (as a percentage from 0% to 100%), thus, the threshold in our approach is of value 0,5 (as 50%).

3.4 Lookup Process

In the following, we explain the steps of the lookup process:

- A search query of resource X is launched in the network.
- The requester (the node that launches this query) sends it to its neighbors.
- Each of these neighbors transmits it to its neighbors except the sender.
- Every time a node receives a lookup request of resource X, it checks if the resource is available or not. If it is found, the node sends a reply to the sender; else, the node checks the popularity of X using the previously defined formula.
- If a node founds that resource X is popular, it keeps information of its index (replication).
- Henceforth, if this node (having the index of X) receives a lookup request for X will be directly reply the requester by this index. So, the lookup process will be achieved in this node without passing to others.
- When the requester finds the X owner, it contacts it directly for download. Here, we are interesting only by the lookup process and not by the download process.

In the proposed formula, we consider: number of received requests, requests distribution over time (over a limited number of periods) and the number of periods (e.g. 3). This allows us to keep a minimum size of the table and to find a more accurate popularity factor. Also, we propose to run the popularity equation periodically.

Scenario: Explanation example:

Suppose we have the following example (Fig. 1):

- Some nodes send lookup requests of resource X.
- The resource X exists in a specific node (N″).
- Node N receives requests in different periods so N calculate the popularity as follows:

Table 4. Explanation example.

Resource/period	P1	P2	P3
X	2	4	7

$$P(X) = \frac{\sum_1^3 Nb\_Req(X).Coeff}{NbP*3} = \frac{2(1)}{13*3} + \frac{4(2)}{13*3} + \frac{7(3)}{13*3} = 0.79 \text{ (Table 4)}$$

- This popularity factor is superior than 0,5, so N considers resource X as popular.
- When N″ receives the request, it responds by a reply, as shown in Fig. 1.

- During the reply, every node receives this reply query checks if this resource is popular or not (popularity > 0.5), like in node N. thus, N keeps information of X index (X, N''). This information does not require other exchanged messages it is just extracted from the received reply query.
- Henceforth, when N receives lookup requests of X (e.g. from N* in Fig. 2), it will respond directly by a reply without forwarding the request to its neighbors, as shown in Fig. 2, without going through many hops or extending latency.

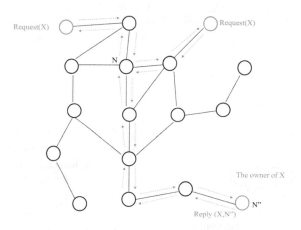

Fig. 1. Lookup process without replication

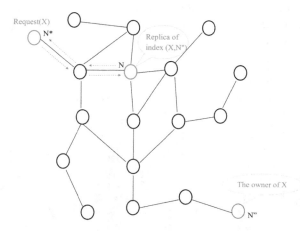

Fig. 2. Lookup process with replication

Discussion. This allows us to conclude the following advantage points:

- Get a popularity factor close to the real one by considering the most important parameters mentioned in the literature.
- Avoid overloading storage space of replica nodes with additional resources that can be heavy and not useful for them.
- Minimize the number of hops in the lookup process by responding directly by the index to the requester.
- Reduce lookup latency (search time).
- Avoid exchanging additional messages between nodes, neither for calculating global popularity nor for replicating data.
- Reduce network overhead by reducing the number of exchanged requests in each lookup process.

4   Experimental Results

We have programed the proposed processes of our approach using JAVA. Our implemented system contains a number of nodes (peers), a set of resources that are associated randomly to each node, and then lookup requests of some resources are launched in the network in order to calculate the lookup hops number from the source to the destination node.

Here, we extracted the number of hops from some launched executions of our implementation (Fig. 3). The number of hops of each case (of one lookup process) is the number of nodes that the lookup request passes through from the requester node until the destination node (the resource owner).

In Fig. 3, we present the results of hops number extracted before and after applying our proposal. From which, we can notice the effect of the proposed approach on reducing the lookup path.

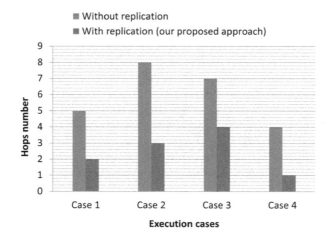

Fig. 3. Hops number

Figure 4 presents the average storage space in each node when replicating resources and when replicating indexes (in our approach) compared to the average number of resources assigned firstly to each node when constructing the network. From the results presented in Fig. 4, we can notice the high storage space avoided by our proposal.

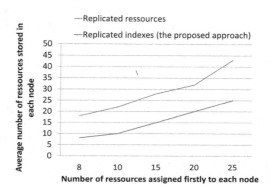

Fig. 4. Storage space

5 Conclusion

In this paper, we are interested by optimizing P2P lookup mechanism based on data popularity. For this, we have started this work by overviewing and reviewing some related works that aim to improve P2P lookup mechanisms and particularly those based on popularity approach. This review allowed us to present our motivations to propose the contribution of this work.

Thereafter, we have presented our proposed contribution of P2P research mechanism based on data popularity. Then, we have explained how the lookup process performs based on the proposed mechanism. For each proposed process, we have mentioned our motivations that conduct us to this proposal and its advantages.

Furthermore, we have presented our proposal implementation. According to some launched executions of this implementation, with different schemes and different lookup processes, we arrived at extract the number of hops and the nodes storage space. The results presented by diagrams prove the effect of the proposed approach on reducing lookup hops number without increasing the storage spaces.

As perspective of this work, we aim to simulate the proposed approach in a large network, to prove its efficiency with more nodes and more requests and also to be able to measure some evaluation criteria like: lookup latency, lookup success ratio and overhead.

References

1. Ripeanu, M., Foster, I., Iamnitchi, A.: Mapping the Gnutella network: properties of large-scale peer-to-peer systems and implications for system design. IEEE Internet Comput. J. **6** (1), 40–49 (2002)
2. Stoica, I., Morris, R., Liben-Nowell, D., Karger, D.R., Kaashoek, M.F., Dabek, F., Balakrishnan, H.: Chord: a scalable peer-topeer lookup protocol for internet applications. In: Proceedings of SIGCOMM, pp. 149–160. ACM Press (2001)
3. Nwebonyi, F.N., Martins, R., Correia, M.E.: Reputation based approach for improved fairness and robustness in P2P protocols. Peer-to-Peer Netw. Appl. **12**, 951 (2019)
4. Ahmad, S., Bouras, C., Buyukkaya, E., et al.: Peer-to-peer live video streaming with rateless codes for massively multiplayer online games. Peer-to-Peer Netw. Appl. **11**, 44 (2018)
5. Chen, F., Li, H., Liu, J., et al.: Migrating big video data to cloud: a peer-assisted approach for VoD. Peer-to-Peer Netw. Appl. **11**, 1060 (2018)
6. Cherbal, S., Boukerram, A., Boubetra, A.: RepMChord: a novel replication approach for mobile chord with reduced traffic overhead. Int. J. Commun Syst **30**, e3285 (2017)
7. Soltani, N., Khaneghah, E.M., Sharifi, M., Mirtaheri, S.L.: A dynamic popularity-aware load balancing algorithm for structured P2P systems. In: Park, J.J., Zomaya, A., Yeo, S.S., Sahni, S. (eds.) Network and Parallel Computing, NPC 2012. Lecture Notes in Computer Science, vol. 7513. Springer, Heidelberg (2012)
8. Seddiki, M., Benchaïba, M.: 2P-lookup: popularity and proximity based P2P lookup mechanism over MANETs. J. Netw. Comput. Appl. **71**, 181–193 (2016)
9. Das, S.K., Naor, Z., Raj, M.: Popularity-based caching for IPTV services over P2P networks. Peer-to-Peer Netw. Appl. **10**, 156–169 (2017)
10. Ganapathi, S., Varadharajan, V.: Popularity based hierarchical prefetching technique for P2P video-on-demand. Multimed Tools Appl. **77**, 15913–15928 (2018)
11. Gnutella 0.4. [En ligne] (2001). http://rfcgnutella.sourceforge.net/developer/stable
12. Gnutella 0.6. [En ligne] (2002). http://rfc-gnutella.sourceforge.net/src/rfc-0_6-draft.html
13. Chawathe, Y., Ratnasamy, S., Breslau, L., Lanham, N., Shenker, S.: Making Gnutella-like P2P systems scalable. In: Proceedings of ACM SIGCOMM (2003)
14. Vishnumurthy, V., Francis, P.: On heterogeneous overlay construction and random node selection in unstructured P2P networks. In: Proceedings of IEEE INFOCOM (2006)
15. Lv, Q., Cao, P., Cohen, E., Li, K., Shenker, S.: Search and replication in unstructured P2P networks. In: Proceedings of the 16th International Conference on Supercomputing, pp. 84–95. ACM (2002)
16. Gkantsidis, C., Mihail, M., Saberi, A.: Random walks in peer-to-peer networks. In: INFOCOM 2004. Twenty-Third Annual Joint Conference of the IEEE Computer and Communications Societies, vol. 1 (2004)
17. Otto, F., Ouyang, S.: Improving search in unstructured P2P systems: intelligent walks (i-walks). In: IDEAL 2006, pp. 1312–1319. Springer (2006)
18. Hamdeni, C., Hamrouni, T., Ben Charrada, F.: Data popularity measurements in distributed systems: survey and design directions. J. Netw. Comput. Appl. **72**, 150–161 (2016)

Dynamic Cluster Based Connectivity Approach for Vehicular Adhoc Networks

Naskath Jahangeer[1]([⊠]), Paramasivan Balasubramanian[1],
B. ShunmugaPriya[1], and Hamza Aldabbas[2]

[1] National Engineering College,
Tuticorin District, Kovilpatti, Tamil Nadu, India
naskath.neccse@gmail.com, bparamasivan@yahoo.co.in,
priyakrishnan.me@gmail.com
[2] Al-Balqa Applied University, Al-Salt, Jordan
aldabbas@bau.edu.jo

Abstract. With a broad scope to communicate and control the roadway of interest, the field of VANET is getting more attention among researchers. Owing to the real-time traffic data monitoring requirements, VANET emerged as an ideal network to a great extent. In VANET, the mobility of vehicles and frequent changes in connectivity of moving on-board units (OBU) along with the deployed (Road side Infrastructures) RSU's necessitated the evolution of several connectivity optimization techniques. In this proposed work, connectivity of network is evaluated using clustering approach and the connectivity probabilities are investigated in both vehicle-to-vehicle (V2V) and vehicle-to-infrastructure (V2I) multilane highway. The simulation results are analyzed using both network and the mobility simulators.

Keywords: Connectivity · Cluster analysis · VANET

1 Introduction

Intelligent transportation system (ITS) is the most promising and emerging technology to achieve safer, rapid, and more ecological transportation [12]. VANETs are a key branch of the (ITS) framework. VANET uses different communication technologies such as 3G, WAVE, WiFi, Dedicated Short Range Communications (DSRC) [10], 4G technologies and others. These communication technologies are varied based on the usage of different communication modes like V2V and V2I and sometimes Infrastructure-to-Infrastructure (I2I) also. V2V is one of the most exigent communications in Highways, but at the same time, it is efficient communication in the urban-related environment. V2I communication is very suitable in low-density highway environment using a single or more than one hop ad-hoc connection. Using infrastructure, vehicle's OBU can also access the Internet facility and other broadband services. When vehicles travel out of the exposure area of infrastructure, it will utilize its nearby vehicles (from both sides) as relays to communicate with the infrastructures [1].

Efficient coverage and connectivity of network is essential for the vehicular communication and its various challenges and applications since it might be complicated to

© Springer Nature Switzerland AG 2020
L. C. Jain et al. (Eds.): ICICCT 2019, LAIS 9, pp. 187–197, 2020.
https://doi.org/10.1007/978-3-030-38501-9_19

broadcast the packets to other vehicles in a network in case of frequent link breakages. Moreover, the high mobility of vehicles and relatively low communication range of DSRC in V2V is a core reason for those frequent disconnections [14]. In this paper, Dynamic Cluster based Connectivity approach for Vehicular Adhoc Networks (DCCV) this link fracture issue is overcome using clustering approaches. In this work, similar property vehicles in terms of velocity and distance can be grouped in to form a cluster to make efficient connectivity in both way scenarios. In VANET, this clustering technique is used to build a dynamic network with stable on the logical level and as long as provide a firm foundation for upper-level layer protocols. Design and developing a good clustering algorithm for efficient communication in VANET is a challenging task due to the high speed OBUs. However, clustering is the primary approach in V2V, V2I and I2I communications depend upon the density of the vehicles in the network.

The proposed clustering algorithm [8] provides better cluster permanence and low overhead. Mobility metrics, like position of OBUs, leading direction, and velocity rates also considered in this work. Moreover, this paper estimated the connectivity probability (CP) of V2V, V2I and I2I using network parameters. This work presents the following contributions. At first, this paper proposes an enhanced V2V and V2I interconnection using clustering approach, then measure the interconnection probability between V2V and V2I in various scenarios with mathematical models. Next, analysis of the performance of proposed works and conclude this paper with some futuristic ideas.

2 Literature Survey

Network connectivity is an essential platform for other vanet related issues. [4] suggested a connectivity improvement strategy in distributed manner to enhance the coverage and connectivity of network with reducing the level of energy utilization and signal variances. Connectivity analysis is categorized based on the environment like urban that is highly dense and the highway that is the low, compact area. To improve the V2V and V2I connectivity Wang et al. [6] proposed infrastructure oriented connectivity approaches for enhancing the intermittent connectivity of the network. They proposed mathematical approaches for reducing the delivery delay of the system. Chen et al. [7] considered user behavior like V2V cooperation has a major impact on connectivity of vehicular network. This cooperation can lead to boosting up of the network performance. The vehicle to anything communication refers that the information exchanges between the vehicles and the various elements of intelligent transport systems. This technology has great potential by enabling applications for increasing road safety, passenger message broadcastings, car firm services, and providing optimized traffic situations. It is enabling to done using two leading communication technologies like DSRC and cellular towers.

Infrastructure oriented communication enhances the communication approaches using platoon or clustering concept in sparse area highways [11, 13]. They designed a MAC protocol for platoon based networks using multi priority Markov model. Clustering is one of the eminent methods for organizing ad hoc networks with high system scalability, increases connectivity, and reduces frequent connectivity breakages [12]. Wang et al. [6] proposed a structure for determining connectivity requirements such as the minimum spatial node density and transmission range for disseminating packets in the network.

Yana et al. [4] provided mathematical models for enhancing the connectivity of the networks with consideration of some network parameters like transmission range of OBUs, and packet size and coverage and mobility parameters like headway distance, acceleration, association time, and the relative speed of vehicles, in V2V communications models. The other communication features such as handshaking time during communication, data transferring, throughput, and its response time, can also considerably impact the communications. Hou et al. [5] stated that the connected vehicles are the main metrics that have a critical impact on the performance of data transmission process. Due to the lack of real-world road traces, connectivity cannot comprehend the realistic large size environment and notably it affects the mobility over-connectivity. They used some metrics, such as the speed of vehicle and component speed to signalize connectivity and mobility. Abuelenin et al. [3] noted that vehicle headway distribution is essential for evaluating the CP of network. They considered the vehicles are coming in the single lane and used the shifted exponential approach to retain a secure distance between them.

3   Proposed Model

Vehicular Ad hoc Networks (VANETs) is a sub part of executing the pervasive environment. In this proposed work, consider [2] the optimal placement of the road side infrastructures. The proposed highway road topology is obtained and trace files are extracted from the Open Street Map OSM database that could be provide as an input to SUMO simulator as shown in Fig. 1. The output of SUMO [10] has been given as input to the NS2 simulator for analyzing of performance of network connectivity.

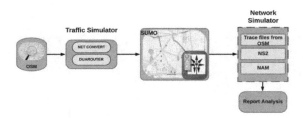

Fig. 1. Modeling of mobility using OSM

3.1   Connectivity Enhancement Using C-THE Clustering Techniques

Connectivity has a wide body of research issue in VANET due to the high mobility of nodes. Physical and technical are two main factors that can affect the connectivity of the vehicles. Coverage barrier between the vehicles, height of heavy duty vehicles and some weather conditions etc. are considered as the physical factors. The technical factors are characterized as the wireless channel randomness such as fading and shadowing. The defects due to the physical factors can be overcome by the deployment of RSU [2, 6] in optimal location with good communication range and enhancing these connectivity using clustering methodologies. The enhancement of connectivity of DCCV can be attained using the clustering modules like, Cluster formation, Cluster head selection, and Cluster Reorganization. Clustering techniques [8, 9] are employed in the VANET to improve the network efficiency. During Cluster formation, the number of vehicles which comes under the RSU's transmission range and it forms a cluster and is referred as Cluster Members (CM), and the vehicles whose distance is less than transmission range of the vehicles in the cluster members, referred as the sub-cluster members. For this, the primary work is to define the transmission range of RSU and OBU. Let I_{tr} and V_{tr} be the notation of transmission range of the RSUs and OBUs. The Algorithm 1 (C-THE) defines the cluster formation, Cluster head selection and reorganization steps.

Algorithm 1. Clustering Techniques for Highway Environment(C-THE)

	Cluster Head selection	Cluster Re-organization

`Read RSU;`
`Rr=(x[r],y[r]);`
`Define I_tr;`
`Read OBU;`
`Vi=(x(i),y(i))`
`i=1....N;`
`Define V_tr;`
`ensure ( N >1 )`
`for ( i=0, i<N,`
`i++ )`
`If (V_tr<I_tr) then`
`        Add Vi as`
`CM;`
`   end if`
`   for ( j=0, j<N,`
`j++ )`
`If (V_tr[j]<V_tr[i])`
`then,`
`Add Vj as sub CM`
`of Vi;`
`end if;`
`end ;`
`end;`
	`M= set of CM ;`
	`ensure ( M >1 )`
	`for ( i=1, i<M,`
	`i++ )`
	`compute di = D[Vi`
	`,Rr];`
	`compute AVi= sum`
	`(ar);      find c =`
	`argmin ( di );`
	`find v = argmin (`
	`Ai );`
	`if (dist of Vi<= c`
	`&&velocity of Vi<=`
	`v ) then`
	`       elect Vi as`
	`the CH;`
	`end if;`
	`end ;`
	`end;`
		`Set i = 1 and x`
		`= X_c ;`
		`Assume x_1c is`
		`perturbed by Δi;`
		`Calculate ḟ =`
		`ḟ(x),`
		`f̈+= ḟ(xi+ Δi),`
		`f̈ -= ḟ (xi -`
		`Δi);`
		`Calculate ḟ,`
		`f̈+, f̈-;`
		`Vmin= min (f, f+`
		`,f -);`
		`Set x <-Vmin;`
		`if ( X=X_c) then`
		`set X_c = X_{c+1};`
		`X_{c+1}=X_c+(X_c-X_{c-1});`
		`set X_{c+1} as BCN;`
		`end if ;`
		`elect (CH(Vi))`
		`end; end;`

   Initially the algorithm defines the coordinates of infrastructure and the vehicle as Rr and Vi respectively. If the range of vehicles falls inside of the RSU's range then, the vehicle Vi is added to the cluster. If the range of the vehicle's falls outside of the cluster but inside of the cluster member vehicles, then add the vehicle add as the sub cluster member. The cluster head (CH) or leader is elected to act as the intermediary between the RSU and the CM. Once the vehicles find their position in the clusters, it is very essential is to elect the cluster lead which can transmit packets of data between the intra and inter-clusters. The cluster members in a single cluster will compete for the election to become the cluster head. In order to select the CH, consider two parameters and they are vehicle's velocity and the distance between RSU. Initially, it calculates the distance from the RSU and the individual cluster members. From that the minimum distance is taken as the optimum vehicle. Next, the Accumulated Velocity (AV) is calculated by aggregating the relative velocity of vehicles which is calculated with respect to all its neighbours. The vehicle which is nearer to RSU and has the minimum AV is elected as the CH. Consider there are M members in the cluster, Xc is current cluster head, N number of vehicles. if there are at least one member in the cluster ($M > 1$), then compute the distance between the CM and RSU (di = D [Vi, Rr]) and compute the Accumulated velocity (AVi = sum (ar)) where ar is relative velocity. Then find the minimum distance and minimum velocity as c = argmin (di) and v = argmin (Ai) respectively. If a CM has the minimum velocity and distance then elect that CM as CH. Algorithm 1. Describes the clustering technique of DCCV.

The maintenance of Cluster longevity or cluster reorganization is main objective of forming the clusters using strong cluster heads. Whenever, the cluster head move left from the current huddle the responsibility of CH changed over to other nodes to maintain the uninterrupted connectivity between and within the cluster. For this it uses the Backup Cluster Node (BCN). The selection of BCN the uses the evolutionary move method, and selects the vehicle which has minimum accumulated velocity. Where $\dot{f}$, $\ddot{f}+$, $\ddot{f}-$ is velocity of vehicles. Once the cluster head comes to the edge of the RSU transmission range, it will hand over all the information to the BCN by the pattern move operation. After that, the cluster head election algorithm is called to elect the next CH. Again by using the pattern move the information from the BCN is transferred to the new CH.

3.2   Analysis of the Connectivity Probability

In this proposed work, cluster based VANET is designed for analyzing the CP. Two different network scenarios like V2V and V2I are considered for this work. The probability of connectivity is analyzed in both one and two way lanes of highway scenario.

3.2.1   Connectivity Probability of V2V Scenario

In V2V based cluster communication scenarios, D_i represents the intervehicle gap between two successive vehicles. Consider V2V clustering approaches, in this cluster the distance between any two successive vehicles is smaller than the cluster size. So the CP of one wayV2V network is, $P' = \prod_{i=1}^{N-1} Pr\{Di \leq R\}$. Vehicles can connect with the cluster member $(1 - p = q)$ or cluster head (p), and the D_i is random variable i = 1, 2.... N − 1 so the connectivity possibility is described as

$$P_{SV} = \prod_{i}^{N-1} q.Pr\{Di \leq R_{CM})\} + p.Pr\{Di \leq R_{CH})\} \tag{1}$$

R_{CM} and R_{CH} transmission range of member and CH vehicles. As per Eq. 1, The CP of V2V in one way scenario is defines as, $P_{SV} = [q.(1 - e^{-\alpha R_{CM}}) + p.(1 - e^{-\alpha R_{CH}})]^{N-1}$. In the two way V2V communication, the arrival rates of vehicles in the two ways are Δ (same in both ways). Assumed that, if the D_i of vehicles on the consecutive road is greater than the transmission range V_{tr} then the connectivity link is not stable to continue so it is consider as broken link B_l. So, the probability that a broken link is represented as P_{Bl}, $P_{Bl} = \{D > l\} = [q.(1 - e^{-\alpha R_{CM}}) + p.(1 - e^{-\alpha R_{CH}})]^{N-1}$. Due to the B_l in the one way road, the CP value is less than 1. So as per binomial distribution the total link probability $P_T(k)$ is defined using broken and healthy links.

$$P_T(k) = \binom{N-1}{k} P_{Bl}^k (P_{Hl})^{N-1-k} \tag{2}$$

Here k of $N - 1$ has broken links and P_{Hl} is healthy connected links in a network.

The link breakages are happened frequently, when the gap between two consecutive nodes is larger than its transmission range. However, in this scenario, the two nodes can connect with each other even if they are placed in coverage barrier then they can be connected with each other using adjacent road lane vehicles. Assume distances between two consecutive vehicles in the adjacent lanes are with its ranges it can possible to communicate and with broken vehicles.

For example, as shown in Fig. 2 the V_{si} and V_{sj} vehicles area in same lane, but out of coverage. Moreover, it communicate with each other using adjacent lane vehicles of V_{ai} and V_{aj}. If V_{ai} can connected with V_{aj}, then the vehicles on the nearby road can communicated between V_{si} and V_{sj}. That is possible when that V_{ai} is situated in the range of V_{si} and V_{aj} is situated in the range of V_{sj}. The probability function of $P_{sa(i)}$ and $P_{sa(j)}$ is,

$$P_{sa(j)} = P_{sa(i)} = Pr\{D \leq R\} = 1 - e^{-\alpha R} \tag{3}$$

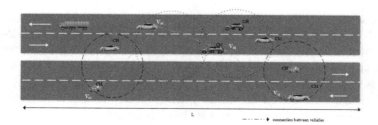

Fig. 2. V2V communication scenario in both one and two way scenario

Consider V_{si} can connect with V_{sj} using n-hop (n connected nodes between of V_{ai} and V_{aj}) communication then, $P_{(n-hop)} = \Pr\{D \le R\}^{n-1}$. So CP of adjacent vehicles is P_a,

$$P_a = \int_R^\infty \sum_1^n \frac{\delta}{\int_R^\infty y^n e^{-\bar{\alpha} \bar{l} y} dy} P_{sa(j)} dl \tag{4}$$

As per Eqs. 3 and 4 the connectivity probability V_{si} and V_{sj} of using healthy links defines, $P_{tV-Hl} = (1 - e^{-\alpha R})^2 . P_a$. Assume possibilities of broken link between of V_{si} and V_{sj} then the connection probability is defined using Eq. 2.

$$P_{AV}(k) = P_{tV-Hl}^k \binom{N-1}{k} P_{Bl}^k (P_{HI})^{N-1-k} \tag{5}$$

3.2.2 Connectivity Probability V2I Scenario

In this communication scenario, vehicles can communicate with other nearby or distanced vehicles through RSU. Assume the subject vehicle can communicate with the RSU using two hop communications. Subject Vehicle to Infrastructure communication without using any relay is termed as one hop with relay is two or multi hop mode. The V2I scenario is depicted in Fig. 3. From this depiction optimal gap between the two adjacent infrastructure is d and its transmission range is I_{TR}. Let, the probability of connectivity of subject vehicle in one way V2I is P_I. This module of the work defined using different scenarios. If the subject vehicle can connect directly under the coverage of two adjacent Infrastructures without any relay or coverage barrier then the total probability of subject vehicle is equal to 1. If the subject vehicle is situated under coverage of one RSU then it can connect using one hop mode. So the CP is, $P_{C'} = \frac{2I_{TR}}{d}$. If the subject vehicle is come under the barrier area of coverage then two possibilities are considered to connect the subject vehicle with RSU. They are, connect Subject vehicle using relays either cluster member or cluster Head.

(a) Connect using Cluster Member: If the subject vehicle travels under the coverage gap of any RSUs then it can try to make a communication with relay. If the relay is member vehicle of Cluster, then the CP is, $P_{CM'} = \Delta q(1 - e^{-\alpha l'})$. Where $\Delta = 1 - P_{C'}$ and $l' = 2(I_{TR} + R_{CM}) - d$.

(b) Connect using Cluster Head: If the subject vehicle travels under the coverage gap of any infrastructures then it can try to make a communication with relay. If the relay is head of Cluster, then the CP is, $P_{CH'} = \Delta p(1 - e^{-\alpha l''})$. Where $l'' = 2(I_{TR} + R_{CH}) - d$. CP of this case is $P_{I'} = P_{CM'} + P_{CH'}$. So $P_{I'} = \Delta \{q(1 - e^{-\alpha l'}) + p(1 - e^{-\alpha l''})\}$.

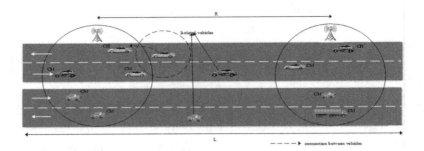

Fig. 3. V2I communication scenario in both one and two way scenario

In 2 way communication scenario, vehicles can communicate with nearby or distanced vehicles through infrastructures and adjacent lane vehicles. In this case, assume that the subject vehicle can communicate with the road side Infrastructures using two or more hop communications. Subject Vehicle can connect with the infrastructure using two or more than two clusters from same and adjacent road lanes. The two way V2I scenario is depicted in Fig. 3. Let, take $P_{I''}$ is connectivity probability of a subject vehicle in two ways V2I. This module of the work defined using different cases of scenarios. If the subject vehicle is situated under coverage of one RSU, then it connects using one hop mode. So the CP is same as $P_{C'}$. If, subject vehicle is come under the breach of coverage then the two possible options are available to connect the subject vehicle with the RSU. They are, connect Subject vehicle using adjacent road relays either CM or CH.

(a) Connect using CM: If the subject vehicle travels under the coverage gap of any RSUs then it can try to make a communication with relay. If the connecting node is from adjacent lane, then the CP is $P_{CM''} = 1 - \Delta q(1 - e^{-\alpha l'}) * P_{CM'}$.

(b) Connect using CH: If the subject vehicle travels under the coverage gap of any infrastructures then it can try to make a communication via relay. If the relay is head of Cluster, then the probability of connectivity is, $P_{CH''} = \Delta p(1 - e^{-\alpha l'})$. Here, $l'' = 2(I_{TR} + R_{CH}) - d$ and CP of this case $P_{I''} = P_{CM''} + P_{CH''}$

$$P_{I''} = \Delta \left\{ q\left(1 - e^{-\alpha l'}\right) + p\left(1 - e^{-\alpha l''}\right) \right\} \tag{6}$$

As per Eq. 6, this connectivity of subject vehicle is simplify using 2×4 decoder structure for two way V2I communication. The subject vehicle can communicate with RSU using one hop or multi hop approach. This system connectivity is designed using decoder with 2 input lines, and 2n output lines. The input lines are connection links is formed using either CMs or CHs or combination of both of different clusters. The connectivity of this two-way vehicle to infrastructure will be active High based on the combination of connected vehicles in the cluster.

4   Simulations and Results Analysis

This work is simulated using two simulators like traffic and network simulators of SUMO and NS2, respectively. The main intention of these simulations [10] is to evaluate the performance of proposed connectivity schemes with and without the presence of roadside infrastructures. A VANET of 45 moving vehicles and infrastructures are deployed into 1500×1000 m^2 area of the highway. The experiment is tested in the presence of various transmission ranges of infrastructures in the proposed area. The efficiency of the proposed schemes is evaluated using the metrics such as end-to-end delay, packet delivery ratio, and connectivity probability (CP). The simulation results show that the proposed study performance is evaluated in different scenarios of V2V and V2I with different transmission ranges.

The CP is defined as, vehicles on multilane highway can access to at least one infrastructure besides the road with minimum number of hops. The CP of proposed multilane highway is analyzed using theoretical as well as traffic and network simulation. In this study first analyze the 4 scenarios of proposed system, they are V2V, V2I (50 m), V2I (100 m) and V2I (150 m) i.e. transmission ranges are varied from 50 m to 150 m for static infrastructures at road side. Figure 4 depicts the comparison chart of Connectivity probability, packet delivery ratio and end to end delay of DCCV.

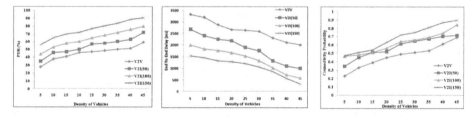

Fig. 4. Comparison graphs of DCCV in terms of PDR, end to end delay and CP

V2I-150 m range of scenario continuously outperforms in packet delivery in the proposed network. The other three scenarios slightly varied from each other based on the density and transmission range of infrastructures. V2I (150 m) scenario performed well than other situations even though the network has dynamically changed its topology. Especially V2I (150 m) continuously minimized the packet delivery delay compared to other proposed schemes due to select the minimum hop cluster leads as infrastructures for preserving stable cluster. The proposed schemes, all are decreases the average delay significantly when the number of vehicles increased. Compared to proposed schemes, V2I (150 m) outperforms with minimum hop communication with end devices to deliver the packets.

5 Conclusions and Future Work

This proposed work discussed and analyzed about the connectivity of the multilane highway topology. The main objective of the work is achieved by formulating the model for connectivity enhancement using clustering techniques. In the proposed work, the cluster is formed using moving vehicles and infrastructures. When the number of vehicles gets increased in the scenario head responsibility is to take over by the infrastructures. The proposed scenarios are compared with each other. The technical factors such as wireless channel randomness, shadowing, fading can also be included as the network parameter while measuring the network connectivity. The above factors can affect the connectivity and it can be considered for providing the quality of service in connectivity for the multilane cross sectional environment and they are left open for the further work. And the performance of proposed algorithms will be compared with other related algorithms.

References

1. Al-Sultan, S., Al-Doori, M.M., Al-Bayatti, A.H., Zedan, H.: A comprehensive survey on vehicular ad hoc network. J. Netw. Comput. Appl. (2013). https://doi.org/10.1016/j.jnca.2013.02.036
2. Naskath, J., Paramasivan, B.: Location optimization for road side unit deployment and maximizing communication probability in multilane highway. Int. J. Heavy Veh. Syst. 25 (3/4), 369–390 (2018)
3. Abuelenin, S.M., Abul-Magd, A.Y.: Effect of minimum headway distance on connectivity of VANETs. Int. J. Electron. Commun. (AEU) 69, 867–871 (2015)
4. Yan, G., Rawat, D.B.: Vehicle-to-vehicle connectivity analysis for vehicular ad-hoc networks. Int. J. Ad Hoc Netw. 58, 25–35 (2017)
5. Hou, X., Li, Y., Jin, D., Wu, D.O., Chen, S.: Modelling the impact of mobility on the connectivity of vehicular networks in large-scale urban environments. IEEE Trans. Veh. Technol. 65(4), 2753–2758 (2016)
6. Wang, Y., Zheng, J., Mitton, N.: Delivery delay analysis for roadside unit deployment in vehicular ad hoc networks with intermittent connectivity. IEEE Trans. Veh. Technol. 65(10), 8591–8602 (2016)
7. Chen, R., Sheng, Z., Zhong, Z., Ni, M., Leung, V.C.M., Michelson, D.G., Hu, M.: Connectivity analysis for cooperative vehicular ad hoc networks under Nakagami fading channel. IEEE Commun. Lett. 18(10), 1787–1790 (2014)
8. Cooper, C., Franklin, D., Ros, M., Safaei, F., Abolhasan, M.: A comparative survey of VANET clustering techniques. IEEE Commun. Surv. Tutor. 19(1), 657–681 (2016)
9. Ahmed, E., Gharavi, H.: Cooperative vehicular networking: a survey. IEEE Trans. Intell. Transp. Syst. 19(3), 996–1014 (2018)
10. Samatha, B., Raja Kumar, K., Karyemsetty, N.: Design and simulation of vehicular adhoc network using SUMO and NS2. Adv. Wirel. Mob. Commun. 10(5), 1207–1219 (2017). ISSN 0973-6972
11. Shao, C., Leng, S., Zhang, Y., Vinel, A., Jonsson, M.: Performance analysis of connectivity probability and connectivity-aware MAC protocol design for platoon-based VANETs. IEEE Trans. Veh. Technol. 64(12), 5596–5609 (2015)

12. Ghosh, R., Pragathi, R., Ullas, S., Borra, S.: Intelligent transportation systems: a survey. In: IEEE International Conference on Circuits, Controls, and Communications (2017). https://doi.org/10.1109/ccube.2017.8394167
13. Pal, R., Prakash, A., Tripathi, R., Singh, D.: Analytical model for clustered vehicular ad hoc network analysis. ICT Express **4**, 160–164 (2018)
14. Baqar, M.A., Aldabbas, H., Alwadan, T., Alfawair, M., Janicke, H.: Review of security in VANETs and MANETs. In: Network Security Technologies: Design and Applications, pp. 1–27. IGI Global (2014)

Transfer Learning for Internet of Things Malware Analysis

Karanja Evanson Mwangi[1(✉)], Shedden Masupe[2],
and Jeffrey Mandu[1]

[1] University of Botswana, Gaborone, Botswana
`sundayfeb29@gmail.com`, `jeffreym@mopipi.ub.bw`
[2] Botswana Institute for Technology Research and Innovation (BITRI),
Gaborone, Botswana
`smasupe@bitri.co.bw`

Abstract. Internet of Things (IoT) environments are characterized by devices that have heterogeneous applications, diverse underlying technologies and most of them are constrained in resources such as low memory, and weak security mechanisms. The synthesis of malware that menaces the Internet of Things environments is an open and evolving research problem. This study proposes use of images and transfer learning to analyze IoT malware. The malware files are converted to three channel images to cull architecture and platform of analysis dependence. The preprocessed malware images are split to training and testing set. The deep neural network is based on pretrained VGG19 Model and adapts the last layer to discriminate the malware images into their malware families. The experimental results on the dataset adopted from IoTPoT dataset shows that our approach can be used effectively to classify malware into families as it attains 89.23% overall accuracy and F measure of 91.3%.

Keywords: Transfer learning · Internet of Things malware · VGG19 model · IoTPoT dataset

1 Introduction

Internet of Things environments have intrinsic features such as heterogeneity in terms of platforms, wide range of central processing architectures, diverse run-time libraries, variant modes of system imaging that makes malware analysis challenging [12]. To provide a solution to the malware menace, malware samples needs to be examined and their capabilities established. Malware analysis involves examining malicious files with an aim of understanding its behavior, evolution, constructs and possible targets. The outcome of the malware analysis process aids in designing defense mechanisms against the studied malware and its family. Previous methods in literature such as static and dynamic analysis approaches are architecture and platform dependent. However, the IoTPoT [16] dataset used in the study demonstrates that one sample of IoT malware, can affect more than four different central processing unit architectures. Therefore, a solution to analyzing IoT malware is either (1) to develop analysis tools for each of the affected architecture and also go to different platforms or operating system which is

© Springer Nature Switzerland AG 2020
L. C. Jain et al. (Eds.): ICICCT 2019, LAIS 9, pp. 198–208, 2020.
https://doi.org/10.1007/978-3-030-38501-9_20

computational costly and might not be feasible or (2) develop an architecture independent and platform independent analysis approach. This work adopts the second form of solution since images can be analyzed independent of the architecture or platform that created them. The overarching distinct contribution in our approach is that live malware can be converted to images and utilize existing image processing algorithms that are robust and well tested to analyses and classify malware into their families.

2 Motivation

Transfer learning offers access to robust and flexible pretrained learning networks for image analysis. Transfer learning schemes do not require intricate malware features to be extracted and engineered through dissembling in target architectures thus reduces the computational complexity in malware analysis. There lacks sufficient IoT based tools for malware analysis or tools with interoperability that can span more than one architecture [4]. Conversion of malware to three channel images and use of pretrained transfer learning models such as VGG19 culls the complexity of malware analysis processes such as malware classification as it reduces the dependency of features reengineering and training on large dataset that might not be available for the study domain.

3 Related Works

Malware has been analyzed in the past either by examining its code in-situ (static analysis); or running it in safe environment or a sandbox (dynamic analysis); or using hybrid methods such as visualization.

When analyzing malware statically, the code structure of the malware is analyzed without executing it. Static analysis involves discerning patterns on static features of the code. These features include string signatures, control flow graph (CFG), operation code (Opcode), byte sequence n-grams and Windows application programming interface (API) calls [24]. The study by Damodaran et al. [5] extracts both API call sequences and opcode sequences from seven malware families. The sequences are used to train the Hidden Markov Models (HMMs) learner. The areas under receiver operating characteristic (ROC) curves and Precision Recall (PR) index area are used as tools of analysis. The shortcoming on their proposed methods is that it is computationally expensive to train the hidden Markov models. Kang et al. [11] develops a system that classifies Android malware using creators information such as API and certificate serial numbers. The system applies Naive Bayes Classifier with 90% classification accuracy. Salehi et al. [23] created API frequencies from binary files. Three set of features (call lists, arguments and joint API argument list) were generated from APIs and using Random Forest classifier their model attained 98.4% in overall accuracy and 3% as false positive rate. In [10], up to ten opcodes features were derived from disassembled files and organized into a feature vector for classification. Each of the feature sets is analyzed using four different algorithms. The results achieve an F-measure of 98%.

Zhang et al. [27] built a graph-based model to analyze Dalvik opcode and their properties. In the model, Dalvik opcode sequences are modelled as directed graph of edges and vertices. Manhattan distance is computed as a similarity measure between pre-labelled opcodes and the target with the model achieving 93.6% classification accuracy.

Dynamic malware analysis involves executing the malware in virtual environment or sandbox and analyzing its behavior. In dynamic analysis, features such as traces of live application programming interface (API) calls, network analysis, system calls, registry changes and memory traces are examined [5]. Various studies work has been done in the area of dynamic analysis of malware, in this subsection a selected few are reviewed. Kim et al. [13] proposes a method that creates a behavior chain to characterize malware based on clustered system calls. Malware classification in their system is undertaken by calculating similarity indexes on clustered malware and the system call of the testing malware. Smith-Waterman (SW) algorithm and the longest common subsequence (LCS) are used as similarity indices. The results show that LCS had better performance at precision rate (95.01%), overall accuracy of 94.89%, false positive rate as 4.04% while the F-measure realized was 93.26%.

Galal et al. [8] proposes analysis of API calls by clustering them based on common shared semantic into sequences that denote actions. Three classification algorithms, that is decision tree, random forests, and support vector machine are used to evaluate the model. Decision tree achieved best performance of sensitivity (97.3%), specificity (96.53%) and overall accuracy (97.19%).

Nari and Ghorbani [19] introduces a dynamic analysis technique that automatically converts network behavior into a graph whose features are used to classify the malware using WEKA library classifiers. The results indicate that J48 decision tree achieved 98% receiver operating characteristic curve (ROC) area for four malware families. Cabau et al. [3] discriminates changes made to the registry keys and file systems in real time using Support vector machines (SVMs). Their model achieves average F-measure of 96.25%.

Although most of the IoT devices are always connected, on-system detection of malware and in vivo malware analysis is hard due to limited computational power caused by battery size, lifetime and memory requirements on most devices. For instance, most smart IoT sensors operate on central processing power (CPU) ranging between 50 and 100 Dhrystone Million Instructions per Second (DMIPS) [2]. Static and dynamic malware analysis requires platform dependent disassemblers and other test tools to extract features from malware code. However, there lacks robust tools for static and dynamic IoT malware analysis [4].

In the recent past visualization-based methods have offered promising results in malware analysis. Visualization systems for malware analysis have shown remarkable results in the recent years. Visualization based analysis has been proposed in few studies for traditional computers malware mainly affecting mainly Windows based platforms [7, 20]. Nataraj et al. [20] converts live malware into grayscale (one channel) images and using gradient information on scales and orientations (GIST) method derives the texture features for analysis. The classification using K-nearest neighbor with Euclidean distance attained 98% accuracy on a Windows based malware dataset consisting of 25 families. Their study is the first work to convert malware binaries to images and proved faster than n-gram method. Makandar and Patrot [15] proposes using feed forward artificial neural networks (ANN) to classify Windows based Mahenhur dataset. Their experiment applies Gabor wavelet transform and GIST for features extraction. The ANN classifier attained 96.35% accuracy on 24 families in the dataset.

In Xiaofang et al. [26], malware visualization methods is proposed. The proposed method uses gray scale image and features extracted using speeded up robust features algorithm (SURF). Liu and Wang [14] converts disassembly files into gray-scale images. In their method, a local mean is used to compress grayscale images to be mapped to a feature vector with 98.2% classification accuracy. Fu et al. [7] method extracts color moments and gray level co-occurrence matrix on Windows platform based malware from 15 families. Experimental classification of malware into their families is conducted using random Forest, K-nearest neighbor and support vector machine. Random forest algorithm achieves better performance with accuracy of 0.9747, precision of 0.9711, re-call of 0.9672 and F-measure of 0.9685. Darus et al. [6] extracts image features using GIST descriptor on Android malware. Three classifiers namely k-nearest neighbor (KNN), random forest (RF), and decision tree (DT) are applied on the features set for classification. Random forest outperformed the rest attaining 84.14% accuracy.

Naeem et al. [18] proposed a malware image classification system that extracts global and local features from malware images. Though their study claims to be for IoT malware, the experiments only handled 25 families of Windows platform malware only. The classification of image features using linear support vector machine attained 99.6%.

Convolution neural network (CNN) is a deep learner that has been used in image analysis [9]. The advantage gained by using CNN is that unlike other features extraction methods such as texture analysis, CNN requires minimal preprocessing and can work directly with raw images. Kalash et al. [9] outlines an approach that converts malware from Malimg and Microsoft malware to grayscale images. A CNN-based architecture was applied to extract features and classify images achieving accuracy of 98.52% and 99.97% on the Malimg and Microsoft datasets respectively.

MalDeep, a malware classification framework using deep learning based on texture visualization is proposed in [28]. The framework uses grayscale images and extracts texture features using CNN with over 90% accuracy on Microsoft BiG 2015 dataset. The model by Ni et al. [21] converts malware codes into gray scale images using sim-Hash and then uses convolutional neural network to classify the malware images into their families. The unique aspect of their model is that performance could be improved using convolutional neural network on images to achieve 98.862% accuracy. Unlike previous studies, our approach does not need use of handcrafted features and uses color images as malware representation.

4 Materials and Methods

The IoT malware used in this work is shared from Pa et al. [16]. The shared live malware are identified using message-digest algorithm (MD5) hash. To establish the ground truth and labels, the executable malware file was scanned using VirusTotal [25]. In the VirusTotal there were 56 active malware scanning engines. For each scanned malware to qualify for further processing, at least 28 malware scanning engines needed to identify it as malware from either Bashlite or Mirai family. The resultant dataset after preprocessing has 133 Gafygt/Bashlite samples, 125 Mirai samples. A set of benign files were extracted from clean system files in various architectures and also scanned through the VirusTotal with zero engine recognition as malware files. Each of the file obtained as dataset sample is converted to its respective grayscale image following a procedure described in [20]. For each byte-plot grayscale image it is further converted to three channel (RGB) image and re-scale it to 64×64 dimensions using image bilinear interpolation algorithm. For detailed understanding of bilinear interpolation algorithm see [1]. The final re-sampled and resized images have class labels as; 0 = Bashlite, 1 = Benign, 2 = Mirai.

Transfer learning is formally defined as: - Let D be a source domain and B the learning task, a target domain E and learning task T. Using predictive function f, transfer learning targets to improve the learning of f in E using the knowledge gained in D and B where $D \neq E$ and $B \neq T$. If the target and source domains $D = E$ and their corresponding learning tasks are similar, that is $B = T$, the learning problem converges to a classical machine learning problem Our malware classification task is defined as inductive transfer learning setting since the target task of classifying malware is different from domain task and also there is no similarity of thee target and the source domains. Transfer learning avails high performance learning experience in various domains such medical applications where there exist small datasets [22].

Transfer learning aids in leveraging on pretrained VGG19 model that has demonstrated high accuracy and state of art performance in object recognition. In the model depicted in Fig. 1, the three-channel dataset of images of malware and benign samples is small and different from the original dataset for VGG19. The baseline model is trained with 64 * 64 * 3. In the training and validation, the dropout is set at 0.5. Dropout is key as it ensures that the model trains in all 20 epochs without over-fitting. The RMSprop optimizer is used and the learning rate is set at 0. 0001. The bottleneck features are collected at block 5 of the VGG19 where the model is truncated to remove the fully connected layers from the baseline VGG19. These bottleneck features form our output layer. The bottleneck features are fed to train our final classifier. The added layers in the classifier are (1) flatten layer (2) Fully connected layer with 512 neurons and rectifier activation (ReLu) and finally a SoftMax layer for classification.

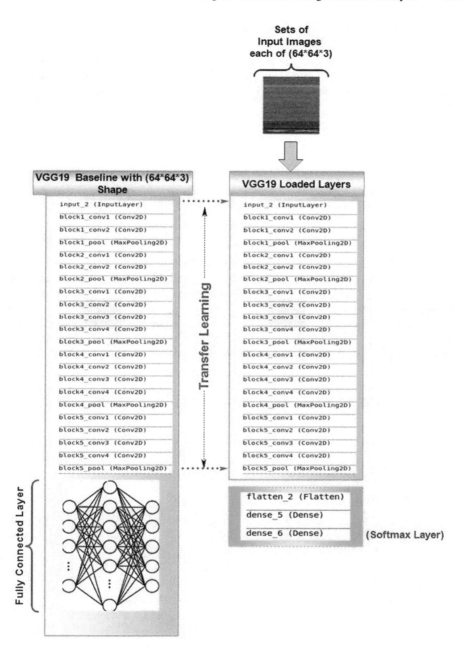

Fig. 1. The proposed model for transfer learning on malware images.

5 Experimental Results and Discussion

The analysis experiments are conducted on the preprocessed dataset and the proposed approach performance is evaluated using metrics such as the precision rate, recall rate, overall accuracy and area under curve on the receiver operating characteristics (ROC) curve.

$$Accuracy = \frac{TrP + TrN}{TrP + TrN + FlP + FlN}. \tag{1}$$

$$Precision = \frac{TrP}{TrP + FlP} \tag{2}$$

$$Recall = \frac{TrP}{TrP + FlN}. \tag{3}$$

Where, True Positive (TrP), True Negative (TrN), False Positive (FlP) and False Negative (FlN). The receiver operating characteristic (ROC) curve is a visualized graph showing the true positive rate(recall) or sensitivity versus the false positive rate (1-Specificity). The area under curve indicates the performance of classifier ranging from 0, random classifier with 0.5 and a perfect classification of 1. Balance accuracy concept introduced by Mohammad et al. [17] is applied using micro-average and macro-average methods for cross validation. The precision recall curve (PRC) is a plot

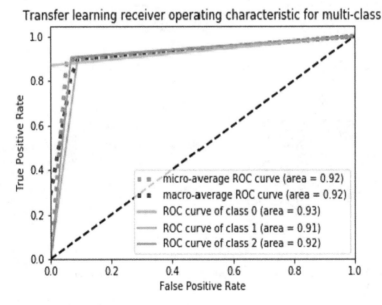

Fig. 2. Transfer learning Receiver Operating Characteristic (ROC) curve

of precision against recall(sensitivity). The recall axis represents the true positive rate. The precision recall curve is key when there is imbalance in the class distribution in the dataset as it is for the dataset in this study. The precision recall curve index from a perfect score of 1 or 100% to zero. The network topology is built using Keras and the neural network API was written in Python with TensorFlow. The model is trained for 20 epochs with RMSprop optimizer and category cross entropy. For multi category classifications, model achieved validation accuracy of 95.6%. The model validation loss of 11.14% at the end of 20th epoch (Fig. 2).

Precision and recall curve are shown in Fig. 3. The average precision score micro averaged over all classes for transfer learning classifier and individual class performance is shown. The Precision Recall curve index of 0.83 across classes is obtained. On individual categories, The PR curve area index on Bashlite is 0.92, benign = 0.78 and Mirai = 0.81 respectively.

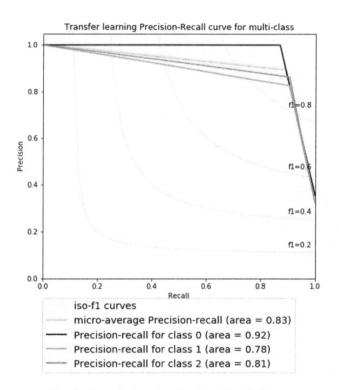

Fig. 3. Transfer learning Precision Recall Curve

6 Future Work

Considering the favorable results obtained in this study, future work will involve testing other pre-trained models for transfer learning with fined tuned vectors. The malware dataset is considerably small, part of the future work will also include exploring use of image data augmentation to increase the features set in the dataset.

7 Conclusion

The results show that usage of images and robust pre-trained models transfer learning in malware analysis is a promising approach that culls the architecture and platform bottlenecks offering promising results for analysis tasks such as classification.

Acknowledgments. This work has been partially supported by METEGA-ACP project at University of Botswana. We also thank the IOTPOT team for giving us access to the malware dataset used in the study.

References

1. Acharya, T., Tsai, P.S.: Computational foundations of image interpolation algorithms. Ubiquity **2007**, 4:1–4:17 (2007). http://doi.acm.org/10.1145/1322464.1317488
2. ARM: A guide to internet of things (IoT) processors (2019). https://www.usenix.org/system/files/conference/woot15/woot15-paper-pa.pdf
3. Cabau, G., Buhu, M., Oprisa, C.P.: Malware classification based on dynamic behavior. In: 2016 18th International Symposium on Symbolic and Numeric Algorithms for Scientific Computing (SYNASC), pp. 315–318. IEEE (2016). https://ieeexplore.ieee.org/stamp/stamp.jsp?arnumber=7829629
4. Costin, A., Zaddach, J.: IoT malware: comprehensive survey, analysis framework and case studies. Black Hat, USA (2018). http://firmware.re/malw/bh18uscostin.pdf
5. Damodaran, A., Di Troia, F., Visaggio, C.A., Austin, T.H., Stamp, M.: A comparison of static, dynamic, and hybrid analysis for malware detection. J. Comput. Virol. Hacking Tech. **13**(1), 1–12 (2017). https://link.springer.com/article/10.1007/s11416-015-0261-z
6. Darus, F.M., Ahmad, S.N.A., Ariffin, A.F.M.: Android malware detection using machine learning on image patterns. In: 2018 Cyber Resilience Conference (CRC), pp. 1–2. IEEE (2018). https://ieeexplore.ieee.org/stamp/stamp.jsp?arnumber=8626828
7. Fu, J., Xue, J., Wang, Y., Liu, Z., Shan, C.: Malware visualization for fine-grained classification. IEEE Access **6**, 14510–14523 (2018). https://ieeexplore.ieee.org/abstract/document/8290767/
8. Galal, H.S., Mahdy, Y.B., Atiea, M.A.: Behavior-based features model for malware detection. J. Comput. Virol. Hacking Tech. **12**(2), 59–67 (2016). https://link.springer.com/content/pdf/10.10072Fs11416-015-0244-0.pdf
9. Kalash, M., Rochan, M., Mohammed, N., Bruce, N.D., Wang, Y., Iqbal, F.: Malware classification with deep convolutional neural networks. In: 2018 9th IFIP International Conference on New Technologies, Mobility and Security (NTMS), pp. 1–5. IEEE (2018). https://ieeexplore.ieee.org/stamp/stamp.jsp?tp=arnumber=8328749

10. Kang, B., Yerima, S.Y., McLaughlin, K., Sezer, S.: N-opcode analysis for android malware classification and categorization. In: 2016 International Conference on Cyber Security and Protection of Digital Services (Cyber Security), pp. 1–7. IEEE (2016). https://ieeexplore. ieee.org/stamp/stamp.jsp?tp=arnumber=7502343
11. Kang, H., Jang, J.w., Mohaisen, A., Kim, H.K.: Detecting and classifying android malware using static analysis along with creator information. Int. J. Distrib. Sens. Netw. **11**(6), 479174 (2015). https://journals.sagepub.com/doi/pdf/10.1155/2015/479174
12. Karanja, E.M., Masupe, S., Mandu, J.: Internet of things malware: a survey. Int. J. Comput. Sci. Eng. Surv. **8**(3), 1–20 (2017). http://aircconline.com/ijcses/V8N3/8317ijcses01.pdf
13. Kim, H., Kim, J., Kim, Y., Kim, I., Kim, K.J., Kim, H.: Improvement of malware detection and classification using API call sequence alignment and visualization. Cluster Comput. 1–9 (2017). https://link.springer.com/content/pdf/10.10072Fs10586-017-1110-2.pdf
14. Liu, L., Wang, B.: Malware classification using gray-scale images and ensemble learning. In: 2016 3rd International Conference on Systems and Informatics (ICSAI), pp. 1018–1022. IEEE (2016). https://ieeexplore.ieee.org/stamp/stamp.jsp?tp=arnumber=7811100tag=1
15. Makandar, A., Patrot, A.: Malware analysis and classification using artificial neural network. In: 2015 International Conference on Trends in Automation, Communications and Computing Technology (I-TACT 2015), pp. 1–6. IEEE (2015). https://ieeexplore.ieee.org/ iel7/7490162/7492634/07492653.pdf
16. Pa, Y.M., Suzuki, S., Yoshioka, K., Matsumoto, T., Rossow, C.: IoTPoT: analyzing the rise of IoT compromises. In: Proceedings of the 9th USENIX Conference on Offensive Technologies, WOOT 2015, p. 9. USENIX Association, Berkeley (2015). https://www. usenix.org/system/files/conference/woot15/woot15-paper-pa.pdf
17. Mohammad, S., Morris, D.: Texture analysis for glaucoma classification. In: 2015 International Conference on BioSignal Analysis, Processing and Systems (ICBAPS), pp. 98–103. IEEE (2015). https://ieeeplore.ieee.org/stamp/stamp.jsp?tp=arnumber=7292226
18. Naeem, H., Guo, B., Naeem, M.R.: A light-weight malware static visual analysis for IoT infrastructure. In: 2018 International Conference on Artificial Intelligence and Big Data (ICAIBD), pp. 240–244. IEEE (2018). https://ieeexplore.ieee.org/stamp/stamp.jsp?tp= arnumber=8396202tag=1
19. Nari, S., Ghorbani, A.A.: Automated malware classification based on network behavior. In: 2013 International Conference on Computing, Networking and Communications (ICNC), pp. 642–647. IEEE (2013). https://ieeexplore.ieee.org/stamp/stamp.jsp?arnumber=6504162
20. Nataraj, L., Karthikeyan, S., Jacob, G., Manjunath, B.S.: Malware images: visualization and automatic classification. In: Proceedings of the 8th International Symposium on Visualization for Cyber Security, VizSec 2011, pp. 4:1–4:7. ACM, New York (2011). https://doi.org/ 10.1145/2016904.2016908
21. Ni, S., Qian, Q., Zhang, R.: Malware identification using visualization images and deep learning. Comput. Secur. **77**, 871–885 (2018). https://doi.org/10.1016/j.cose.2018.04.005. http://www.sciencedirect.com/science/article/pii/S0167404818303481
22. Pan, S.J., Yang, Q.: A survey on transfer learning. IEEE Trans. Knowl. Data Eng. **22**(10), 1345–1359 (2010). https://ieeexplore.ieee.org/stamp/stamp.jsp?tp=arnumber=5288526
23. Salehi, Z., Sami, A., Ghiasi, M.: Using feature generation from API calls for malware detection. Comput. Fraud Secur. **2014**(9), 9–18 (2014). https://doi.org/10.1016/S1361-3723 (14)70531-7
24. Sihwail, R., Omar, K., Ariffin, K.A.Z.: A survey on malware analysis techniques: static, dynamic, hybrid and memory analysis. Int. J. Adv. Sci. Eng. Inf. Technol. **8**(4–2), 1662–1671 (2018). https://doi.org/10.18517/ijaseit
25. Virus Total: VirusTotal search (2019). http://www.virustotal.com

26. Xiaofang, B., Li, C., Weihua, H., Qu, W.: Malware variant detection using similarity search over content fingerprint. In: The 26th Chinese Control and Decision Conference (2014 CCDC), pp. 5334–5339. IEEE (2014). https://ieeexplore.ieee.org/stamp/stamp.jsp?tp= arnumber=6852216tag=1
27. Zhang, J., Qin, Z., Zhang, K., Yin, H., Zou, J.: Dalvik opcode graph based android malware variants detection using global topology features. IEEE Access **6**, 51964–51974 (2018). https://ieeexplore.ieee.org/stamp/stamp.jsp?arnumber=8466776
28. Zhao, Y., Xu, C., Bo, B., Feng, Y.: MalDeep: a deep learning classification framework against malware variants based on texture visualization. Secur. Commun. Netw. **2019** (2019). http://downloads.hindawi.com/journals/scn/2019/4895984.pdf

Hybridization of Local Search Optimization and Support Vector Machine Algorithms for Classification Problems Enhancement

Raid Khalil[1,2(✉)] and Adel Al-Jumaily[1]

[1] University of Technology Sydney, UTS 81-113 Broadway,
Ultimo, NSW 2007, Australia
`Raid.M.Khalil@student.uts.edu.au`,
`R.M.Khalil@bau.edu.jo`, `Adel.Al-Jumaily@uts.edu.au`
[2] Al-Balqa Applied University, Al-Salt, Jordan

Abstract. The classification technique is defined as a task of supervised learning aims to identify the set for which a new input element is classified. SVM is a classifier, which is simple in structure, easy for training and often applied in classification problems. In the paper introduced, we improved SVM model that employs local search (LS) to enhance the accuracy of the classification. The proposed approach was experimented by eleven standard benchmark datasets from UCI repository. The results show enhancement in classification accuracy of proposed SVM model in comparison with that of the conventional SVM model.

Our proposed new technique combines local search optimizer and support vector machine classifier as a reasonable solution to handle classification problems. SVM was utilized and then optimized by investing the LS optimization technique. Results of experiments drove using 10 benchmark datasets showed that the proposed LS-SVM outperformed the original SVM on all datasets. Also, when comparing the accuracy of LS-SVM to the accuracy of other approaches detailed in the literature, LS-SVM's classification accuracy was the highest on most of eleven datasets.

Keywords: Machine learning · Support vector machine (SVM) · Local search (LS) · Classification · Optimization

1 Introduction

In machine learning algorithms, the instance is described by a group of features. Instances are divided into two categories. These include supervised instances which are characterized by known labels or functions and unsupervised which lack labeling. Scientists anticipate that unsupervised algorithms would assist in exploring unidentified, but beneficial, classes of items [1, 2]. The current study focuses on the mechanisms required to examine the relationships or associations between unknown items [6]. Machine learning is a methodology for classifying a set of rules from instances, or generally, generating a classifier to postulate from new instances [3]. Applying

© Springer Nature Switzerland AG 2020
L. C. Jain et al. (Eds.): ICICCT 2019, LAIS 9, pp. 209–217, 2020.
https://doi.org/10.1007/978-3-030-38501-9_21

supervised machine learning to actual real problem is described in Fig. 1. Such settings, a satisfactory selection of a specific learning algorithm as a classifier is considered critical since it would be accessible for use in routine investigations [4]. The classifier assessment usually based on prediction accuracy which is percentage of correct predictions divided by total number of predictions [8, 9].

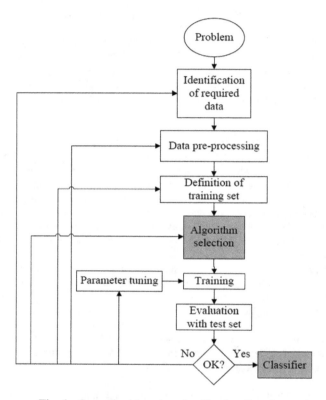

Fig. 1. Supervised learning classification flowchart

In solving optimization problems, a local search optimization algorithm improves one solution. They can be imagined as "walks" within neighborhoods or the search routs over the search area of the problem. The walks are accomplished by repeated steps that shift from the current solution to another solution in the search area. LS algorithm efficiency is shown in manipulating different optimization problems within different domains [5, 10]. LS algorithms proceed from primary solution and repeatedly attempt to exchange the present solution by a superior solution in a properly defined neighborhood of the present solution. The basic LS algorithm is repetitive improvement which only exchanges the current solution with a better one and stops as soon as no better, no more neighbored solutions can be found [7, 11, 12].

2 The Proposed Technique

Our proposed method of classification based on SVM classifier and local search optimizer. To solve a problem, the local search algorithm restricts the local optimization; our work introduces a feedback mutation local search algorithm, so that the classification model parameters is optimal. Taking in consideration the effect of parameters training respecting the SVM classifier performance, the LS algorithm is inserted to explore the parameters training in global area. The classifier weights have significant impact on the strength of results of the SVM model [3]. With reference to the kernels generate various outputs for the problem itself. Figure 2 illustrates LS-SVM flowchart.

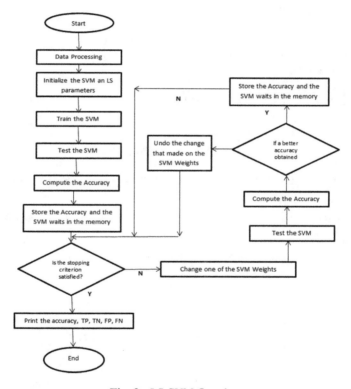

Fig. 2. LS-SVM flowchart

As mentioned previously, in this research the authors propose a LS-SVM approach in which LS is employed to optimize the classification results obtained from using the original SVM model. As shown, the LS is employed to optimize the weights of the standard SVM model. After generating a population of N candidate solutions, the SVM is applied. The quality of each solution in the population is then evaluated. The solutions are then sorted in an ascending order, and the best value is stored.

3   Experiments

The proposed method was applied to a set of different datasets taking into account the results of previous studies. The superiority of the proposed method was observed, with better results than other previous methods. Table 1 shows a group of selected datasets available on UC Irvine Machine Learning Repository; we compare their previous results with the result of LS-SVM method.

Table 1. The datasets characteristics

Datasets	# of attributes	Training set	Testing set
Breast Cancer (BC)	10	193	72
German Credit Data (GCD)	20	675	250
Haberman Surgery Survival (HSS)	3	206	77
BUPA Liver Disorders (LD)	6	233	86
PIMA Indian Diabetes (PID)	8	518	192
Statlog (SH)	13	182	68
Appendicitis (AP)	7	71	27
Australian Credit Approval (ACA)	14	465	173
SPECTF	45	180	67
Parkinsons (PA)	23	131	49

To compare the past studies with our proposed technique, Table 2 summarizes that.

Table 2. Comparison results between LS-SVM and others approaches

Dataset	Approach name	Accuracy	Rank
PID	Flexible Neural Fuzzy Inference	78.6	–
	Fuzzy Neural Networks	81.8	–
	CBA	76.7	–
	SVM	77.08	–
	SVM-LS	84.38	1st
HSS	SVMs using linear terms	71.2	–
	"Proximal SVMs"	72.5	–
	CBA	72.7	–
	SVM	76.62	–
	SVM-LS	83.12	1st
AP	Predictive Value Maximization approach	89.6	–
	CBA	91.4	–
	SVM	92.59	–
	SVM-LS	92.59	1st

(*continued*)

Table 2. (*continued*)

Dataset	Approach name	Accuracy	Rank
BC	Hybrid simplex-GA	79.687	–
	BP	73.5	–
	SVM	69.40	–
	SVM-LS	81.94	1st
LD	FAIRS	83.4	–
	PSOPRO-RVNS	70.97	–
	CBA	77.5	–
	SVM	63.95	–
	SVM-LS	81.40	2nd
SH	Attribute weighted artificial immune	87.4	–
	CBA	77.9	–
	PSOPRO-RVNS	84.36	–
	SVM	88.24	–
	SVM-LS	95.59	1st
GCD	SVMs	77.9	–
	CBA	86.4	–
	SVM-LS	79.20	2nd
PA	ANNs	81.3	–
	SVMs	91.4	–
	CBA	85.7	–
	SVM	75.51	–
	SVM-LS	91.83	1st
SPECTF	CLIP 3	77	–
	CBA	83.6	–
	SVM	76.12	–
	SVM-LS	86.57	1st
ACA	C4.5	85.7	–
	SVMs	85.5	–
	CBA	69.9	–
	SVM	88.43	–
	SVM-LS	92.49	1st

As the results shown in Table 2, the proposed method was superior to those of previous studies. Figures 3, 4, 5, 6, 7, 8, 9, 10, 11 and 12 show graphically the comparison between the proposed method with some of the methods used previously.

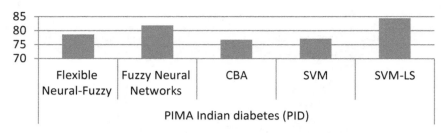

Fig. 3. LS-SVM vs. past studies based on PIMA dataset

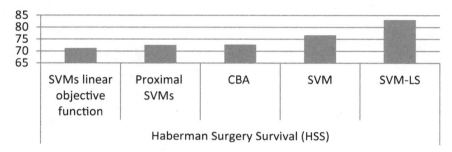

Fig. 4. LS-SVM vs. past studies based on HSS dataset

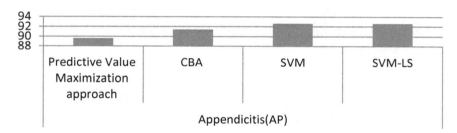

Fig. 5. LS-SVM vs. past studies based on AP dataset

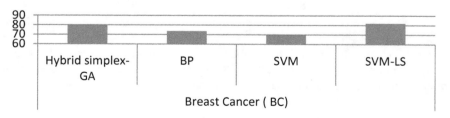

Fig. 6. LS-SVM vs. past studies based on BC dataset

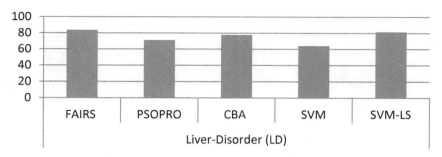

Fig. 7. LS-SVM vs. past studies based on LD dataset

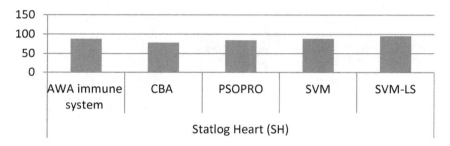

Fig. 8. LS-SVM vs. past studies based on SH dataset

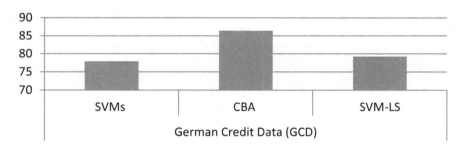

Fig. 9. LS-SVM vs. past studies based on GCD dataset

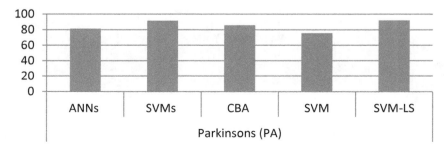

Fig. 10. LS-SVM vs. past studies based on PA dataset

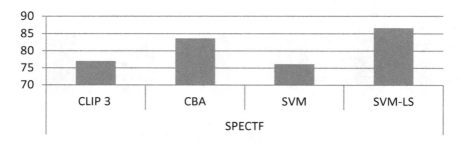

Fig. 11. LS-SVM vs. past studies based on SPECTF dataset

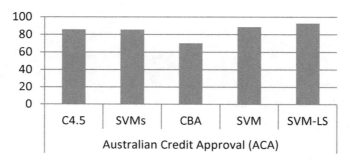

Fig. 12. LS-SVM vs. past studies based on ACA dataset

To evaluate the performance, the overall confusion matrix shown in Table 3, the TP "True Positive", TN "True Negative", FP "False Positive", and FN "False Negative" were calculated. To provide a measure for quality of classification, the classification accuracy as shown in Eq. (1) was calculated.

$$Accuracy = \frac{TP + TN}{TP + TN + FP + FN} \tag{1}$$

Table 3. Confusion matrix

		Predicted class	
		Yes	No
Actual class	Y	*TP*	*FN*
	N	*FP*	*TN*

4 Conclusion

In this research, the authors propose the hybridization of LS and SVM as a possible solution to address classification problems. Initial solutions were randomly generated; SVM was applied and then optimized by exploiting the LS. Results of experiments carried using 10 benchmark datasets showed that the proposed LS-SVM outperformed the original SVM on all datasets. Furthermore, when comparing the accuracy of LS-SVM to the accuracy of other approaches detailed in the literature, LS-SVM's classification accuracy was the highest on eight out of ten datasets.

References

1. Zhang, Y., Wu, L.: Classification of fruits using computer vision and a multiclass support vector machine. Sensors **12**, 12489–12505 (2012). https://doi.org/10.3390/s120912489
2. Gao, L.F., Zhang, X.L., Wang, F.: Application of improved ant colony algorithm in SVM parameter optimization selection. Comput. Eng. Appl. **51**, 139–144 (2015). https://doi.org/10.3778/j.issn.1002-8331.1307-0281
3. Frank, A., Asuncion, A.: UCI Machine Learning Repository. University of California, School of Information and Computer Science, Irvine, Calif, USA (2010). http://archive.ics.uci.edu/ml
4. Marcano-Cedeño, A., Quintanilla-Domínguez, J., Andina, D.: WBCD breast cancer database classification applying artificial metaplasticity neural network. Expert Syst. Appl. **38**(8), 9573–9579 (2011)
5. Chalup, S., Maire, F.: A study on hill climbing algorithms for neural network training. In: Congress on Evolutionary Computation, University of York, UK, pp. 2014–2021 (1999)
6. Kuhn, M., Johnson, K.: Applied Predictive Modeling, vol. 26. Springer, New York (2013)
7. Kotsiantis, S.B., Zaharakis, I.D., Pintelas, P.E.: Machine learning: a review of classification and combining techniques. Artif. Intell. Rev. **26**(3), 159–190 (2006)
8. Alpaydin, E.: Introduction to Machine Learning. MIT Press, Cambridge (2004)
9. Bishop, C.M.: Pattern Recognition and Machine Learning. Springer, New York (2006)
10. Hastie, T., Tibshirani, R., Friedman, J.: The Elements of Statistical Learning: Data Mining, Inference, and Prediction. Springer, New York (2001)
11. Mitchell, T.M.: Machine Learning. McGraw Hill, New York (1997)
12. Vapnik, V.N.: Statistical Learning Theory. Wiley, Hoboken (1998)

The Analysis of PPG Time Indices to Predict Aging and Atherosclerosis

Yousef K. Qawqzeh[✉]

Computer Science Department, College of Science, Majmaah University,
Riyadh, Kingdom of Saudi Arabia
y.qawqzeh@mu.edu.sa

Abstract. This paper aims to investigate the effects of age on systolic and diastolic peaks times. Two indices from PPG waveform were extracted to analyze the variations of systolic and diastolic peaks time as we age. The study focused on two main groups (less than or equals to 40 years, and more than 40 years old). The study showed an inverse relationship between age and diastolic peak time (DpTime) index in which it tends to be decreased as we age. On the other hand, the systolic peak time (SpTime) index tends to be increased as we age. These findings declared the progress of arterial stiffness and atherosclerosis with aging. The more we age, the more our arteries become stiffen.

Keywords: Arterial stiffness · PPG · Systolic peak · Diastolic peak · Aging

1 Introduction

The photoplethysmogram (PPG) waveform consists of pulsatile (AC) waveform contributing to changes in blood volume with each heart beat [1]. All measurements were obtained using PPG's sensor at right hand index finger. PPG technique is used normally in medicine due to its easy recording procedure [2, 3]. Many research works have emphasized the important information embedded in PPG waveform that can be used for arterial stiffness assessment and cardiovascular activities monitoring [4–6]. The use of PPG technique to studying endothelium dysfunction, arterial stiffness, atherosclerosis, arteriosclerosis, and erectile dysfunction is highly appreciated [7]. The disturbances of atherosclerosis play an essential role in the loss of elastic properties in arteries [7]. Aging represents a critical parameter that seems to be highly affecting the morphological changes of PPG contour. The more we age, the more the roundness in PPG waveform will be, which makes PPG's dicrotic notch less pronounced [6, 7]. PPG can be used to measure arterial stiffness noninvasively. PPG waveform reflects the changes in blood volume with each heartbeat, which in turn makes the PPG analysis a vital tool to track vascular changes and to study aging effects on its morphology. One of the most exciting applications that depends on the analysis of PPG waveform is providing a non-invasive, rapid, and cost effective measure for aging and diseases [8, 9]. Figures 1 and 2 show a single_pulse PPG waveform for a young person (20 years old), and a single_pulse PPG waveform for an old person (66 years old) respectively. The use of

© Springer Nature Switzerland AG 2020
L. C. Jain et al. (Eds.): ICICCT 2019, LAIS 9, pp. 218–225, 2020.
https://doi.org/10.1007/978-3-030-38501-9_22

PPG's time indices is used to facilitate the investigation and tracking of arterial stiffness and aging effects [7].

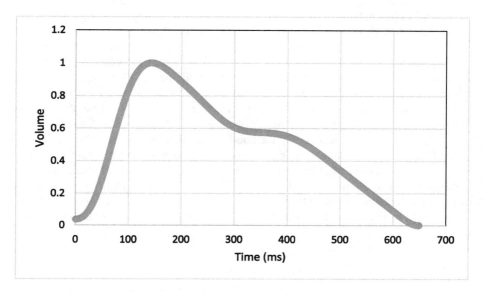

Fig. 1. Single PPG pulse 20 years old male

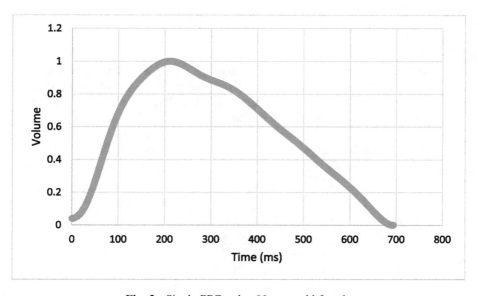

Fig. 2. Single PPG pulse 66 years old female

2 Research Methods

A total of 190 participants were enrolled in this work. A PPG recording system is used to collect PPG signals from the right hand index finger. Subjects are requested to fast for at least 6 h before the recordings. While patient lying on a supine position, breathing normally and remaining quite, the recording ran for 1 min. The recording was repeated again for 1 min to allow the selection of best recording and best clear PPG pulses. Data preprocessing is achieved by digitizing PPG signal, and then transmitting the data to a local computer. The probes functioned on a transmission type sensor. A written consent is collected from each participant before the start of recordings. Participants represents Arab people from both Jordan and Saudi Arabia in which their participating lunched by the coordination between the study team, Al Zulfi general Hospital, and college of science in Zulfi, Riyadh-KSA. All measurements have been recorded in a clinical environment (Hospital room ± 25). A customized algorithm is developed using Matlab (MathWorks MATLAB R2013a) to visualize, analyze and extract PPG features. Finally, statistical analysis is performed using Matlab, SPSS, and Excel software's. The descriptive analysis of the collected data is described in Table 1 below.

Table 1. Descriptive statistics

	N	Minimum	Maximum	Mean	Std. Deviation
Age	190	18.00	72.00	40.2000	21.86810
Sp_time	190	125.00	255.00	168.4000	30.24115
Dp_time	190	260.00	420.00	345.0000	50.18042
Valid N (listwise)	190				

Table 2 examines the correlation between age, SpTime and DpTime indices. It shows that age is inversely correlated with DpTime (r^2 value = −94.8% while it correlates positively with SpTime (r2 value = 81.3%). These findings strengthen the associations between age and PPG time indices.

Table 2. Pearson correlations (Age Vs. Sp time and Dp time)

		Age	Sp_time	Dp_time
Age	Pearson correlation	1	.813[**]	-.948-[**]
	Sig. (2-tailed)		.000	.000
	N	190	190	190
Sp_time	Pearson correlation	.813[**]	1	-.753-[**]
	Sig. (2-tailed)	.000		.000
	N	190	190	190
Dp_time	Pearson correlation	-.948-[**]	-.753-[**]	1
	Sig. (2-tailed)	.000	.000	
	N	190	190	190

3 Results and Discussions

The findings of this work showed that PPG time indices (SpTime index and DpTime index) are statistically significant with age. Figure 3 explains the association between systolic peak time and age. As the boxplot visualizes, the SpTime index tends to be increased as we age. This means that the PPG pulses of young and healthy people will take less time to reach the systolic peak in compare to elder and unhealthy subjects where their PPG pulses will take longer time to reach the systolic peak. This phenomenon in fact mimics the elastic properties of arterial wall. The more the elasticity of arterial wall, the less time it takes to arrive to the systolic peak. Figure 3 draws the variations on systolic peak time index based on two age groups.

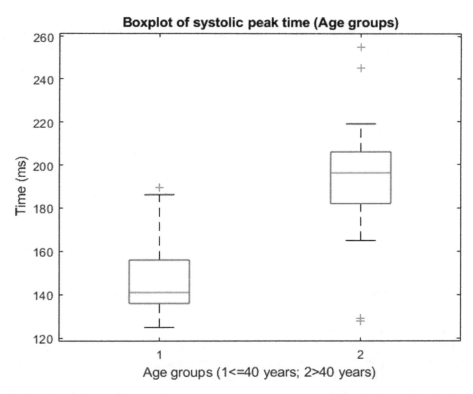

Fig. 3. Boxplot of systolic_peak time index based on two different age groups

On the other hand, Fig. 4 demonstrates the association between diastolic peak time and age. As the boxplot shows, the DpTime index tends to be decreased with age. This means that young and healthy people will obtain a PPG waveform in which it will take more time to reach the diastolic peak in compare to elder and unhealthy subjects where their PPG waveforms will take shorter time to reach the diastolic peak. These findings indicate the acceleration of the diastolic phase of PPG's waveform due to arterial stiffness and atherosclerosis. Figure 5 and 6 also describe the associations between age and SpTime index, and between age and DpTime index. The scatter plot in Fig. 5 demonstrates the positive relationship between age and SpTime index. The more the age, the more it takes to arrive to the systolic peak. Systolic peak represents the highest peak of PPG's pulse. While diastolic peak represents the second highest peak of PPG's pulse. Figure 6 shows the inverse relationship between age and diastolic peak time index. The more we age, the less time it takes to arrive to the second highest peak of PPG's pulse (DpTime).

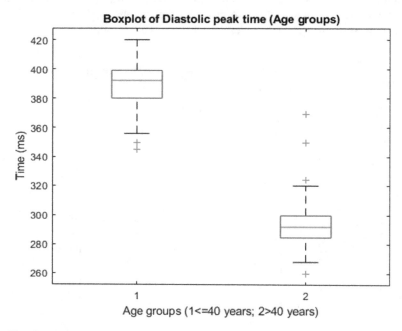

Fig. 4. Boxplot of diastolic_peak time index based on two different age groups

The difference between PPG's SpTime index values and PPG's DpTime index values is given in Fig. 7 below. The scatter plot draws to distinct groups of values in which the young and healthy subjects belong to the first group in the left of the graph having their (SpTime – DpTime) value ranges between 150 ms and 280 ms approximately. While the second group of difference elaborates the (SpTime – DpTime) value ranges between 23 ms and 195 ms approximately.

Fig. 5. Scatter plot of systolic_peak time index based on two different age groups

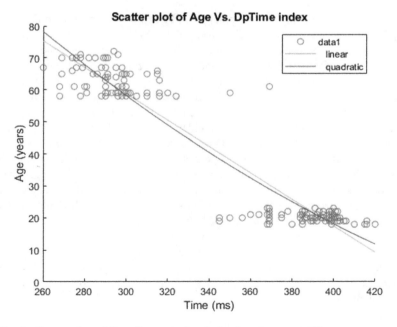

Fig. 6. Scatter plot of diastolic_peak time index based on two different age groups

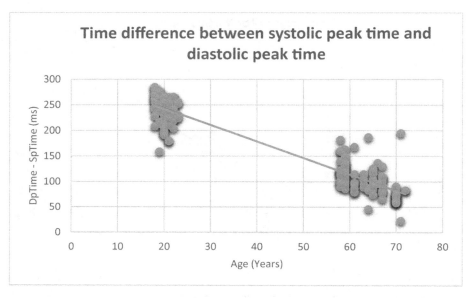

Fig. 7. Scatter plot of the difference between systolic peak time and diastolic peak time based on two different age groups

4   Conclusions

This paper focused on analyzing the changes in PPG's waveform time indices (SpTime and DpTime) among two different age groups. The SpTime represents the time needed to arrive to the systolic peak of PPG's pulse. This is very important to be studied since it indicates the compliance of the artery and reflects the ability of arterial wall to absorb blood pressure. The results showed that, in the young subjects, it takes lesser time for PPG pulse to reach the systolic peak compared to elder and unhealthy subjects. Moreover, the obtained results showed that DpTime index is higher in young subjects. The findings of this work showed that in elder and unhealthy subjects there are small difference between the SpTime and DpTime compared to the young subjects. The stiffening in small to medium arteries opens the way to start the investigation of early atherosclerosis in sub-clinical settings. investigating the effects of aging, atheroscle-rosis, arterial stiffness on PPG contour represents a useful method for disease pre-vention and risk prediction. The findings of this work showed that SpTime and DpTime indices are strongly associated with age. The decrement of DpTime index as we age might indicate the present of atherosclerosis and the start of arterial stiffness. Bring-ing PPG waveform as a diagnostic tool to clinical settings represents an advancement in science and therapy [5]. Therefore, the DpTime and SpTime indices are counted to represent a new way for atherosclerosis prediction and risk management rapprochement.

References

1. Mohamed, S., Mahamod, I., Zainol, R.: Artificial neural network (ANN) approach to PPG signal classification. Int. J. Comput. Inf. Sci. **2**(1), 58–65 (2004)
2. Allen, J., Frame, R., Murray, A.: Microvascular blood flow and skin temperature changes in the fingers following a deep inspiratory gasp. Physiol. Meas. **23**(2), 365–373 (2002)
3. Allen, J.: Photoplethysmography and its application in clinical physiological measurement. Physiol. Meas. **28**, R1–R39 (2007)
4. Hong, K.S., Park, K.T., Ahn, J.M.: Aging index using photoplethysmography for a healthcare device: comparison with Brachial-Ankle pulse wave velocity. Healthc. Inform. Res. **21**(1), 30–34 (2015)
5. Elgendi, M., Liang, Y., Ward, R.: Toward generating more diagnostic features from photoplethysmogram waveforms. MDPI Dis. **6**, 20 (2018)
6. Jermana, M., et al.: Advances in photopletysmography signal analysis for biomedical applications. MDPI Sens. **18**, 1894 (2018)
7. Qawqzeh, Y.K., Uldis, R., Alharbi, M.: Photoplethysmogram second derivative review: analysis and applications. Sci. Res. Essays **10**(21), 633–639 (2015)
8. Qawqzeh, Y.K., Reaz, M.B.I., Maskon, O., Chellappan, K., Ali, M.A.M.: Photoplethysmogram reflection index and aging. In: Proceedings of SPIE 8285, International Conference on Graphic and Image Processing (ICGIP), vol. 82852R (2015)
9. Peralta, E., Lazaro, J., Bailon, R., Marozas, V., Gil, E.: Optimal fiducial points for pulse rate variability analysis from forehead and finger photoplethysmographic signals. Physiol. Meas. **40**(2), 025007 (2019). https://doi.org/10.1088/1361-6579/ab009b

Hybrid Particle Swarm Optimization with Science Cosine Algorithm and Mathematical Equations for Enhancing Robot Path Planning

Hussam N. Fakhouri[✉], Amjad Hudaib, and Azzam Sleit

King Abdullah II School of Information Technology,
The University of Jordan, Amman, Jordan
{h.fakhouri,ahudaib,asleit}@ju.edu.jo

Abstract. This paper introduces a hybrid metaheuristic algorithm that combines Particle Swarm Optimization (PSO) algorithm with Sine Cosine algorithm and Mathematical equations. The algorithm makes a contribution to optimization field by providing better strategy for finding the global minimal value, enhancing exploration and exploitation features, speeding up the converge rate over the tested benchmark optimization problems. The results show that combining SCA and ME with PSO in a new hybrid algorithm called PSE. However, the new algorithm overcome the drawbacks of PSO and it effectively solve high dimensional optimization problems. PSE algorithm is being applied in order to enhance robot path planning, robots can find a high efficiency specification objective when powered by hybrid algorithm. The objective of optimization is to reduce the path lengths to target. Autonomous path planning is necessary to prevent obstacles during the motion of robot. The result show that the algorithm proposed is able to give better performance in reaching targets.

Keywords: Optimization · Meta heuristic · Particle Swarm · Sine Cosine Optimizer · Mathematical equations of contradiction · Reflection and shrinking

1 Introduction

Optimization is currently one of the most applicable approaches that solve real world problems by finding the best solution from all feasible solutions [1]. It solves problems by determining the best value of a set of particles by mimicking biological or physical phenomena [1–4]. The problem of optimization can be solved using metaheuristic algorithms; one of the most popular metaheuristic algorithms is particle swarm optimization (PSO) algorithm, which is part of a wider class of the swarm intelligence algorithms that aim to solve global optimization problems [5].

Metaheuristic algorithms are needed for the optimization process in various fields of science including computer, medicine, Math, and engineering etc. In order to find optimum solutions, they try to satisfy the convergence needs. Fast and successful convergence is needed especially for online real time applications [6]. Convergence time has been shortened by developing new or hybrid optimization algorithms over

© Springer Nature Switzerland AG 2020
L. C. Jain et al. (Eds.): ICICCT 2019, LAIS 9, pp. 226–236, 2020.
https://doi.org/10.1007/978-3-030-38501-9_23

time. In order to do this, two ways of achieving better results were used: developing new optimization techniques, and hybridization between optimization algorithms. The method of (hybridization) is being followed in this paper. Further, an effective benchmarking and testing is being examined. PSO is being used in several disciplines and hybridized with many newly developed optimization algorithms [1, 7]. Combining an algorithm or formulas with PSO can however increase its efficiency. These modifications are necessary to solve the PSO drawbacks for optimization problems. The hybrid algorithm, however, is usually assessed [8] by a set of benchmarking functions which examine the efficiency of the developed hybrid optimizer.

This paper proposes a combination of the PSO with both the sine cosine optimization algorithm (SCA) and mathematical Equations (ME) of orthogonal, slope, and derivative position update equations. The proposed hybridization algorithm (PSE) significantly has speeds up the convergence rate over the tested benchmark optimization problems. It maintains a suitable balance among exploration and exploitation. In the proposed PSE algorithm, each new solution shall be calculated the results first by using sine and cosine functions, then by mathematical Equations of orthogonal, slope and derivatives. These Mathematical Equations and SCA allows for a further search in the search space to find an optimal solution. PSO first determine the position of the particles, and then SCA update the position of the particles if the SCA best founded solution is better than the value found by PSO. Otherwise, ME will update the particle's position using the three equations of contradiction, expansion and reflection.

2 Particle Swarm Optimization (PSO)

PSO is inspired by the works by C. Reynolds and Heppner and Grenander. These researchers were interested in the movement of schools of fish or birds, and created mathematical models to simulate them.

In PSO every particle has two characteristics (velocity vector V and position vector X). It moves in the search space based on the experience of the particle and the experience of the all particle. The speed and position of the particles are updated mathematically according to Eqs. 1 and 2 [13]:

$$Vid(t+1) = w * vid(t) + c1 * r1 * (pid(t) - xid(t)) + c2 * r2 * (pgd(t) - xid(t)) \quad (1)$$

$$Xid(t+1) = xid(t) + vid(t+1) \quad (2)$$

Where Vid(t + 1) and Vid are particle velocities at iterations t and t + 1. Pid is particle's best position. Pgd's neighborhood's best position at iteration t. C1 and c2 are acceleration coefficients [17].

3   Sine Cosine Algorithm (SCA)

Mirjalili [12] Proposed Sine Cosine algorithm (SCA) to solve problem related to optimization. SCA produces a variety of random solutions and then approaches them with the optimum solution using equations such as sine and cosine functions. The sine and cosine equations (Eqs. 3 and 4) are used to update solutions for exploration and exploitation [12]:

$$Xi\,(t+1)\; =\; Xi\,(t)\, +\, r1\, \sin\,(r2)\,|\,r3Pt - Xi\,(t)| \tag{3}$$

$$Xi(t+1)\; =\; Xi(t)\, +\, r1\, \cos(r2)\,|\,r3Pt - Xi(t)\,| \tag{4}$$

where $Xi(t)$ is the position of the current solution in i th dimension at t th iteration, r1, r2, and r3 are random numbers.

4   Mathematical Equations (ME)

A very popular multi dimensional optimization search method is using mathematical equations such as orthogonal, slope, and derivative to find the solution directly. The vertices are sorted by the value of the objective function. The algorithm tries to replace the worst solution with a new point [74, 75].

The orthogonal trajectory of straight lines defined by the Eq. 5, is any circle having center at the origin (Eq. 6):

$$y = kx \tag{5}$$

where k is a parameter (the slope of the straight line)

$$x2 + y2 = R2 \tag{6}$$

where R is the radius of the circle.

Similarly, the slope of line between two points is defined in Eq. 7.

$$m = (y2,\, y1)/(x2,\, x1) \tag{7}$$

$(y2, y1)$ = coordinates of first point in the line, $(x2, x1)$ = coordinates of second point in the line.

5   The Proposed PSE Algorithm

In this paper, the equations of position updating of SCA (Eqs. 3 and 4) is combined in equation (Eq. 6) and added to PSO. After that, the update position of Mathematical Equations of contradiction, reflection and shrinking (Eqs. 5 to 7). [10] is secondly cascaded after SCA (Eqs. 3 and 4). These two enhancements will modify the PSO particles movement strategy to be able for better search the search space, detecting the global optima and speed up the converge rate. After that the algorithm will apply ME Eqs. (5 to 7) which will change the movement direction of the particles.

In the new proposed hybrid algorithm, at each iteration the particles position is first updated according to PSO equations. The particles at this point will move toward the global optimum point (gbest). Then its motion will be affected by SCA (Eqs. 3 and 4) and ME Eqs. (5 to 7) respectively. However, if the solution found by any of the two algorithms gives better optimum solution than PSO then the particles will move towered the better solution. However, if the particle position does not improve at the end of every iteration of the optimization process, then the combination of SCA and Mathematical Equations of contradiction, reflection and shrinking will generate a new position for every particle in the search space.

The PSOSCNSM algorithm's pseudo code is shown in Fig. 1. The populations are represented in the same set as the actual parameter vectors xi = (x1,..., xD), I = 1,..., N, in which D is a problem dimension and NOP the population size. At the beginning of the search, the single vectors xi is randomly initialized. MaxIt is the maximum iteration, the lower and upper bounds in search space for Xmin and Xmax. At the start of the proposed algorithm are those parameters set. The position and velocity of particles are initially generated randomly.

Before updating the vector for the velocity of each particle the limit value of the particle is checked. Firstly, Particle velocity and position are updated by the simple PSO Eqs. 1 and 2, Secondly, Eqs. 3 and 4 in SCA algorithm shall determine the next position of the particle. If the fitness value of applying the equations is better than the value obtained from Eqs. 1 and 2 [19].

Thirdly, the algorithm then applies Eqs. 5 and 6 respectively which constitute of mathematical equations of contradiction and expansion. Then, the global value is chosen among the three equations if the founded solution is better than the previous founded solution. The algorithm then saves the best solutions, assigns them to the destination and updates further solutions. In the meantime, the upper and lower search space range is updated so that search space is used to achieve the global optimal point at each iteration.

When the termination condition is archived, the PSE algorithm terminates the optimization process and the iteration number reaches its maximum limit by default.

6 Applying PSE on Robot Path Planning

PSE algorithm is being applied in order to enhance robot path planning; robots can find a high efficiency specification objective when powered by PSE algorithm. The hybrid PSE algorithm reduces the path lengths to target. Autonomous path planning is necessary to prevent obstacles effects during the movement of robot. The result show that the algorithm proposed is able to give better performance in reaching targets.

Particles in the search space are moving in iteration to get to new positions, but firstly it keep checking to see whether the new positions outside the obstacle boundary, the avoidance of obstacle functions is being initiated.

Realization how to move robot within search space and where the new position of the robot is determined by the hybrid PSE algorithm. Robot will move and update its

a.	Initialize the parameters (NOP, MaxIt, Xmin, Xmax, dim)
2.	Generate particles vectors xi = (X1, ..., XD)
3.	i = 1, ..., NOP, D is the dimension, and NOP is the population size.
4.	Initialize the particles with random positions (Xi) and random velocity (Vi)
5.	Fore I =1 to NOP do
6.	Calculate fitness for Xi to Xn
7.	End loop
8.	Set Xi to be XpBest , initialize the gbest value to the best fitness value (Fi) of (X1 to Xn)
9.	Set the particle with best fitness to be XgBest
10.	// Main loop
11.	for i = 1 to NOP do
12.	update the velocity and position in regard to equation (1)
13.	for X1 to Xn do
14.	update Xi position and fitness value according to equation (2)
15.	Calculate fitness (Fi)
16.	End loop
17.	for X1 to Xn do
18.	update Xi position and fitness value
19.	Calculate fitness (Fi)
20.	End loop
21.	for i = 1: NOP
22.	if trial(i)< 0.5
23.	Update the position Xi of particle using Eq. (3)
24.	else
25.	Update the position Xi of particle using Eq. (4)
26.	end if
27.	Evaluate fitness value for new particle Xi
28.	for X1 to Xn do
29.	update Xi position and fitness value according to equation (5)
30.	Calculate fitness (Fi)
31.	for X1 to Xn do
32.	the updated velocity and position according to equation (6)
33.	Calculate fitness (Fi)
34.	End loop
35.	the updated velocity and position according to equation (7)
36.	Calculate fitness (Fi)
37.	End loop
38.	if f(XpBest) < f(XgBest)
39.	XgBest = XpBest
40.	end if
41.	keep new position within range of (Xmin, Xmax)
42.	End loop

Fig. 1. Pseudocode of the PSE algorithm

positions according to mathematical equation in PSE algorithm, convergence of PSE with the best solutions by changing every individual particle's path to its optimal position based on that robot target position is first updated according to PSO equations. The robot at this point will move toward the global optimum point. Then its motion will be affected by SCA and ME equations respectively. However, if the solution found by any of the two algorithms and gives better optimum solution than PSO then the robot will move towered the better solution.

Changes to the position of the robot are achieved by position and speed data. Cartesian coordinates, represent the robot position and the speed of the robot are changed by PSE algorithm.

A random position (M1, M2... MD) is being defined as the robot initialization position by the PSE algorithm that distribute consistently around (Md, Fd) start position. Particles is described in the obstacles topology context within the population, its neighbors can be performed depending on the robot position range.

For each particle, fitness value is being calculated based on the target region coverage over obstacles. When robot finds a neighbor, the coordination of the new optimal positions is stored while a particle detects a pattern which is better than the global minimal stored value, so that in order to reduce potential obstacle collisions the robot rotates and changes its path.

The position of the robot is randomly initialized firstly (Md, Fd) at certain time (T), at different time (Time + 1) it will then go to an intermediate location (Md', Fd'). Until it reach target position (Mg, Fg), We determine the robot between position (Md', Fd') with PSE algorithm.

We choose the particle with the optimal position at the end of the execution of the PSE algorithm, to be reach robot's next position as shown in Fig. (2).

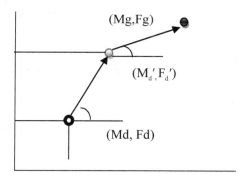

Fig. 2. Position of the xth particle of the robot

6.1   Objective Function

The problem of optimization involves an objective function to reduce the distance of Euclidean's distance between particles with their corresponding robots target positions from the presents robot position to the target position of robots. This objective function determines the length of the robot's path. The path connects the particles pairs {(Md, Fd), (Md', Fd')} and {(Md', Fd'), (Mg, Fg)} Euclidean distance is calculated by functions so that the total length of the path is minimized without touch the obstacle, as shown in Fig. (3).

In Fig. 3 robot select (Md', Fd') to avoid collision with obstacle and determine the next robot position. Replacing Mx and Fx with better values, the next position is calculated replacing the values of Fx.

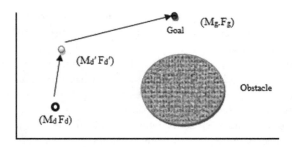

Fig. 3. Robot select of (Md′, Fd′) to avoid collision with obstacle.

6.2 Obstacles Functions

In this paper the obstacles functions (OB) is being used according to state of the environment. That defines the enhancement of robot path planning directly from first position to target avoiding obstacle collision to happen. When robot sees obstacles it must develop the objective function by adding the obstacle function to it. That is why we suppose that every obstacle function has a different value. Robot selects the shortest path to reach the target, when the robot sees the obstacles it decides to turn right or left far away from this obstacle. This turn can be any movement degree.

$$\text{Radius } 0 \leq \alpha \leq 2 * 3.14$$

The neighboring obstacles are influenced by the length of the arc for that part. We suppose the robot has a direct path to target and a vertical line from the obstacle center to this line, allow crossover of M, suppose that four triangles have a right angle in Fig. 4 for all triangles.

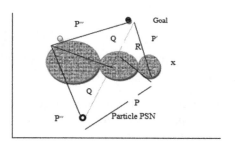

Fig. 4. Four Triangles with right angle.

Clarifying the robot continuous path planning is by PSE algorithm, the suggested algorithm supposes robot present position and its speed, then determine the robot's next position by using the hybrid algorithm that restricted objective function.

We apply a method for the robot path planning with PSE algorithm by Add the present robot position to the path, when a distance is less than or equal to a preset limit between the present robot position and the target then move the robot to next position f. Relocate obstacles and targets position according to the respective probability of relocation. Start the PSE swarm around the robot's present position. Evolve the algorithm of the PSE.

Pick the robot's best global position to take the robot's current position and go back to the path and stop at the target position.

6.3 Improvement

Robots have a limitation for the path, vstart, where robots will go forward in each stage from this starting point. There's also a path to goal. vtarget, its value supposed to be less than robot's current value. If not, even if the target does not move or moves according to a probability objective ptarget. It might escape the robot all the time in the worst case. Every environmental obstacle has relocation probability's that pg = 0 for static obstacles. With regard to this probability, Vg represent the changes with all the obstacles that begin before the planning phase. Obstacle is supposed to be circular; no two obstacles' have overlap in the setting, but can be contiguous.

Taking into account static and dynamic environmental obstacles, to the objective function, we add obstacle function, and calculate the distinct computer simulation results from those approached in Table 1.

R1 results in improved obstacles functions and R2 results in 15 different environments, the time and time of the path is less than R1.

Environmental parameters are: probability of relocation goal's (PRG), step of moving Goal's (MG), step of Robot's moving (RM) and step Obstacles' moving (OM).

Experiment and computer simulation of applying PSE algorithm is applied in four different environments, where maximum velocity (step size) and relocation probability of the goal and obstacles has been tested as in Figs. 5 and 6. In Figs. 7 and 8 the shadow obstacles are the static obstacle and goal is the uncolored obstacles in Table 1.

For robot Path planning, we apply PSE algorithm, the latest location of the robot is [0, 0] and that the target is [1000, 1000]. The obstacles and the goal are being fixed as shown in Figs. (5, 6, 7 and 8). Figure 8 shows the varying of numbers, size and position of the obstacle, while in Figs. (5, 6, 7 and 8) it has different goal velocity, goal relocation probability and dynamic relocation probability. Table 1 shows each figure have different results in regards to obstacle number, static obstacle number, dynamic obstacle number.

Table 1. Four environments

Figure	Obstacle number	Static obstacle number	Dynamic obstacle number	Goal relocation probability	Goal velocity	Robot velocity
5	9	9	0	0	0	50
6	4	4	0	1.5	10	30
7	11	11	0	0.7	10	20
8	14	7	7	0	0	40

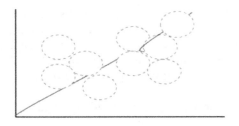

Fig. 5. Visualization of different goal velocity and goal relocation for the path planning

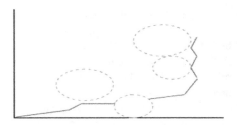

Fig. 6. Visualization of goal relocation

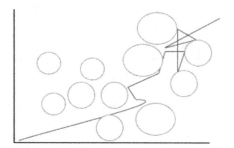

Fig. 7. Visualization of dynamic relocation probability

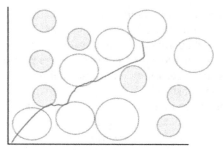

Fig. 8. Visualization of obstacle varying of numbers, size and position

7 Conclusion

In this paper a hybrid evolutionary algorithm that combines Particle Swarm Optimization (PSO) with Sine Cosine Optimizer and mathematical equations of contradiction, reflection and shrinking have been proposed in order to solve the drawbacks in PSO algorithm in locating local minima rather than global minima, low converge rate and low balance between exploration and exploitation. The solutions are updated by the proposed algorithm with the best position found by the three cascaded hybrid algorithms.

The convergence of the PSE algorithm is ensured as the electrons always tend to update their positions with regard to the best solution discovered among all equations that are the most promising solutions acquired in the course of iterations so far. Furthermore, The PSE algorithm is capable of solving actual difficult issues with unknown and restricted search spaces and solving the issue of robotic path planning. Moreover, there is no optimization algorithm to solve all optimization problems, according to the NFL theorem. Since the PSN algorithm in this research has been able to outperform other algorithms on most sample instances, it can be regarded as an alternative optimizer to solve optimization problems among the present well known algorithms. PSE algorithm is being applied in order to enhance robot path planning, robots can find a high efficiency specification objective when powered by hybrid algorithm. The objective is to reduce the path lengths to target. Autonomous path planning is necessary to prevent obstacles during the motion of robot. The result show that the algorithm proposed is able to give better performance in reaching targets.

References

1. Al Sayyed, R.M., Fakhouri, H.N., Rodan, A., Pattinson, C.: Polar particle swarm algorithm for solving cloud data migration optimization problem. Mod. Appl. Sci. **11**(8), 98 (2017)
2. Altay, E.V., Alatas, B.: Performance comparisons of socially inspired metaheuristic algorithms on unconstrained global optimization. In: Bhatia, S., Tiwari, S., Mishra, K., Trivedi, M. (eds.) Advances in Computer Communication and Computational Sciences, pp. 163–175. Springer, Singapore (2019)
3. Amodeo, L., Talbi, E.G., Yalaoui, F.: Recent Developments in Metaheuristics. Springer, Cham (2018)
4. Arora, S., Singh, S.: Butterfly optimization algorithm: a novel approach for global optimization. Soft. Comput. **1**, 20 (2018)
5. Krawiec, K., Simons, C., Swan, J., Woodward, J.: Metaheuristic design patterns: new perspectives for larger scale search architectures. In: Handbook of Research on Emergent Applications of Optimization Algorithms, pp. 1–36. IGI Global (2018)
6. Stützle, T., López Ibáñez, M.: Automated design of metaheuristic algorithms. In: Gendreau, M., Potvin, J.Y. (eds.) Handbook of Metaheuristics, pp. 541–579. Springer, Cham (2019)
7. Hudaib, A.A., Fakhouri, H.N.: Supernova optimizer: a novel natural inspired meta heuristic. Mod. Appl. Sci. **12**(1), 32 (2017)
8. Ong, P., Chin, D.D.V.S., Ho, C.S., Ng, C.H.: Metaheuristic approaches for extrusion manufacturing process: utilization of flower pollination algorithm and particle swarm optimization. In: Handbook of Research on Applied Optimization Methodologies in Manufacturing Systems, pp. 43–56. IGI Global (2018)
9. Trivedi, I.N., et al.: A novel hybrid PSO–WOA algorithm for global numerical functions optimization. In: Bhatia, S., Mishra, K., Tiwari, S., Singh, V. (eds.) Advances in Computer and Computational Sciences, pp. 53–60. Springer, Singapore (2018)
10. Chegini, S.N., Bagheri, A., Najafi, F.: PSOSCALF: a new hybrid PSO based on Sine Cosine Algorithm and Levy flight for solving optimization problems. Appl. Soft Comput. **73**, 697–726 (2018)
11. Mirjalili, S.M., et al.: Sine cosine algorithm: theory, literature review, and application in designing bend photonic crystal waveguides. In: Mirjalili, S., Song, D.J., Lewis, A. (eds.) Nature Inspired Optimizers, pp. 201–217. Springer, Cham (2019)
12. Kennedy, J.: Particle swarm optimization. In: Sammut, C., Webb, G.I. (eds.) Encyclopedia of Machine Learning, pp. 760–766. Springer, Boston (2011)
13. Salcedo Sanz, S.: Modern meta heuristics based on nonlinear physics processes: a review of models and design procedures. Phys. Rep. **655**(1), 70 (2016)
14. Sörensen, K.: Metaheuristics—the metaphor exposed. Int. Trans. Oper. Res. **22**(1), 3–18 (2015)
15. Glover, F.: Tabu search – part I. ORSA J. Comput. **1**, 190–206 (1989)
16. van Laarhoven, P.J., Aarts, E.H.: Simulated annealing. In: van Laarhoven, P.J.M., Aarts, E. H.L. (eds.) Simulated Annealing: Theory and Applications, pp. 7–15. Springer, Dordrecht (1987)
17. Mladenović, N., Hansen, P.: Variable neighborhood search. Comput. Oper. Res. **24**(11), 1097–1100 (1997)

18. Krishnanand, K., Ghose, D.: Glowworm swarm optimization: a new method for optimising multi modal functions. Int. J. Comput. Intell. Stud. **1**, 93–119 (2009)
19. Yang, X.S.: A new meta heuristic bat inspired algorithm. In: González, J.R., Pelta, D.A., Cruz, C., Terrazas, G., Krasnogor, N. (eds.) Nature Inspired Cooperative Strategies for Optimization (NICSO 2010), pp. 65–74. Springer, Heidelberg (2010)
20. Beni, G., Wang, J.: Swarm intelligence in cellular robotic systems. In: Proceedings NATO Advanced Workshop on Robots and Biological Systems, Tuscany, 26–30 June 1989. https://doi.org/10.1007/978-3-642-58069-7_38

Setting Up Process of Teaching Software's on Human Anatomy and Physiology in University of Vietnam

Mai Van Hung[1]([⊠]) and Nguyen Ngoc Linh[2]

[1] VNU University of Education, Hanoi, Vietnam
hungmv@vnu.edu.vn
[2] National College for Education, Hanoi, Vietnam

Abstract. To set - up teaching software, there is no absolute need for a teacher to grasp the capacity of computer's programming. Enough to establish a cooperation between a specialist of the given science and a specialist of informatics, to set-up successfully a teaching software of quality, meeting satisfactorily any requirements of a didactic aid.

The cooperation to set - up a teaching software begins with the teacher's ideas, formulating the contents of lessons on computer, for organizing the teaching procedure, aiming at the objectives of active teaching, to guide students to implement self-learning. The given lesson will be composed of how many parts, and of what questions, tables, diagrams, figures, pictures, scientific films (movies), modeling experiments...? Let students know how to use these aids, self-observe and self - answer the questions.

Keywords: Teaching software · Human · Anatomy · Physiology

1 Introduction

Nowaday, to set - up a teaching software is very importance at every universities in Vietnam, teaching software is getting more and more popular. So it becomes an important method for students to learn. The Ministry of Education and Training in Vietnam also was encouraged the teachers build many teaching softwares for teaching. The teacher presents ideas of lesson with 4 parts: objectives; new knowledge; registering and memorizing; control and evaluation. The objectives are merely of a written channel. It's the study of knowledge that integrates the written and voice channals (audio - visual channel), scientific films, modeling experiments...for the learners to observe, then answer the various questioning forms, such as "questions - and - answers", "objective tests with multiple choices", "filling gaps with beforehand given words" exercises. The control - evaluation part consists chiofly of objective tests...

Basing upon these suggestions, the informatics specialist will design a computer's programme with appropriate terms and forms, with the capacity to meet satisfac - toryly the teacher's requirements. At the end, the informatics specialist will provide the teacher with a programme - frame, to enter the information from the scenario furnished by the teacher. The processes of scenario - setting and programme - frame designing can be implemented at the same time.

© Springer Nature Switzerland AG 2020
L. C. Jain et al. (Eds.): ICICCT 2019, LAIS 9, pp. 237–246, 2020.
https://doi.org/10.1007/978-3-030-38501-9_24

2 Setting up Process of Teaching Software

2.1 Step 1: Definition of the Teaching Objectives

The objectives of the manual "Human Anatomy and Physiology " are to provide university. Students with scientific knowledge on the structural characteristics and the living activities of a human being. Basing on these data, the progamme proposes adequate measures of hygiene; corporal training; increase of efficiency and achievements in studies; in contribution to the implementation of the objectives in training dynamic and creative workers, meeting with success the requirements of socio - economic development of the country.

To reach these objectives, besides the traditional didactic auxiliaries (as tables, photos, plastic models, samples, equipment for experiments and practical exerciser...), our university are in great and pressing need of didactic aids in applies informatics technologies, to renovate our teaching methodology.

2.2 Step 2: Analysis of the Teaching Content

The programme of "Human anatomy and physiology" includes knowledge on the human body structure and functional activities happening in various organs and organ systems. In recent years, the programme of this object has also introduced some basic knowledge on human development, under the integrative form, in particular some practical advices on human reproduction and contraceptive methods, to safeguard the reproductive health.

The physiological concepts reflect the typical activities of organs, organ systems and of the whole body. The concepts on physiology include its phenomena and processes. Physiological phenomena display the exterior aspect of the organ and organ system's functional activity. They show only the beginning and ending of this activity.

Physiological processes investigate the interior mechanism of this activity, discover the inter-actions between the various components of the mechanism, for example the coagulation of the blood, the different phases of digestion...

The programme does not quote concrete physiological laws, but in some paragraphs, it indicates some processes suggesting the presence of physiological laws, such as the regulation processes, ensuring the equilibrium and stability of the human body; some cyclic phenomena. All these data must be exploited in the teaching process.

Though these data are divided into chapters, lessons, paragraphs, ... they are in reality associated, and closely realted.

The programme of "Human anatomy and physiology" incluses 10 parts:

Part 1. Generalities on the human - body.

This chapter introduces in a general way the basic knowledge on the human body; and asscripts that the unit - component of all tissues, organs and organ - systems of this body is the "cell"; the functional unit - component of all human activities is a "reflex".

Part 2. The system of movements.

This organ - system is the first to be introduced, because all activities typical of the animal kingdom, including humans, are displayed into movements.

The system of movements is the most easily observed and studied system in the human body. It is relatively simple and easily recognized, in comparison with other systems.

This part includes knowledge on muscles and bones (skeleton); on their structure, properties and functional activities; as well as on their respective places in the body. It helps us to differentiate various muscles and bones and get.

It helps us to differentiate various muscles and bones and get acquainted with their mechanism of activity.

Part 3. Circulation system: include various acknowledge on the blood and internal environment; the white blood cells and immunity; the coagulation of blood; the principle of blood transfusion; the main components of blood circulation and lymph movement; their role; the external and internal structure of the heart; different blood vessels: vein, artery and capillary; different phases of contraction, dilatation, and pause of the heart; the mechanism of blood transport in vessels; heart and blood vessel disfunction and disorders.

Part 4. Respiration system: includes various knowledge on the structure, functions and role of respiratry organs in a living organism. The main characteristics in the mechanism of ventilation of lungs. Exchange of gas in the lungs and cells. Harmful consequences of air pollution, scientific basics of the respiratory system training, to maintain it healthy. Active struggle against air-polluting agents.

Part 5. Digestion system: Includes various knowledge on the structure of the digestive system and different digestive acts, as deglutition, salivation, conduction of food along the digestive track; the digestion of glucids, lipids and proteins; the absorption of the digested food in the small intestine and the discharge of faces through the retum and anus. Chemical action of digestive enzymes from the salivary, stomach, intestinal glands; the pancreas and liver on different kinds of food. The passage of digested food from the small intestine into the blood and to all the body cells. Action of harmful eaten agents and bad eating habits on digestive tract. Sanitary measures to protect the digestive system and ensure its good, effective functioning.

Part 6. Exchanges of matter and energy. The coordination and combined action of different systems and organs of circulation, respiration and digestion to ensure the implementation of an essential living activity, that is the permanent exchanges of matter and energy between the environment and the organism, as well as between different organs of the human body. This process is displayed by an uninterrupted breaking-down and building-up of matter and energy, going-on in pair in the intra-cellular environment and ensured by a ceaseless exchange of matter and energy.

The metabolism, including the two opposed but intimately related phenomena of assimilation and disassimilation, is a basic and essential living condition and process of the body.

Part 7. Excretion system: Includes various knowledge on the structure and functioning of the excretory-urinary system, composed essentially of 2 kidneys, urinary conducts, urinary bladder and an excretory canal. It describes how the urine is made and dismissed from the human body. It says too some words about urinary stones, and sanitary measures to keep the urinary system in good conditions.

Part 8. The nervous system and sense organs: The nervous system receives all the time excitations from the body interior environment as well as from the surrounding environment, through sense-organs; and reacts to these excitations by controlling, promoting and coordinating the living activities of different organs, systems of the human body. The global result is to constantly ensure the correct adaptation of the body to the environmental conditions, to coexist and survive.

The structure and functioning of the nervous system and different sense- organs, as well as about their functioning mechanism. The sense-organs include the organs of smell, audition, vision, and touch. The chapter describes their functioning mechanism. The unconditional, conditional and operative reflex; the process of formation of new reflexes; the conditions necessary for the training of new conditional reflexes. The physiology of "higher nervous activity" in humans, with capacity of abstract thinking, and mastering the ability to speak and write "words".

Part 9. The endocrine system: Includes various knowledge about the structure and functioning of endocrine glands. Learners must get acquainted with some chief endocrine glands. They must get able to indicate their place on and in the human body; describe shortly their functioning mechanism; say some words on the nature and effects of the importance of endocrine glands in human life. Regulation and coordination of endocrine glands.

Part 10. Reproduction system: Includes various knowledge on the structure and functioning of the male and female reproductive systems. Process and mechanism of the fecondation, fertilization, and development of the embryon-foetus. Scientific basics of contraceptive methods. Sexually transmitted Diseases. AIDS, the terrible humanity calamity.

2.3   Step 3: Compilation of Materials; Collecting and Selection; Setting-up and Pedagogic Treatmant; Technical Treatment of Pictures, Photos, Illustrations, Films; Experiment Modeling... Compatible with the Teaching Requirements

Gathering, selection and setting-up of pictures, scientific films episodes, experiment models:

–   Some difficult experiments (requiring much time; many living samples not easily found or the numerous collecting of which is able to influence badly on the nature ecological equilibrium) can be done for demonstration by capable teachers in well piped schools, then filmed by qualified cameramen and provided to teachers to use, as modern means of teaching and self learning or to show in class at schools to children to watch and learn.
–   With regard to knowledge on anatomy, the teacher must search and collect or made pictures, photos, tables and scientific films.

- With regard to knowledge on physiology and hygiene, the teacher must search and do experiment-modeling or/and scientific films.
- With regard to knowledge on hygiene, the teacher must gather appropriate pictures, photos, propaganda leaves and posters, scientific films and experiment-modeling pieces. Pedagogic processing and technical treatment of illustrating aids and teaching softwares, appropriate bits of scientific films, flash, experiment-modeling.
- To lead the pedagogic processing and technical treatment of illustrating materials, pictures use software tools, as the automatic photo-improving. Picas a system; the Photoshop software; the CorelDraw software, the 3-dimensional technology software...
- For the pedagogic processing and technical treatment of bits of films and experiment modeling, use the software tools, as Multimedia technology, Windows movie maker, Nero photos, Nero Vision, Audio...

2.4 Step 4: Designing of Tool-Software: To Format and Display Data by Programming Language (by the Informatics Experts)

Softwaves with capacity to display and project word-chanel, pictures, bits of film, experiment-models, can integrate the combined use of picture and film treatments.

2.5 Steps 5: Entering of Information from Scenarios into Designed Teaching Softwaves

The source of information can come from word, picture, film or experiment models channels.

2.6 Step 6: Test-Playing of Softwaves; Correction and Setting-up of Programmed CD-Rom

These can be errors, shortcomings and mistakes to be corrected while entering data from scenario into softwaves.

2.7 Step 7: Writing of the Users Guide

Includes instructions and advices to fully use the properties and capacities of teaching-softwaves, as well as to fully exploit any information, as word, picture, scientific films, experiment-models... channels, aiming at organizing true "activation of learning activities".

3 Results and Discussion

After the putting - up in concrete form of the procedure of teaching - software, we have made an enlarged inquest or investigations, to collect the opinions of 94 teachers in 8 universities, the obtained results are as follows (Figs. 1, 2 3, 4, 5, 6 and 7):

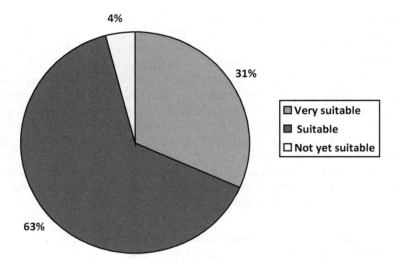

Fig. 1. Definition of the teaching objectives

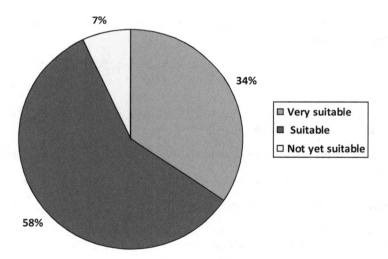

Fig. 2. Analysis of the teaching contents

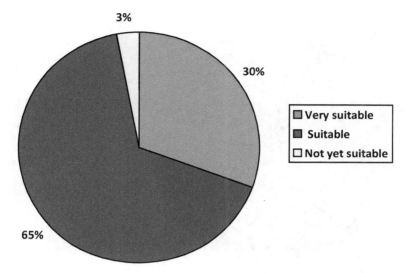

Fig. 3. Gathering and selection of teaching materials; didactic setting –up and consolidation of drafts; technical treatment of figures and illustrations; scientific, films; experiment modelling…all in conformity with the teaching contents

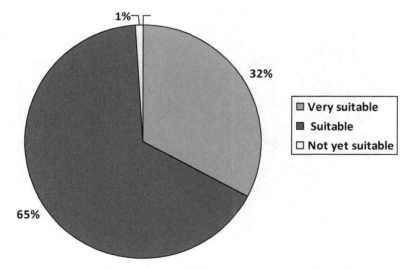

Fig. 4. Formatting and expression of software data in informatics programming language, by the informatics specialist.

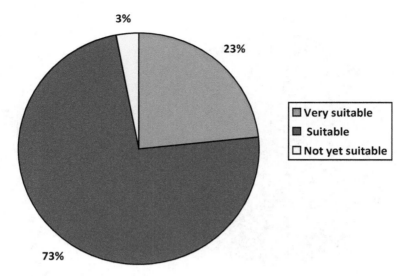

Fig. 5. Entering of information from the scenario into the designed software

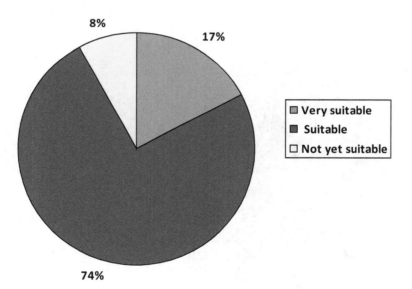

Fig. 6. Testing and correcting the programmed draft; setting – up of the programmed CD-Rom

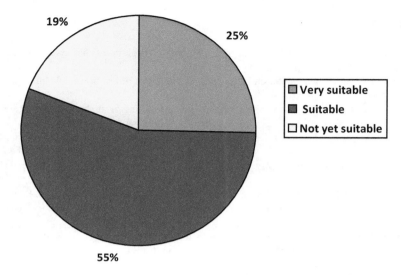

Fig. 7. Writing of the teaching – software Users' guide

The inquest among teachers shows that the above procedure is highly valued. Applying this procedure, we have determined the objectives and analyzed the knowledge contents of the programme and text - book on the "Human Anatomy and Physiology" at the universities of Vietnam.

4   Conclusion

With the built-up procedure of teaching-software making in general and in the field of Biology Teaching in particular, we have succeeded in producing 8 softwaves for the teaching of the 8th Form Programme of Biology, namely, Blood Circulation and Lymph Flowing; Heart and Vessels; Respiratory activities; Sympathetic and Parasympathetic Nervous System; Auditory Sense Organ.

These softwaves have been tested and use in earnest to teach in 8 Universities in Vietnam.

The use of these softwaves has proved to be of great help in renovating the teaching methodology and also in improving the quality of teaching biology in these universities. What is more, they have been highly valued by the teachers using them as well as by the teachers and educational cadres and responsible witnessing these softwaves use in Universities.

References

1. Cat, T.D., et al.: Practical Exercises of the Manual of Biology of the 8th Form. Education Publishing House, Ha Noi (2005)
2. Hung, D.M., et al.: Document on Biology of the 8th Form. Education Publishing House, Ha Noi (2004)
3. Phan, C.X.: Confection of softwaves, serving in the teaching of the lessons "Structure of the cell". Master-of-Sciences dissertations, in Ha Noi University of Education (2001)
4. Hoat, D.V., et al.: Theory of Teaching in University. Education Publishing House, Hanoi (2010)
5. Huan, D.: Some insights into modulized vocational training worldwide. J. Educ. Technol. (1992)
6. Hung, M.V.: Biology Development. Vietnam National University of Education Press (2018)
7. Lan, T.T., Hung, M.V.: Human Annatomy and Physiology. Vietnam National University Press (2018)
8. Quyen, N.T.: Setting-up and use of softwaves to teach the Chapter II. In: Biology, 7th Form, Secondary school, Master-of-Sciences dissertation in Education, Ha Noi University of Education, Ha Noi (2004)
9. Trang, N.T.T.: Setting–up of softwaves, to teach some contents of the part "natural and Social Sciences" on Multimedias echnology in Primary Education, Master-of-Sciences dissertation on Education. Ha Noi University of Education, Ha Noi (2004)
10. Vinh, N.Q., et al.: Manual of Biology of the 8th Form. Education Publishing House, Ha Noi (2004)
11. Guide for Self-Study by Photoshop Ver.2. Education Publishing House, Ha Noi (2006)

Feature Selection Models for Data Classification: Wrapper Model vs Filter Model

Asma'a Khtoom[(⊠)] and Mohammad Wedyan

Prince Abdullah Ben Ghazi Faculty of Communications and Information
Technology, Al-Balqa Applied University, Salt, Jordan
{asmakhtoom,mwedyan}@bau.edu.jo

Abstract. Data Classification is a managed learning process, which sort and organize the data into various categories to be used in more effective and efficient way. Feature selection (FS) is a noteworthy theme for the advancement of data classification process. It is regularly a basic information handling step preceding applying a learning algorithm, where the choosing of the most fitting subset of features that portrays a given classification task and the removing of irrelevant and repetitive information is occurred, hence this increases the performance of machine learning algorithms. Usually, two public methods of feature selection are used: a wrapper method in which the proposed learning algorithm itself is used to estimate how much the features are helpfully and a filter method, which assesses features heuristically, depend on the data general characteristics. In this work, the goal is to make a performance and effectiveness analysis of feature selection methodologies. This analysis handles a comparison between wrapper and filter method, by applying them on seven datasets prior to using these datasets in three classification algorithms. The results explains a set of important issues related to FS methods such as the selection taken time and the accuracy of the classification algorithms when apply FS methods to their testing datasets. The experiments were tested on seven standard datasets downloaded from multiple resources such as Machine Learning Repository and Weka tool. The experimental results have revealed that different feature selection methods can effectively enhance classification problems, but not all the time. No single method is the best for all datasets.

Keywords: Feature selection · Wrapped model · Filter model · Naïve Bayes · Logistic Regression · C4.5 tree

1 Introduction

Data classification is known as a process of data management that take in consideration the demands for different business or individual objectives when it classify and separate the involved data. In the classification process different approaches, algorithms and criteria's are used for scanning, sorting and classifying data within a database or repository. The ability to examine, identify, select and separate data is generally done through an intelligence software, which use many algorithms for classification such as the Probabilistic Neural Network (PNN) [10, 26], Naive Bayes (NB) [6] and Logistic Regression (LR) [1].

© Springer Nature Switzerland AG 2020
L. C. Jain et al. (Eds.): ICICCT 2019, LAIS 9, pp. 247–257, 2020.
https://doi.org/10.1007/978-3-030-38501-9_25

In this work, three classification algorithms representing different ways to learning were utilized: Naive Bayes, Logistic Regression (LR) and a Decision Tree (DT).

Normally, raw data that used in the classification algorithms can be in various formats as it originates from different sources, it might comprise of corrupted or missing data, irrelevant and redundant attributes, meaningless data and so forth. For these reasons, data needs to be preprocessed before applying any type of data classification algorithms and this accomplished by following a set of pre-steps such as features selection (FS).

In machine learning, pattern recognition and statistics fields, researchers have been focused in using features selection different methods, that allow selecting the most proper subset of features and eliminating the irrelevant and repetitive features for a given classification task. Generally, there are two main categories of feature selection methods: The initial one is *filter method* [5, 20] that works as a filter to sift the repetitive, noisy and irrelevant attributes in a way that is independent of a learning procedure. The second process is called *wrapper method* [9] where the selection of features depends on which the relevant attributes are determined by a specific classifier and it deals as a wrapper around a learning algorithm. Filter method does not depend on any classifiers when selecting the features. Instead, wrapper method uses a particular classifier to evaluate the quality and the nature of selected features; furthermore, it offers a clear and staggering way to deal with addressing the issue of features selection, paying little respect to the picked learning machine [8].

In wrapper methods, the main three steps to use a typical wrapper method in the training process, given a predefined classifier, which works as a black box, are:

First Step Search: looking through the features to create a subset of features from all conceivable element subsets.

Second Step Evaluation: estimating and evaluating the exhibition and performance of the picked classifier on the previously chosen features subset and returned them back to the feature search component (step1) for the following round of feature subset selection.

Third Step: repeating first and second steps until the desirable quality is reached, so the feature set with the highest evaluated worth will be picked as the last set to learn the classifier.

Fourth Step Classifier Testing: During testing process, an assessment process for the resulting classifier is made using new independent testing group that is not used in the procedure of training. Figure 1 shows a general framework for features selection wrapper method.

As you see in Fig. 1, the framework consists of three main parts: Feature search, Feature evaluation and Classifier Induction Algorithm.

In this work, we analyzed how feature selection methods (FS) affect classification process performance and efficiency. This analysis includes a comparison between wrapper and filter approaches, explanation of an important issue of the features selection (the required time to extract features) and a discussion of a data classification issue (data classification accuracy). The features selection is employed first, using filter model then wrapper model, to obtain better classification accuracy and later three classification algorithms: Naive Bayes (NB), The Logistic Regression (LR) and Decision Trees (C4.5) were applied to classify the given data.

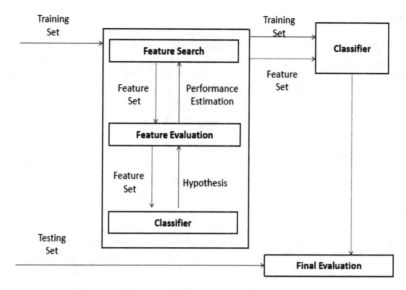

Fig. 1. A standard framework for wrapper methods.

The paper is composed of the next sections: Sect. 1 introduces an introduction about data classification and FS methods. Section 2 presents the background and literature. The work strategy is explained in Sect. 3. The experiment results and analysis are discussed in Sect. 4. Section 5 presents the study results in this work.

2 Literature Review

Data classification is defined as the operation of sorting and categorizing data into different groups, sets or any other separated class. Data separation enables the classification of data to agree to data group requirements for different trading or human objectives. It is fundamentally a data and information management method. That comprises into binary-phase stages in which the first stage is the training step where the classifier process builds a classifier with the training group of tuples and the next step is categorization step where the model is utilized for categorization and its evaluation is analyzed with the testing group of tuples [15, 19].

The pre-process of applying any classification algorithm is filter irrelevant attributes. Irrelevant attributes are removed by using different feature selection techniques like linear discriminant analysis a wrapper, filter, and embedded technique. In addition, it plays a very significant role in machine learning by reducing processing and analysis inputs and extracting for the significant attributes. Feature extraction term means to manipulate the raw data, to produce new attributes. In the other hand, feature selection select features from many existing features [22, 23].

2.1   Wrapper Model

In the wrapper method is a features subgroup selection, the seeking for a best group of features is prepared utilizing the induction method as a black box. The evaluated future accomplishment of the method is the heuristic guiding the treatment. Statistical algorithms for selecting a subset features including backward elimination, forwarding selection, and their stepwise variations can be observed as simple hill-climbing methods in the space of feature subgroups [12, 15, 17].

2.2   Filter Model

BFS Filter (Correlation-based Feature Subset Selection) [4, 7, 18].

2.3   Naive Bayes Classification Algorithm

Naive Bayes (NB) is one of the famous and generally utilized methods for categorization. It was named by Thomas Bayes, who proposed the Bayes Theorem.

Naive Bayes is a supervised machine-learning algorithm that uses the Bayes' Theorem (from Bayesian statistics), which assumes that features are statistically strong (naive) separated. In basic terms, a naive Bayes classifier accepts that the nearness (or nonappearance) of a specific feature of a class is disconnected to the nearness (or nonattendance) of some other component. It regularly performs shockingly well in numerous genuine applications, in spite of the solid presumption that all highlights are restrictively free given the class.

In the learning process of this classifier with the identified construction, group probabilities and contingent possibilities are determined to utilize making information, and afterward, estimations of these possibilities used to order new perceptions [21, 24, 25].

2.4   Logistic Regression Classification Algorithm

Logistic regression is a robust measurable technique for indicating a binomial outcome with minimum one illustrative factors. It determines the relation between the absolute ward variable and at minimum one free inputs by evaluating probabilities using a strategic capacity, which is the aggregate calculated circulation [3, 14].

2.5   Decision Trees Classification Algorithm

A decision tree means backing apparatus that utilizes a tree-like diagram or model of choices and their possible outcomes, counting chance-event results, resource costs, and utility [2, 16].

3   Work Strategy

In this study, firstly the data classification algorithms: Naive Bayes, Logistic Regression and Decision Trees (C4.5) were employed to classify the datasets without using any feature selection models (using All Features).

Figure 2 shows the mechanism of employing classification algorithm to classify datasets without using feature selection methods. Here the accuracy of the classified data was measured by Eq. (1) for all used classification algorithms.

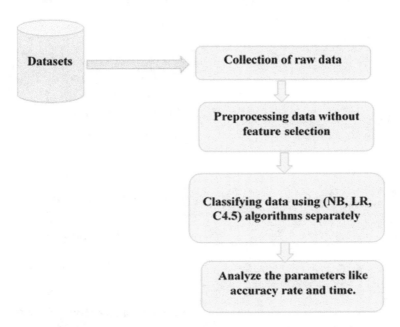

Fig. 2. The mechanism of data classifying without feature selection.

Later the filter (Correlation-based Feature Subset Selection (CFS) filter) and the wrapper (WrapperSubsetEval) were executed to the full training set to select features. The same search strategy (Breadth-First Search) and stopping criterion were employed for both methods. Then separated subsets of features selected by both methods were created to contribute as training and testing sets for the previously mentioned classification algorithms which are tested on these selected feature datasets for 10 times. In each run, training and testing sets were randomly selected from the given dataset. Table 1 shows the sizes of the training and testing dataset. The results of accuracy rate were registered and calculated by using Eq. (1) for all of the classification algorithms and the feature extraction time was noticed.

Figure 3 shows the mechanism of employing classification algorithms to classify datasets preceded by filter/wrapper feature selection.

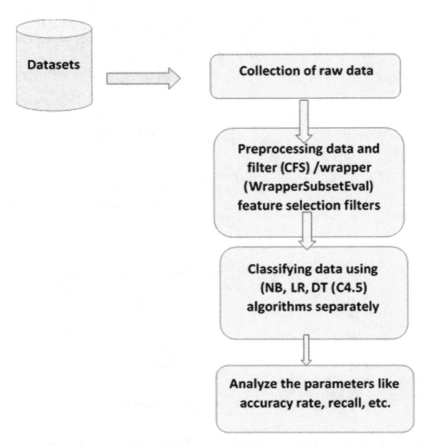

Fig. 3. The mechanism of data classifying using filter/wrapper feature selection methods.

Here the dataset was preprocessed and features were selected using a filter method algorithm called Correlation-based Feature Subset Selection (CFS) filter. In addition, WrapperSubsetEval filter was used as a wrapper method. The excluded features were thrown and the classification algorithms were applied ten times on every reduced feature dataset as previously mentioned.

4   Experiment Analysis

4.1   Weka Data Mining Tool

The experiments in this work were carried out using a simulator written in java used for data mining and classification that allow users to classify and measure the accuracy of algorithms based on datasets by applying different algorithms, this simulator is called Waikato Environment for Knowledge Analysis (Weka) program. Simulations were done on an Intel(R) Core(TM) 2, 1.8 GHz computer. Eshwari Girish Kulkarni, Raj B. Kulkarni provided a reference for more explanation about Weka tool in [13].

4.2 Benchmark Datasets

Dataset is a collection of raw data where every attribute of data represents instance, each instance has its own description, and each dataset consists of different numbers of attributes. This work used seven datasets that can be downloaded from this website "https://archive.ics.uci.edu/ml/index.php" and free. The characteristics of the contributed datasets are outlined in Table 1.

Table 1. Characteristics of datasets

Datasets	No. of attributes	Training set size/Test set size	No. of classes	Dataset file format
Chronic_Kidney_Disease	25	240/160	2	.arff
phishing	31	6633/4422	2	.arff
Diabetic Retinopathy Debrecen	20	691/460	2	.arff
Breast-Cancer	10	172/114	2	.arff
Weather. Symbolic	5	8/6	2	.arff
German_Credit	21	600/400	2	.arff
Vote	17	261/174	2	.arff

4.3 Evaluation Criteria

Normally, the accuracy value of the classification algorithm is used to measure its performance and effectiveness. It is calculated by Eq. (1), that can be referred to as True-Positive (X), True-Negative (Y), False-Positive (Z), and False-Negative (W) as defined in Table 2 [11].

Depending on the classification outcomes, the accuracy is measured as shown in Eq. (1):

$$\text{Accuracy} = \frac{A + B}{A + B + C + D} \tag{1}$$

Table 2. Classification results of a two-class problem

	Predicted Class: Yes	Predicted Class: No
Actual Class: Yes	True-Positive (A)	False-Negative (D)
Actual Class: No	False Positive (C)	True-Negative (B)

4.4 Experimental Results

The results presented in Tables 3, 4 and 5 show the comparison of the classification accuracy (%) of naive Bayes(NB), Logistic Regression(LR) and C4.5 decision tree classification algorithms using the selected features and also the accuracy results for these algorithms without using feature selection (All features) are shown. The average of the accuracies over the ten runs gives the percentage of correct classifications of the specific algorithm. Results for the best-achieved accuracy are shown in italic format.

In Table 3 below, the accuracy values of Naïve Bayes classification algorithm are registered and compared together, at the first time NB used datasets with filter model selected features (using CFS as filter). On the second time NB used a datasets with wrapper model selected features (using Naïve Bayes as evaluator) and on the last time all features of the datasets are contribute in the classification algorithm.

Another accuracy comparison between the different outcomes of Logistic Regression classification algorithm that used filter model selected features datasets and then used wrapper model selected features datasets as a testing datasets for the classification process. This is repeated for each dataset, as seen in Table 4, the last column shows the accuracy of LR classification algorithm when using all features of the datasets.

The last comparison between the accuracy values of C4.5 decision tree classification process using datasets with filter feature selection, wrapper feature selection and finally using datasets without any feature selection prior the classification process(All Features), as shown in Table 5.

A whole comparison between the accuracy outcomes of the three classification algorithms that used the attributes selection models before their data classifying was made and discussed later in this paper.

Table 3. Naive Bayes (NB) accuracy (%) outcomes.

Dataset	Using selected features datasets by Filter model (CFS)	Using selected features datasets by Wrapper model	No feature selection
Chronic_Kidney_Disease	96.25	*99.38*	95
Phishing	92.27	92.65	*92.94*
Diabetic retinopathy debrecen	53.91	*63.91*	53.04
Breast-Cancer	71.05	*75.44*	71.93
Weather. Symbolic	*100*	*100*	66.67
German_Credit	75.5	75.75	*77*
Vote	*97.70*	*97.70*	91.38

As shown in Table 3 wrapper outperformed the CFS five times for Naive Bayes, while they performed equally for two times. Furthermore, comparing the two methods (CFS and wrapper) to all features we found that wrapper enhanced accuracy five times and declined accuracy two times, while the CFS enhanced accuracy four times but declined accuracy three times. In the same way, Table 4 shows Logistic Regression (LR) classification outcomes comparisons.

Table 4. Logistic regression accuracy (%) outcomes.

Dataset	Using selected features datasets by Filter model (CFS)	Using selected features datasets by Wrapper model	No feature selection
Chronic_Kidney_Disease	95.63	*98.75*	96.88
phishing	93.17	92.38	*94.08*
Diabetic Retinopathy Debrecen	72.61	64.35	*72.83*
Breast-Cancer	66.67	*75.44*	66.67
Weather. Symbolic	66.67	66.67	*83.33*
German_Credit	74.75	75.25	*76.25*
Vote	97.70	97.70	*98.28*

Table 4 shows that wrapper outperformed the CFS three times for Logistic Regression classification algorithm, CFS outperformed the wrapper two times, while they performed equally for two times. In addition, by comparing the two methods (CFS and wrapper) to all features we found that the accuracy was improved two times and degraded five times when using the wrapper, while the CFS did not improve the accuracy for any time. Finally, Table 5 represents the accuracy outcomes of the decision tree (C4.5) classification algorithm.

Table 5. DT (C4.5) accuracy (%) outcomes.

Dataset	Using selected features datasets by Filter model (CFS)	Using selected features datasets by Wrapper model	No feature selection
Chronic_Kidney_Disease	*100*	97.5	*100*
phishing	93.87	92.70	*95.03*
Diabetic Retinopathy Debrecen	61.30	59.57	*62.17*
Breast-Cancer	65.79	65.79	*70.18*
Weather. Symbolic	*66.67*	*66.67*	50
German_Credit	71.5	*73.75*	71.25
Vote	97.70	97.70	*98.28*

Table 5 shows that CFS outperformed the wrapper three times, for one time only wrapper outperformed the CFS, while they performed equally for three times. Moreover, by comparing the two methods (CFS and wrapper) to all features we found that's wrapper enhanced accuracy two times and declined accuracy five times, while the CFS enhanced accuracy two times and declined accuracy four times and the accuracy was not affected by the CFS feature selection once only.

It appears that the wrapper works better than CFS in both Naïve Bayes and Logistic Regression. Furthermore, CFS works better in decision tree (C4.5) algorithm than the wrapper. Unfortunately, it appears that feature selection does not improve the accuracy all the time; it works well in some classification algorithms such as Naïve Bayes algorithm where a significant improvement on the accuracy can be noticed, but in both Logistic Regression and decision tree (C4.5) algorithms it seems that using all feature will give a better accuracy results.

According to the time taken for selecting the features, wrapper took more time than CFS especially in the datasets that contain a large number of attributes such as phishing and German Credit datasets. The running of CFS model is faster than the wrapper and in many times, the number of features is reduced to the half by both methods.

In addition, the time taken in the classification process is differs from one algorithm to another, thus Logistic Regression take more time than Naïve Bayes when classifying the phishing dataset.

5   Conclusion

The behaviors prediction helps businesses to make knowledge decisions, this direct to enhance business queries were consumed too time to be resolved, one of the common ways to accomplish this is by using data classification tools and feature selection approaches.

This paper shows performance and effectiveness analysis of feature selection methodologies on the outcomes of different classification algorithms. This analysis handles a comparison between wrapper, which utilizes a target-learning algorithm to guide their search for good features, and filter method, which is a correlation, based model, by applying them on seven datasets before using those datasets in three classification algorithms.

In many cases, the outcomes of the experiments show that wrapper gives results that are similar or better than the Filter model. Filter model can be applied to larger datasets because it is much quicker than the wrapper. Applying feature selection models on the datasets of the classification algorithms removes these unwanted features from the vast size of the data, which makes it easy to learn good classifiers. Unfortunately, feature selection is not necessary can improve the data classification all the time; this depends on the used datasets and the classification algorithms that have been used.

Embedded Model is the third model of feature selection, which has the advantages of wrapper model and filter model. Future work will aim to handle feature selection using the embedded model, compare the results with both wrapper, and filter models.

References

1. Azoff, E.M.: Neural Network Time Series Forecasting of Financial Markets. Wiley, New York (1994)
2. Biggs, D.: A method of choosing multiway partitions for classification and decision trees. J. Appl. Stat. **18**(1), 49–62 (1991)
3. Caigny, A.D.: A new hybrid classification algorithm for customer churn prediction based on logistic regression and decision trees. Eur. J. Oper. Res. **269**(2), 760–772 (2018)
4. Campos, A.R.: Selection of Environmental Covariates for Classifier Training Applied in Digital Soil Mapping. Revista Brasileira de Ciência do (2018)

5. Dietterich, H.A.T.: Learning boolean concepts in the presence of many irrelevant features. Artif. Intell. **69**(1–2), 279–305 (1994)
6. Friedman, N., Geiger, D., Goldszmidt, M.: Bayesian network classifiers. Mach. Learn. **29**, 131–163 (1997)
7. Isaev, I.I.: Melif+: optimization of filter ensemble algorithm with parallel computing. In: Iliadis, L., Maglogiannis, I. (eds.) Artificial Intelligence Applications and Innovations. Springer, Cham (2016)
8. John, R.K.G.: Wrappers for feature subset selection. Artif. Intell. **97**(1), 273–324 (1997)
9. John, R.K.G., Eger, K.P.: Irrelevant feature and the subset selection problem, pp. 121–129. Morgan Kaufmann Publisher (1994)
10. Kamber, J.K.H.M.: Data Mining: Concepts and Techniques. Morgan Kaufmann Publishers Inc., San Francisco (2008)
11. Kohavi, R.: Wrappers for performance enhancement and oblivious decision graphs (1995)
12. Kohavi, R.D.: Feature subset selection using the wrapper method: overfitting and dynamic search space topology. In: KDD (1995)
13. Kulkarni, E.G., Kulkarni, B.R.: WEKA powerful tool in data mining. Int. J. Comput. Appl. **975**, 8887 (2016)
14. Kurt, I.: Comparing performances of logistic regression, classification and regression tree, and neural networks for predicting coronary artery disease. Expert Syst. Appl. **34**(1), 366–374 (2008)
15. Mitchell, T.M.: Artificial neural networks (1997)
16. Otukei, J.T.: Land cover change assessment using decision trees, support vector machines and maximum likelihood classification algorithms. Int. J. Appl. Earth Obs. Geoinf. **12**, 27–31 (2010)
17. Peng, H.: Feature selection based on mutual information criteria of max-dependency, maxrelevance, and min-redundancy. IEEE Trans. Pattern Anal. Mach. Intell. **27**(8), 1226–1238 (2005)
18. Picek, S.: The secrets of profiling for side-channel analysis: feature selection matters. Cryptology ePrint Archive. Report (2017)
19. Ramana, B.V.: A critical study of selected classification algorithms for liver disease diagnosis. Int. J. Database Manage. Syst. **3**(2), 101–114 (2011)
20. Rendell, K.K.L.: The feature selection problem: traditional methods and a new algorithm, pp. 129–134. AAAI Press/The MIT Press (1992)
21. Taheri, S., Mammadov, M.: Learning the naive Bayes classifier with optimization models. Int. J. Appl. Math. Comput. Sci. **23**, 787–795 (2013)
22. Weber, R.: A wrapper method for feature selection using support vector machines. Inf. Sci. **179**(13), 2208–2217, (June 2009)
23. Wedyan, M., Jumaily, A.: Upper limb motor coordination based early diagnosis in high risk subjects for Autism. In: 2016 IEEE Symposium Series on Computational Intelligence (SSCI), pp. 1–8 (2016)
24. Wedyan, M., Al-Jumaily, A., Crippa, A.: Using machine learning to perform early diagnosis of autism spectrum disorder based on simple upper limb movements. Int. J. Hybrid Intell. Syst. Preprint, 1–12 (2019)
25. Wu, J.: Self-adaptive attribute weighting for Naive Bayes classification. Expert Syst. Appl. **42**(3), 1487–1502 (2015)
26. Wu, Z.: Passive indoor localization based on CSI and naive bayes classification. IEEE Trans. Syst. Man Cybernet. Syst. **48**(9), 1566–1577 (2018)
27. Zhang, G.P.: Neural networks for classification: a survey. Trans. Syst. **30**, 451–462 (2002)

The Effect of Teaching Using the E-Learning Platform on Social Communication Among Ninth Grade Students in Computer Subjects

Mamon Saleem Alzboun[(✉)], Lina Talal Aladwan,
and Wafaa Sleman Abu Qandeel

The University of Jordan, Amman, Jordan
m.alzboon@yahoo.com, lena.aldwan@gmail.com,
wafaakandel@hotmail.com

Abstract. The study aimed to identify the effect of teaching using the e-learning platform on social communication among ninth grade students in computer. The study sample consisted of two sections chosen intentionally, one of which formed the experimental group of (24) students, while the second formed the control group of (23) students. A social communication skills gauge consisting of (20) items was prepared to measure the impact of teaching using the e-learning platform on social communication among ninth grade students in computer subjects for the sample members. To answer the study questions, the arithmetic averages, the standard deviations of the study sample and the associated variance analysis (ANCOVA) were calculated. The results showed the following: There are statistically significant variations between the mean results on the dimensions of the third scale of social communication abilities: communication with others, communication with the teacher, communication with peers, and on the scale as a whole because of the technique of teaching and for the advantage of the experimental community studied using integrated learning. The study concluded to disseminate the experience of using the e-learning platform that was applied to the students of the subject of computer material on various subjects, and benefit from the use of the e-learning platform in the development of social communication skills of the ninth students, and to conduct other new studies with different designs and measurement tools to discuss the impact of using the education platform Electronic in a variety of materials and levels of study.

Keywords: E-learning platforms · Social networking · Ninth grade students

1 Introduction

Technological advances have contributed to activating the role of electronic platforms in the educational process in all educational stages. Since the advent of the Internet, educators have sought to use it in the educational process. As a result, e-learning has emerged characterized by its participatory interactive techniques and tools, which moved the learner from being a passive recipient of knowledge to being a participant, as the learner collaborates, interacts and communicates with colleagues. Therefore, one

© Springer Nature Switzerland AG 2020
L. C. Jain et al. (Eds.): ICICCT 2019, LAIS 9, pp. 258–270, 2020.
https://doi.org/10.1007/978-3-030-38501-9_26

of the features of e-learning is that it is social learning and interactive learning, and this led to the emergence of the so-called social e-learning, which indicates that knowledge is built socially; through conversations about the content provided and interaction on problems and educational tasks and procedures followed to solve them [1]. Subsequently, the so-called e-learning platforms have become increasingly popular because of their ease of use, free participation, low costs, their ability to make the learner the focus of the learning process, the abundance of information resources, and simultaneous and asynchronous communication between people [2].

Therefore, calls began to rise in order to keep pace with educational methods rapid technological advancement and benefit from the Web and its software and interactive tools that share the interactive transfer of the learner from being a passive recipient of knowledge to being a participant and collaborate and interact and communicate with his colleagues [3, 4]. the most prominent of what technology has created is the e-learning platforms Ease of use, free participation, low costs, ability to make the learner the center of the educational process, the abundance of information sources, and simultaneous and asynchronous interpersonal communication [5, 6].

One of the objectives of these platforms in the educational process is that they help to provide a variety of educational experiences and attitudes with a variety of visual and auditory stimuli, and work to create an integrated interactive learning environment through diversity in the sources of exciting and attractive electronic information that overcome the problem of the mental stray of the learners and their attention to the subject of learning, and support interaction Through the exchange of views and educational experiences, dialogues and meaningful discussions, and overcome the problem of time and space that are opposed to the teacher and the learner, students and teachers gain technological skills [7, 8]. For modern, modeling and standardizing instructional lessons through the optimal use of sound, image, motion and related multimedia techniques, and shifting towards the way of research and exploration rather than teacher presentation and indoctrination [2].

One of the benefits of e-learning platforms in the educational process: The use of learning platforms has facilitated and developed a system of communication between each of the clients of this platform of teachers, students and parents, and education across platforms increases the effectiveness of students and helps to Developing the concept of continuing education outside the classroom, facilitating the process of teachers access to educational resources In addition to the large number of references that help teachers in the selection of references and methods of education attractive and renewable, the student through this platform has a greater ability to learn by himself and assess his level of education, the platform contributed Finally, the platform has contributed to enhancing the ability of schools to meet the needs of students who have greater learning difficulties than most of their peers [3, 4].

In this context, the emergence of e-learning platforms is the beginning of changing e-learning environments to become more effective between learners and teachers at the same time, as they help in the development of social e-learning environments because it supports cooperation and sharing, it enables learners to communicate with colleagues and participate in effective learning experiences It also creates and exchanges content among learners [14].

In the context of the need to prepare learners who are able to practice their learning outside the classroom and deal with the requirements of the current information age, and qualified to integrate with the technology of modern devices and social networks it becomes clear the importance of acquiring social communication skills, which is defined as the ability to work with the group to achieve a specific goal through the organization of behavior, Which develops through the process of learning and interaction [5].

Educational platforms work to acquire the skill of social networking so that the learner interact and communicate with other learners and exchange educational content and discuss with each other about the content provided. This communication has developed rapidly in the modern century due to the multiplicity of channels of communication, which has become better services and features of others, began the process of communication in writing, and then voice, then audio and video, and the participation of more than one person at the same time, then appeared forums and blogs and finally appeared Social networking sites that provided greater opportunities for communication between different peoples and cultures and the exchange of experiences and information, so social communication is now a science in itself with its own techniques, methods and forms for it [6].

There are many advantages of e-learning platforms such as increasing student interaction, developing their scientific and cognitive abilities, in addition to increasing their motivation towards learning, facilitating the role of teacher, improving the level and quality of learning, increasing interaction during classes between students and subject, and the possibility of opening dialogue and discussion frameworks. Among the students and the teacher about the subject to be taught, in general, these platforms work to develop the educational process, and to achieve positive cope with technological developments [9–11].

In this regard, Al-Enezi (2017) conducted a study aimed at knowing the effectiveness of the use of educational platforms Edmodo for students of mathematics and computer at the College of Basic Education in the State of Kuwait [7] Students benefit from the applications of the programs offered and help them to exchange experience among colleagues in solving their duties and contribute to education through participatory cooperation between students. One of the main difficulties that limit the application of developer programs weaken the use of applied courses taught in applied materials to be applied in the rest of the theoretical curricula prescribed to them. The number of students in one laboratory is also accumulated, as well as the failure of the library in the college to meet the number of students [12, 13]. The three dimensions of social communication skills and the whole scale are attributed to the teaching method. Wright, Williams and Gedera (2013) conducted a study aimed at learning about the reality of students 'use of e-courses through the e-course management system (Moodle) and students' perspectives on their use of the course through a system at a university in New Zealand [19]. The researchers conducted interviews with students and monitored them when they used the course and activities through the system. The study sample was divided into three groups by one teacher for each group. Each group had three students from the university. Load scheduled on their computers through the management of electronic courses system (Moodle) and the difficulty of downloading material from the machine and took a lot of time during loading, but for the views of the students consider the communication process, the majority of students have demonstrated their desire to communicate the process and the discussions that take place across the system.

2   Study Problem and Questions

In light of the rapid technological developments that touched the various areas of the educational process, especially the methods and strategies of teaching, and in light of this development has emerged many strategies and methods of distance learning and e-learning, which posed many challenges to all educational institutions, including schools.

The problem of research comes from the urgent need to investigate the impact of teaching e-learning platforms on the development of social communication skills of ninth grade students in the subject of computer because of the lack of study concerned with the use of e-learning platforms in schools, despite the interest of studies the importance of using e-learning platforms in the educational process Among them is a study [18] which emphasized the need to adopt platforms to improve the educational process. The study [17] said that electronic communication will benefit many learners, especially those who have difficulties in expressing their thoughts and opinions face to face, and study of Munoz and Towner, (2009) pointed to the benefits that social networking platforms can bring to both the teacher and the learner. By offering modern learning methods rather than traditional methods and creating e-learning communities. However, it has not been applied in schools and used effectively and take advantage of its features and raise the level of educational learning [16].

Although scientific studies emphasize the importance of using e-learning platforms in the educational process, and the need to adopt platforms to improve the educational process, given the benefits that platforms can bring to both the teacher and the learner through the possibility of employing modern methods and methods of learning and teaching instead of traditional methods and the establishment of learning societies [16], however, the reliance on e-learning platforms and their application in the teaching process in universities and their actual use and benefit from their advantages in raising the level of educational learning is still limited. In Jordanian schools.

Thus, the problem of the study is to know the effect of teaching e-learning platforms in the development of social networking skills of ninth grade students in the subject of computer because these platforms are a new system used little in educational institutions, and rarely in schools.

Specifically, this study attempts to answer the following question:

– What is the effect of teaching in e-learning platforms on the development of social communication skills of ninth grade students in computer?

3 Study Hypotheses

– There is no statistically significant difference at ($\alpha = 0.05$) in the scores on the social communication skill scale of the ninth grade students in computer subject attributed to teaching in e-learning platforms.

4 The Importance of Studying

The importance of the current research is as follows:

The importance of this study is highlighted by the impact of teaching on e-learning platforms on the development of skill and social communication among ninth grade students in computer. The importance of the study comes through:

5 Theoretical Importance

1. It is in line with modern educational trends that emphasize the importance of modern educational strategies that help increase and develop social communication and make the student's role positive in the educational situation.
2. Contribute to draw the attention of teachers in schools to the importance of using e-learning platforms in education fully or complementary to traditional education.
3. This study will provide information about educational platforms and their use in the educational process.
4. Focus on the learner and his needs and desires, and work to take into account individual differences.

5.1 Practical Importance

1. Contribute to help students to perform their tasks and achieve their goals in promoting e-learning to achieve a high quality in education compatible with contemporary and future needs.
2. Contribute to the creation of a social e-learning environment emphasizes the interactive and active learner in the educational situation.
3. Communication through groups of various media and methods, which gives the ability to access and exchange information in various ways.
4. May benefit those who develop the educational process and the preparation of courses and teachers and students in schools.
5. Objectives of the study:
   This study aims to identify the effect of teaching using e-learning platforms on the development of social communication skills of ninth grade students in computer.

6 Terms and Procedural Definitions

The study adopts the following terms:

– E-learning platforms: Mei (2012, Mei) defines them as web-based remote-learning platforms, which are the arenas where business is displayed, and all that is related to e-learning, including e-courses and activities. Learning by using a set of communication tools that enable the learner to obtain the required courses, programs and information.

It is known as: Interactive educational sites that help ninth grade students in computer courses in the exchange of information and discussion, sharing content, distributing roles among students, and conducting tests and assignments electronically.

Social Networking: Yusuf (2011, 26) are the skills required to transmit and share information and ideas online through some practical applications.

It is procedurally defined as: The skills that ninth graders should acquire in computer courses that help them communicate and interact with other students in order to achieve a particular goal. Students' performance will be measured on the social media scale prepared for study purposes.

7 Study Limits and Limitations

The researchers conducted this study, within a variety of limits that limit its generalization process:

– Human limits: The study sample was limited to computer subjects. Of the ninth grade student's in
– Time limits: The study was conducted in the second semester of the academic year 2018/2019.
– Spatial boundaries: Students of the University of Jordan Model School.
– The study was limited to the use of e-learning platforms in teaching.
– The results of the study were determined according to the psychometric characteristics that will be extracted for the study instruments.

8 Method and Procedures

8.1 Study Approach

The study aimed to address the impact of teaching in e-learning platforms on the development of social communication skills of ninth grade students in computer subjects. To achieve the objectives of the study, researchers will use the quasi-experimental approach, and the information will be collected through a questionnaire that will be prepared by reference to previous studies and research.

8.2   Study Individual

The study members consisted of all students of the ninth grade computer in the model University of Jordan. A class was randomly selected using lots and taught using e-learning platforms in the second semester of the academic year 2017/2018, and the other class will be taught in the traditional way.

8.3   Study Variables

In this study, researchers will study two types of variables:

- Independent variables: e-learning platform.
- Dependent variables: social communication.

8.4   Social Communication Skills

A tool to measure social communication skills has been developed to answer study questions. Using the tools of the previous studies such as the study of the customer (2015) and the study of Ali (2016), this tool has been developed, as the final scale of (24) paragraphs included social communication skills divided into three main dimensions: After social communication with the teacher (6) paragraphs, and after social communication with colleagues and included (6) paragraphs, the researchers also used a five-step scale (always, often, sometimes, rarely, absolutely). The scale of social communication skills was corrected as follows: always (5°), often (4°), sometimes (3°), rarely (2°), never (1°).

8.5   Validity and Stability of the Scale

To verify the validity of the measuring instrument, it was presented to a group of (10) arbitrators with competence in curriculum and teaching, technology and education, metrology and evaluation, and sociology, for their views on the soundness of the language of the paragraphs, the degree of clarity, and how they relate to the dimension. To be measured, and any observations, amendments or additions they deem appropriate, and adjusted the scale according to the consensus of the majority of arbitrators, and the final instrument consisted of (20) paragraphs.

8.6   Construction Validity

To extract the significance of the building validity of the scale, the correlation coefficients of the paragraphs of the scale with the total score were extracted in a survey sample from outside the study sample consisting of (20) students. Table 1 shows that.
   Study tool stability:
   To ensure the reliability of the tool, the internal consistency was calculated on a survey sample from outside the study sample of (20) students according to the Kronbach Alpha formula. The table below shows these coefficients (Tables 2 and 3).

Table 1. The correlation coefficients between the paragraphs and the total score

Paragraph number	The correlation coefficient with the tool	The correlation coefficient with the tool
1	.51(*)	.62(**)
2	.55(*)	.75(**)
3	.33(*)	.49(*)
4	.51(*)	.41(*)
5	.50(*)	.51(*)
6	.60(**)	.57(**)
7	.85(**)	.68(**)
8	.56(**)	.67(**)
9	.50(*)	.44(*)
10	.39(*)	.42(*)
11	.63(**)	.60(**)
12	.80(**)	.63(**)
13	.75(**)	.60(**)
14	.71(**)	.50(*)
15	.51(*)	.50(*)
16	.42(*)	.34(*)
17	.50(*)	.48(*)
18	.55(*)	.35(*)
19	.45(*)	.63(**)
20	.63(**)	.73(**)

*Statistically significant at the significance level (0.05).
**Statistically significant at the level of significance (0.01).

Table 2. Correlation coefficients between fields and the total score

The dimension	Skills as a whole
My social networking	.805(**)
Social communication with the teacher	.701(**)
Social networking with colleagues	.653(**)

*Statistically significant at the significance level (0.05).
** Statistically significant at the level of significance (0.01).

Table 3. Coefficient of internal consistency Alpha Kronbach

Fields	Internal consistency
My social networking	0.78
Social communication with the teacher	0.75
Social networking with colleagues	0.76
Total marks	0.8

9 Discussion of the Results

Question: There is no statistically significant difference at ($\alpha = 0.05$) in the scores on the social communication skill level of the ninth grade students in computer subject attributed to teaching in e-learning platforms.

To answer this question, the arithmetic mean and standard deviations of the social communication skills of the ninth grade students were calculated for the pre- and post-secondary measurements according to the teaching method (e-learning platform, regular), as shown in Table 4:

Table 4. Arithmetic averages and standard deviations of social communication skills among ninth grade students for pre and post measurements by teaching method

Way	Number	Tribal measurement		Telemetry	
		Mean	Standard deviation	Mean	Standard deviation
Experimental (e-learning platform)	24	2.8	0.21	4.45	0.177
Officer (regular)	23	2.05	0.198	2.48	0.212
Total	47	2.4	0.21	3.47	0.922

Table 4 shows that there are obvious variations between the arithmetic circles in the pre- and post-measurement of social communication abilities in fifth grade learners according to the teaching technique (integrated learning strategy, ordinary) and whether these obvious variations are statistically important, the analysis of the associated variance was used (One way ANCOVA) for the telemetry of the social communication skills of the ninth grade students according to the teaching method (e-learning platform, regular) after determining the impact of their pre-measurement.

Table 5. Results of the one way ANCOVA analysis for dimension measurement of social communication skills among ninth grade students according to the teaching method (E-Learning Platform, Standard) after neutralizing their pre-measurement impact

Contrast source	Total squares	Degrees of freedom	Average sum of squares	Values F	Significance level	Square ETA η^2
Tribal measurement	0.07	1	0.07	1.885	0.174	0.037
Teaching method	40.664	1	40.664	1104.883	0	0.958
Total error	1.802	45	0.037			
	43.379	46				

Table 5 shows that there are statistically significant differences at the level of ($a = 0.05$) in the social communication skills of ninth grade students according to the teaching method (e-learning platform, regular), the value of (F) (1104.883) in terms of statistical (0.000), which is a statistically significant value, which means there is an effect of the teaching method.

As shown in Table 5, the effect of teaching method was large; the value of the ETA square ($\eta2$) explained (95.8%) of the explained (predicted) variation in the dependent variable, social communication skills.

To determine the benefit of the differences, adjusted averages and standard errors were extracted according to the teaching method, as shown in Table 6.

Table 6. Adjusted arithmetic averages and standard errors for social communication skills among ninth grade students by teaching method

The group	Arithmetic mean	Standard error
Experimental (e-learning platform)	4.45	0.038
Officer (regular)	2.501	0.038

The results in Table 6 indicate that the differences were favorable for those who studied using the e-learning platform compared to the control group.

The results showed that for the experimental group studied using the e-learning platform, there is a important effect on the growth of social communication abilities among learners ascribed to the teaching method. This finding is consistent with several previous studies that have pointed to the role of the use of modern technology and techniques in general and the use of integrated learning in particular in the development of social communication skills, such as Ali (2016) and the client (2015).

This can be explained by the fact that the use of the e-learning platform has enabled students to perform skills well than their parents who have studied in the normal way, because the presence of students together at the same time during the study has helped to create interdependence and lead to an atmosphere of cooperation Constructive and meaningful, and to identify among themselves on things that others may not know the group students, through the atmosphere of intimacy, love, affection and social relations prevailing among them, which reflected in one way or another on the development of their social communication skills, also led to a positive interaction between the Filleting while working with each other to raise students' motivation towards mastery of the material and skill. In addition, their presence at the same time, working in groups and cooperating with each other to solve duties and activities led to a state of meaningful cooperation and how to dialogue remotely and develop positive behaviors, reduce tension and reduce anxiety among them and a sense of self-confidence and self-fulfillment.

This result may also be attributed to the fact that learning through the e-learning platform led to communication between them and the teacher and them and each other through the agreed time among them, through the discussion room, comment and write new posts through a special page on Facebook, which had Effective impact in improving their skill level performance, where they were able to inquire about any information they need at any time, which promoted their social communication, and education through the e-learning platform helped students use it to create an atmosphere of cooperation between students and constructive dialogue, Coordination of their colleagues follow blogs and comment on them.

The result can also be attributed to the fact that the diversity in the forms of assignments given to students of duties and activities, and the freedom to work during the implementation, has contributed to the student's interaction with himself, and take responsibility for each student to learn from himself, and may prompt the student to pay attention to what he does during the work It also pushes him to keep doing homework at best.

The change of the role of the teacher from the role of the initiator of knowledge, to the role of guidance and participation and listening, and cooperation with students, may be reflected in the way students interact with each other, and enhance the confidence of the student himself.

Researchers have noted that teamwork can contribute to students getting to know each other more, and giving the opportunity to express themselves, or prove their presence in the group, which in turn may push the learner to develop good relations with his colleagues, and seeks to stick to his role during the discussion, and respect for others and self-control, Which may develop ways of communication with his colleagues, and try to raise his competence to reach the best level, whether academically or socially.

10   Recommendations

The researchers concluded a set of recommendations and suggestions:

1. Disseminate the experience of using the e-learning platform that has been applied to the students of the computer subject to various subjects.
2. Utilize the use of the e-learning platform in the development of social communication skills of the ninth basic students.
3. Conducting new studies with measurement tools and different designs to investigate the impact of using the e-learning platform in a variety of subjects and levels of study.

References

1. Al-Judeibi, R.B.M.: Develop social communication skills in the work environment from the perspective of Islamic education. Thought Creativity: Egypt **10**(1), 158–256 (2016)
2. Hijazi, T., Abdulmonem, M., Hindawi, S.: Quality standards for virtual classrooms from the viewpoint of faculty members at King Saud University. Paper presented to the 6th Arab International Conference on Quality Assurance in Higher Education, Sudan University of Science and Technology, Sudan (2016)

3. Hamdan, M.A.: The effect of using embedded technology in the Arabic language curriculum on the acquisition of concepts and developing social communication skills among fifth grade students in Saudi Arabia. Master Thesis, University of Jordan (2016)
4. Al-Dossari, M., Al-Omari, A.: The reality of the use of electronic teaching platforms by faculty members in teaching English at King Saud University. Unpublished Master Thesis, Yarmouk University, Irbid (2016)
5. The Customer, Mamoun Selim: The effect of teaching using the electronic course system (MODEL) on the achievement of students of the University of Jordan with computer skills and in the development of self-learning skills and social communication. Unpublished doctoral thesis, University of Jordan (2016)
6. Ali, S.M.: The impact of some gloomy social e-learning environments on social media platforms on the development of educational e-communication skills among students of the Faculty of Education. Arab Stud. Educ. Psychol. Saudi Arabia **69**, 87–156 (2016)
7. Al-Enezi, Y.A.M.: The effectiveness of the use of educational platforms Edmodo for students of mathematics and computer, Faculty of Basic Education in the State of Kuwait. J. Fac. Educ. Assiut Univ. **33**(6) (2017)
8. Yousef, A.A.: The proposed educational design for an interactive website in social studies and its impact on the development of critical thinking and some electronic communication skills of seventh grade students of basic education. The Arab Bureau of Education for the Gulf States (2011). https://www.abegs.org/aportal/article/showDetails?id=51298. Accessed 11 Nov 2017
9. Homanova, Z., Prextova, T.: Educational networking platforms through the eyes of Czech primary school students. In: Academic Conferences International Limited, European Conference on e- Learning; Kidmore End, pp. 195–204. Kidmore End (2017)
10. Jewitt, C.G., Hadjithoma Clark, Wilma Banaji, Shakuntala Selwyn
11. Neil: School use of learning platforms and associated (2010)
12. Technologies, British Educational Communications and Technology
13. Agency (BECTA), London, UK
14. Kaplan, A., Haenlein, M.: Users of the world, unite! The challenges and opportunities of Social Media. Bus. Horiz. **53**(1), 59–68 (2010)
15. Mei, H.: The construction of a web-based learning platform from the perspective of computer support for collaborative design. (IJACSA) Int. J. Adv. Comput. Sci. Appl. **3**(4), 105–112 (2012)
16. Munoz, C., Towner, T.: Opening Facebook: how to use Facebook in the college classroom. In: Gibson, I., Weber, R., McFerrin, K., Carlsen, R., Willis, D. (eds.) Proceedings of Society for Information Technology & Teacher Education International Conference 2009, pp. 2623–2627. Association for the Advancement of Computing in Education (AACE), Chesapeake (2009)
17. Shi, L., AlQudah, D., Cristea, A.: Social e-learning in Topolor: a case study. Paper presented at the International Association for Development of the Information Society (IADIS) International Conference on e-learning, pp. 57–64 (2013)
18. Shihab, M.: Web 2.0 tools improve teaching and collaboration in English language classes. In: National Educational Computing Conference, San Antonio, TX (2008). Retrieved from ProQuest database. (UMI No. 3344829)
19. Gedera, D., Williams, P.J., Wright, N.: An analysis of Moodle in facilitating asynchronous activities in a fully online university course. Int. J. Sci. Appl. Inf. Technol. (IJSAIT) **2**(2), 6–10 (2013)
20. Baldonado, M., Chang, C.-C.K., Gravano, L., Paepcke, A.: The stanford digital library metadata architecture. Int. J. Digit. Libr. **1**, 108–121 (1997)

21. Bruce, K.B., Cardelli, L., Pierce, B.C.: Comparing object encodings. In: Abadi, M., Ito, T. (eds.) Theoretical Aspects of Computer Software. Lecture Notes in Computer Science, vol. 1281, pp. 415–438. Springer, Heidelberg (1997)
22. van Leeuwen, J. (ed.): Computer Science Today. Recent Trends and Developments. Lecture Notes in Computer Science, vol. 1000. Springer, Heidelberg (1995)
23. Michalewicz, Z.: Genetic Algorithms + Data Structures = Evolution Programs, 3rd edn. Springer, Heidelberg (1996)

Blended Learning and the Use of ICT Technology Perceptions Among University of Jordan Students

Jehad Alameri$^{(\boxtimes)}$, Haifa Bani Ismail, Amal Akour,
and Hussam N. Fakhouri

The University of Jordan, Amman, Jordan
{j.alameri, h.baniismail, a.akour, h.fakhouri}@ju.edu.jo

Abstract. This research focuses on students' attitudes towards Blended learning and the use of information and communication technology (ICT) at the University of Jordan. A sample of 150 students who have taken blended course have been chosen to complete a questionnaire that includes aspects of knowledge and attitude towards blended learning. The results of the study showed that Blended learning is beneficial to students, and most students understand the goals of e-learning by Blended learning fully. The students generally showed a positive attitude towards Blended learning. In addition, they have shown sufficient information on the area of Blended learning and to determine when and how to use the tools they are provided with in combination. It also found that blended learning is more effective to develop and improve knowledge and skills than conventional methods of teaching. The findings also showed that Blended research online tools help students to gain insight and improve their skills and that their roles in their own learning process have a huge impact on student lives.

Keywords: Blended learning · E-Learning · University · Education

1 Introduction

Blended Learning incorporates the benefits of distance learning and traditional methods of learning and has many advantages: Face-to-face training allows for communication between the instructor and the student. In the process of joint study, a favorable educational environment is created, the mood to achieve results, the motivation to study. The students receive immediate feedback, chat about the details and ask questions [1].

In addition to increasing cognitive capabilities and student experiences, teamwork also improves the emotional intelligence of students. With regard to online learning, a Blended model allows students greater freedom: they are able to choose the subject matter, speed, time and place of study [2].

Using ICT and blended learning in education has many features, the most important of which is the education of a large number of learners without the constraints of time and space, and in a short time using thousands of sites and many interactive multimedia that encourage self-learning and provide immediate feedback. It also allows to take into

© Springer Nature Switzerland AG 2020
L. C. Jain et al. (Eds.): ICICCT 2019, LAIS 9, pp. 271–280, 2020.
https://doi.org/10.1007/978-3-030-38501-9_27

account the individual differences between learners through multiple sources of knowledge as a result of connecting to various sites on the Internet and ease of use of tools and equipment and the use of virtual classrooms and exchange of experiences between educational institutions and the ease and speed of updating and development of programs compared to video systems and CDs. In addition, the e-learning system gives education a global character and exits the local context [3].

1.1 Models of Blended Learning

1.1.1 Inversion Class

The best way to implement models is to limit frontal function to a minimum (the educator demonstrates that children listen). Students work at home in an online learning environment using their own Internet-connected electronic devices: get to know the content or do what they read. The class consolidates and interacts with the curriculum, which can be done in activities, workshops or in other interactive ways. The Blended learning method can be used in college for 3–5 grade students [4].

1.1.2 Station Rotation

Rotation of the primary and secondary school Blended learning model. Includes the use of computers and tablets in classrooms, and the use of learning management systems. All students are categorized according to their level of academic activity. In a separate class-a station-each group works. Positions have different purposes: instructor work–educator input; digital education–capacity building, personal accountability, self-regulation, learning ability; task work–information implementation in the resolution of practical issues, the improvement of communication skills and the receipt of class-mate's feedback [5].

During the course, students move from one station to another to visit each one. The groups vary according to pedagogical task from lesson to lesson [2].

There could be two stations rather than three: a teacher work and online work; or four stations: work with an educator, online work, collective work and self-employment.

1.1.3 Laboratory Rotation

The Blended learning model for the "Rotation of Laboratories" suggests that part of the lessons are kept in a daily lesson and are moved for the one lesson to a computer class (laboratory) for the individual training in an on-line environment. Students can learn new material in the online environment, build up their knowledge, develop diverse skills and also work on their own projects. If this model is implemented in the class-room, teachers create an online space for children common to different subjects can allow the most valuable learning. This blended learning model is ideal for students of all ages, given that their age is met by the online environment.

1.1.4 Dynamic Model

The basis for a dynamic Blended learning system is that one form of learning activity is not restricted by students in time. Students draw up a work schedule independently; choose a subject, and how quickly the topic will be studied. This model uses an internet interface mainly. The teacher works with small groups or with students individually who need assistance [3].

For secondary school students, students, and adults, this model is most successful as it requires an advanced self-organizational ability blended teaching, for example, has demonstrated many benefits for effective verbal and written communications through various technology-based teaching materials [6, 7] carried out a study to show that BL was effective in redesigning the contact module of nursing graduates in enhancing their level of satisfaction and self-efficacy. The students had positive attitudes to communicative reading. The students reported a significantly higher level of satisfaction in the blended level of training, communication skills and communication skills of the self-performance stage of the nursing students [17].

Harona announced that academics in Malaysia have embraced blended learning. The results of the study showed that the rate of acceptance of Blended learning is small, as shown by the result in only 13% of academics taking the method. The study shows that The utility of the program, learning goals and educational technology preferences are among factors that have driven the acceptance of Blended learning [15]. Hudaib and Fakhouri (2017) emphasized that Blended learning has more quality than traditional eLearning and is more effective [17].

Studies have shown however that scholars are worried about Blended-learning teaching [13]. Zaghoul explores what is Blended learning, its advantages and disadvantages, and eventually gives teachers some hands on interactions and class exercises to incorporate Blended learning into higher education [22]. Garrison and Kanuka (2004) propose that combined teaching could promote cooperative and independent learning experiences. Blended learning provides both a research community and a forum for open and collaborative dialogue. Nazarenko also talk about how students communicate for Blended learning. Students become digitally literate and develop opportunities to expand their classroom lessons and discussions [21].

Hudaib and Fakhouri [16] experiments shows how the quality of e-learning systems can be evaluated by two quantitative methods: one when the systems already in place are at stake; another when it concerns voicing the desires of students or their views as regards the ideal system to be established in the future [17].

The Robison research studied the perspectives of ten faculty members of the Brigham Young University in the creation and teaching of blended training courses. The results of the study showed that three significant advantages were viewed by the participating faculty in the combined experience [19]. Coynea conducted a study which includes online video support resources and can be a useful tool to advise health students, including nursing, on their clinical skills [18]. Blended learning not only improves the knowledge and skills of students, but is often preferred by students because of its versatility. Ersoy conducted a study with 65 graduate students enrolled in the "Programming languages II" course to demonstrate how much web-based teaching is adding to conventional face-to-face learning [14]. The course was guided with the conventional face-to-face approach and a website was added. The quantitative data revealed that students have a positive perception of web-based training and online teaching, but online collaboration perceptions are negative. In addition to the research questions, other development issues are raised in the Blended learning scenario.

The observation of Hien (2017) indicates that BL is substantially more closely associated with students in STEM than with traditional classroom learning. Discussions on the outcomes and consequences for future research are therefore created. The results show, however, that STEM courses have an average effect size that is lower than that of non-STEM. Ironically, in Schmid et al. (2014), it was noticed that the topic was moderating, as technical rates were used to subdivide the control state, i.e. no technology and some use of technology. Graham (2006) points out that BL allows for more immersive and reflective comprehension as a mixture of conventional face-to-face teaching and online learning. In this learning mode, technology has increased the use of multi-format resources and archived discussions, as well as the changing role of instructors and more time to discuss and reflect [11, 12].

This research aims to investigate student's attitude and Perceptions towards Blended Learning and the use of ICT among University of Jordan Students.

2   Study Problem and Questions

The problem of the study stems from the importance of education using electronic resources because of its important role in increasing students' understanding of the teaching material. The problem of the study is also that "there is an urgent need to improve the methods currently used in the teaching to raise the level of achievement of students where the trend towards the use of modern teaching methods in the educational process and invested to develop higher scientific and intellectual skills of students and increase their educational attainment, the most important use of technology Education. In addition to the need to prepare students with skills and experience that enable them to deal with the data of the times and its challenges and to employ technological innovations and invest their potential in the field of education, and this requires to identify the most important features of e-learning technology and its various programs where this technology is one of the modern applications of computer and Internet networks It requires identifying the possibility of using them in educational institutions in order to achieve the directions related to preparing individuals capable of dealing with the variables of this era.

3   Objectives of the Study

Study the Perceptions and attitude of students towards Blended Learning and the use of information and communication technology (ICT) among University of Jordan Students.

4   The Importance of Studying

This study comes in line with the changes brought about by technology in education, and the explosion of knowledge, and focus on that the learner is the focus of the educational process of learning, and guide education to build a knowledge society, to become a process of human development, which is concerned with the design of environments and conditions according to scientific knowledge about human behavior, in order to build Individual personality or composition of the desired social composition, and provide the learner cognitive experiences. Accordingly, the importance of the study can be explained in the following points:

(1) Introducing new methods that may contribute to the sustainable development of students.
(2) Attempting to keep pace with contemporary global trends and in response to many research recommendations and conferences of the need to employ technological innovations in the educational process to improve the teaching and learning processes.
(3) Address the stage of the most important stages of education, which has the responsibility to prepare individuals who will lead the wheel of development in the field of educational process.
(4) The importance of diversity in teaching methods, the use of modern technological techniques in blended learning, and the use of information provided by the Internet.

5   Methodology

The research was attended by students of the University of Jordan's bachelor and Master Degree from the academic years 2018–2019. The initial sample used in the research methods was a random sample of 150 students over the duration of (18–35) years as shown by Table 1.

Table 1. Demographic characteristics

Demographic variables		Number of students	Percentage
Age	18-21	85	56.7
	21-26	35	23.3
	26-29	17	11.3
	Above 29	13	8.7
Major	Bachelor	130	86.6
	Master degree	19	12.7
	PhD	11	7.3

The demographic characteristics of the participants are shown in Table 1. The table shows that the age group was between 18 and 21 years, 56.7% and 23.3% between 21 and 26 years of age, 20% above 26 years of age. Approximately 86.6% of the sample is in a bachelor's degree, 12.7% master's degree and 7.3% PhD.

5.1   Method of Data Collection and Evaluation

Google type on-line questionnaire was employed as the method of data collection. The questionnaire was developed by Birbal et al. (2018). The groups of college and graduate students filled out 150 questionnaires. Students who completed the survey received information about the completion of the survey questionnaire; questionnaires for students who enrolled in Blended courses were filled out during Blended learning in university computer laboratories. For final results and conclusions, the findings of the questionnaire were evaluated using the SPSS software. According to Chisnall (1997), non-response is a vital restriction of study, so, in order to reduce the non-response rate, at the time of distribution of the sample online questionnaire, the researcher was accessible and the participant completed an online questionnaire. The rating of the SPSS questionnaire was 0.78, which suggests a high degree of internal consistency. According to Cronbach's Alpha A values of more than 0.7, reliability is reasonable in the range of 70% and reliability is perfect for those over 80% (Sekaran 2003). Experts from the School of art and design and the Educational sciences school in the University of Jordan evaluated the questionnaire. SPSS v.20 has been used in statistical analyses after collection of data from the sample of the research and accuracy analyzes as well as moderate and standard deviations have been used (Table 2).

It can be observed from Table 6 that 73.2% of the Sample like to have their lessons online. However, 84.9% of the participants see that Blended learning and online learning provides richer instructional content. And it has privacy, so the students have the ability to express themselves without any boundary through the online learning discussion and internet tools which is considered private. Most of the responses indicated that when using online learning there is also a need for attendees to traditional lectures and that blended online learning also provide different types of learning techniques which is not possible in face to face traditional lectures. Despite that they also see that lecture time in the classroom to be reduced with percent of 80.8%. The results showed high rate from the Sample whom believed that they organize my time better when studying online, furthermore the results showed that Online learning motivates them to prepare well for their studies with percent of 82.7%. Most sample results showed makes them more responsible for their studies.

Results in Table 5 showed that a percentage of 44.3 believe that face-to-face learning is more effective than online learning, 75.6% can learn in a group, as well as they can learn on my own as well and they have sense of community when they meet other students in the classroom with percent of 78.6% think it is important for an online counseling provider to meet with in-person before providing online counseling services. The percentage of the sample that find learning through online resources is more effective than face-to-face classroom lectures is 85.5% and a percentage of 71.6% learn better through lecturer-directed classroom-based activities and a sample of 73.5% prefers that someone guides them personally in a face-to-face setting. The results showed that online learning is available and easily to be accessed from everywhere anytime they want by using the internet with percentage of 85.9% and they feel that the information they get from online learning is useful and is very helpful for them with a percentage of 88.7%. The results of this study also indicate that students like to interact with my lecturer online. A percentage of 87.1% are able to communicate effectively

Table 2. perception toward blended learning

Statement	Yes	No
Online learning		
I do not resist having my lessons online	73.2%	26.8%
I like online learning as it provides richer instructional content	84.9%	15.1%
Blended learning have privacy	88.1%	11.9%
Blended learning provide different types of learning techniques which is not possible in face to face traditional lectures	90.5%	09.5%
I "would like lecture time in the classroom to be reduced	80.8%	19.2%
I would like to have my classes online rather than in the classroom	14.3%	85.7%
I organize ray time better when studying online	63.4%	36.6%
Online learning motivates me to prepare well for my studies	72.7%	27.3%
Online learning makes me more responsible for my studies	44.7%	55.3%
Classroom learning		
I believe face-to-face learning is more effective than online learning	44.3%	55.7%
I like to learn in a group, but I can learn on my own as well	75.6%	24.4%
I have a sense of community when I meet other students in the classroom	78.6%	21.4%
I find learning through online resources is more effective than face-to-face classroom lectures	85.5%	14.5%
I learn better through lecturer-directed classroom-based activities	71.6%	28.4%
I learn better when someone guides me personally in a face-to- face setting	73.5%	26.5%
Online interaction		
Online learning is available and easily to be accessed from everywhere anytime they want by using the internet	85.9%	14.1%
I feel that the information they get from online learning is useful and is very helpful for them	88.7%	11.3%
I would like to interact with my lecturer online	84.2%	15.8%
I am able to communicate effectively with others using online technologies (e.g. email, chat, discussion board.)	87.1%	12.9%
I appreciate easy online access to my lecturer	90.8%	9.2%
I am able to express myself clearly online through my writing	87.7%	12.3%
I can collaborate well with a virtual team in doing assignments	83.5%	16.5%
I respect opinions and information provided by others m online communities.	89.2%	10.2%
Technology		
I am comfortable using my computer	73.4%	26.6%
I believe the Web is a useful platform for learning	82.7%	17.3%
I am comfortable using Web technoloaies	84.7%	15.3%
I think we should use technologies in learning	86.3%	13.7%
I am comfortable in using Web technologies to exchange knowledge with others	81.2%	18.8%

(*continued*)

Table 2. (*continued*)

Statement	Yes	No
I feel that there is difficulties in using blended learning such as low Internet connection speed; Network and lack of live communication.	72.6%	27.4%
It difficult to use electronic resources. I need more practice, skills and training courses to use online learning	73.7%	26.3%
Learning flexibility		
Using blended learning I can get unlimited access to lecture materials	89.5%	10.5%
Using blended learning I can decide where I want to study the amount of knowledge gained is high	88.2%	11.8%

with others using online technologies (e.g. email, chat, discussion board.) and they appreciate easy online access to my lecturer with percentage of 90.8%.

Most sample results showed that they are able to express them self clearly online through my writing with a percentage of 87.7%. However, a percentage of 83.5% say that they can collaborate well with a virtual team in doing assignments and 89.2% of the participant's respect opinions and information provided by others in online communities. Regarding the technology 83.4% of respondents feel comfortable using my computer and 82.7% believe the Web is a useful platform for learning and 84.7% feel comfortable using Web technologies and they are using Web technologies to exchange knowledge with others with percentage of 81.2%. Further, most of respondents about 92.9% say that they should use technologies in learning. For the learning flexibility questions we can see that most of the students see that using blended learning they can get unlimited access to lecture materials and decide where they want to study with percentage of 89.5% and 88.2% respectively.

6   Discussion

The goal of this study was to explore the attitude of students of the Jordanian University in Blended learning and their needs for Blended learning. The results of the study indicated that students had positive attitude about blended learning. And the combination of online learning and traditional in-class learning is better than a one-way stream of information answers, they prefer and believe, shows that many students are more positively attuned to this type of learning, which confirms the views of many researchers. Nevertheless, it is quite natural that few students communicate their unhappiness to engage in the Blended learning process. The findings overall indicate student satisfaction with the program as it enhanced their learning ability and interacted, engaging and interesting the learning process. It has also been verified that it facilitates contact between teachers and students outside the classroom.

A brief look at the results reveals that some of the students' drawbacks from the Blended learning process are clear. The noted drawbacks to Blended learning include low Internet connection rate, difficulties using a blended learning model. Live contact network. Many students find it difficult to use digital resources; the amount of knowledge acquired is not insignificant, but training, skills and guidance for use are missing. Social isolation is also part of the minuses, not good material organization. Students also need more time to communicate, trick students, and other moral behavioral activities. One of the major drawback is that communications through internet require more time than live communications in the classroom. Students also noticed that online video, audio, chat and electronic materials are useful to them for better learning. And so we speak of the need for a deeper analysis. The course materials are appealing to students. The use of many computer programs, the use of smartphone, phone, laptop and more knowledge from the electric E-Learning Platform lead to self-development and to success.

7   Conclusion

This study shows that blended learning, the use of information and communication technology (ICT), mixing online learning and classroom learning, is more successful than conventional learning. It also indicates that more than one way is obtained with Blended learning. Students agreed that Blended learning activities give them the chance to read and learn more.

In addition, the results showed Blended learning is useful for students and most students are fully aware of e-learning goals through Blended learning. The students generally showed positive attitudes through Blended learning. They showed however sufficient information about the field of Blended learning and decided when and how to use the resources provided by Blended learning. Therefore, the study suggests more research to explore different approaches to teaching, including online learning, remote learning and composite learning; further fields of research should be considered in the study to understand how Blended learning attitudes contribute to some variables such as age, sex and disability if the student has a disability.

References

1. Al Adwan, F.E., Al Awamrah, A.F., Al Adwan, F.E.: The extent to which students have sufficient awareness of e-learning and its relation to self-studying and academic achievement. Modern Appl. Sci. **12**(1) (2018)
2. Aladwan, F., Fakhouri, H.N., Alawamrah, A., Rababah, O.: Students Attitudes toward Blended Learning among students of the University of Jordan. Mod. Appl. Sci. **12**(12) (2018)
3. Alshehri, F.: Student satisfaction and commitment towards a blended learning finance course: a new evidence from using the investment model. Int. Bus. Finan. **41**, 423–433 (2017)
4. Behnke, C.: Blended learning in the culinary arts: tradition meets technology. In: Glazier, F. (ed.) New Pedagogies and Practices for Teaching in Higher Education: Blended Learning: Across the Disciplines (2011)
5. Birbal, R., Ramdass, M., Harripaul, M.C.: Student teachers' attitudes towards blended learning. J. Educ. Hum. Dev. **7**(2), 9–26 (2018)
6. Blended Learning: Global Perspectives, Local Designs, pp. 3–21

7. Brinkert, R.: A literature review of conflict communication causes, costs, benefits and interventions in nursing. J. Nurs. Manag. **18**, 145–156 (2010)
8. Brooks, L.: An Analysis of Factors that affect faculty attitudes toward a blended learning environment. Ph.D. dissertation, Faculty of the College of Education, TUI University, California (2008)
9. Bunyarit, M.: E-learning systems: an evaluation of its effectiveness in selected higher learning institutions in Malaysia. M.S thesis, International Islamic University Malaysia (2006)
10. Dearnley, C., McClelland, G., Irving, D.: Innovation in Teaching and Learning in Health Higher Education. The Higher Education Academy, London (2013)
11. Elisabeth, C., Hazel, R., Valda, F., Victoria, K., Melanie, P., Marion, M.: Nurse Educ. Today **63** (2018)
12. Garrison, D.R., Kanuka, H.: Blended learning: uncovering its transformative potential in higher education. Internet High. Educ. **7**, 95–105 (2004)
13. Ha, J.F., Longnecker, N.: Doctor-patient communication: a review. Ochsner J. **10**(1), 38–43 (2010)
14. Ersoy, H.: Blending online instruction with traditional instruction in the programming language course: a case study (Doctoral dissertation, METU) (2003)
15. Hamtini, T.M., Fakhouri, H.N.: Evaluation of open-source e-Learning platforms based on the Qualitative Weight and Sum approach and Analytic Hierarchy Process. In: Clark, R.C., Mayer, R.E. (eds.) Proceedings of the International Multi-Conference Society, Cybernetics and Informatics, pp. 1–7. E-learning and the Science of Instruction. Jossey−Bass, San Francisco (2012)
16. Hudaib, A.A., Fakhouri, H.N.: An automated approach for software fault detection and recovery. Commun. Netw. **8**(03), 158 (2016)
17. Hudaib, A.A., Fakhouri, H.N.: Supernova optimizer: a novel natural inspired meta-heuristic. Mod. Appl. Sci. **12**(1), 32–50 (2018)
18. Coyne, E., Rands, H., Frommolt, V., Kain, V., Plugge, M., Mitchell, M.: Investigation of blended learning video resources to teach health students clinical skills: an integrative review. Nurse Educ. Today **63**, 101–107 (2018)
19. Robison, R.A.: Selected faculty experiences in designing and teaching blended learning courses at Brigham Young University. The University of Nebraska-Lincoln (2004)
20. McCutcheon, K., Lohan, M., Traynor, M., Martin, D.: A systematic review evaluating the impact of online or blended learning versus face-to-face learning of clinical skills in undergraduate nurse education. J. Adv. Nurs. **71**(2), 255–270 (2015)
21. Nazarenko, A.L.: Blended learning vs traditional learning: what works? (A case study research). Procedia Soc. Behav. Sci. **200**, 77–82 (2015)
22. Zaghoul, F.A., Rababah, O., Fakhouri, H.: Website search engine optimization: geographical and cultural point of view. In: 2014 UKSim-AMSS 16th International Conference on Computer Modelling and Simulation, pp. 452–455. IEEE, March 2014

Improved Crypto Algorithm for High-Speed Internet of Things (IoT) Applications

Mohammed Gouse Galety[1]([✉]), Chiai Al Atroshi[2], N. Arul Kumar[3], and Saravanan[4]

[1] Department of Computer Network, Lebanese French University,
Erbil, KR, Iraq
galety.143@gmail.com
[2] Department of Information Technology, Lebanese French University,
Erbil, KR, Iraq
[3] Department of Computer Science, Christ (Deemed to be University),
Bangalore, India
[4] CMR Institute of Technology, Bangalore, India

Abstract. Modern technologies focus on integrated systems based on the Internet of Things (IoT). IoT based devices are unified with various levels of high-speed internet communication, computation process, secure authentication and privacy policies. One of the significant demands of present IoT is focused on its secure high-speed communication. However, traditional authentication and secure communication find it very difficult to manage the current need for IoT applications. Therefore, the need for such a reliable high-speed IoT scheme must be addressed. This proposed title introduces an enhanced version of the Rijndael Cryptographic Algorithm (Advanced Encryption Standard – AES) to obtain fast-speed IoT-based application transmission. Pipeline-based AES technique promises for the high-speed crypto process, and this secure algorithm targeted to fast-speed Field Programmable Gate Array (FPGA) hardware. Thus, high-speed AES crypto algorithms, along with FPGA hardware, will improve the efficiency of future IoT design. Our proposed method also shows the tradeoff between High-Speed communications along with various FPGA platforms.

Keywords: Crypto algorithm · Advanced Encryption Standard · Pipeline stages · High-speed hardware · Field Programmable Gate Array

1 Introduction

Modern systems and technologies are seeking for the fast-speed Internet of Things (IoT) based secure design. Special care needs to be taken for data security in High-speed communication channel, data encryption/decryption and in real time secure applications. Secure information needs to be prevented more from malicious attacks. These attacks will breakdown the devices physically and collapse the internet communication. Finally, it leads to malfunction and unauthorized access to the IoT system. Earlier systems were adopted by the software-based security method. Due to more malicious attack, pure software-based security systems are not able to full fill the need for secure systems. So, there is an essential requirement for hardware-based secure

© Springer Nature Switzerland AG 2020
L. C. Jain et al. (Eds.): ICICCT 2019, LAIS 9, pp. 281–289, 2020.
https://doi.org/10.1007/978-3-030-38501-9_28

systems, which will be better to protect secure data from various malicious attack. Thus, this paper focuses on hardware-based secure systems with an improved mechanism in Rijndael cryptographic algorithm (Advanced Encryption Standard – AES) to achieve High-Speed, secure communication. Our proposed method also targeted to fast-speed Field Programmable Gate Array (FPGA) based hardware design.

2   Literature Review

This section provides the literature survey done by various researchers in the research arena of AES security algorithms and its hardware design. In paper [1, 6], provides the AES algorithm's easy features along with its specification. Federal Information Processing Standard (FIPS) provides the standard for the AES algorithm. FPGA based AES was proposed in paper [2, 4, 5] for embedded systems-based design. This paper also shows the efficient implementation of AES in FPGA hardware. In paper [3, 17–19], it shows how AES has targeted to VLSI based hardware with high-speed design. Pipeline-based AES [8, 9, 20, 21] was considered with various techniques like unrolling, tiling etc. It also shows how a pipeline processor on FPGA hardware.

In paper [10], a new strengthened view of AES was proposed along with new logical topology. In paper [11, 12, 15], S-box was considered with higher order mask scheme and also showed its optimized design. This paper also targets AES in non-volatile based FPGA. In paper [13, 14], reconfigurable hardware was focused on cypher security with low power, high throughput and low-cost VLSI design. Lightweight AES algorithm was proposed in paper [16], which gives an optimal approach to AES method. Thus, our proposed method observes the need for a hardware-based secure algorithm to achieve more efficiency.

3   Block Cypher-Based Modes of Operation

Rijndael Cryptographic Algorithm's mode operation is classified into five modes. They are listed below.

Electronic Codebook Mode (ECB). This ECB mode converts primary input of plaintext into a single output of cipher text for every given block [19, 20]. To transform shorter texts into that block, its same key has been used. For longer message, it needs to break down into various small blocks. If there is a need for message bit, padding can be used. Thus, this mode is well suitable for a small amount of data [17, 18].

Cipher Block Chaining (CBC) Mode. Separate text blocks of cipher are generated from the same plain text blocks as required by the user. XORing logic is used for cypher block chaining, which is functioning with previous rounds along with the given key [21, 22].

Cipher Feedback (CFB) Mode. This CFB mode is used to convert block cypher into various stream cypher. Padding concept is also used depends upon the need of last message bit. Left most of the bits are handled together with the 1st segment of plain text with XORing activities. Shift register concept is also used to convert plain text into cypher text [23, 24].

Output Feedback (OFB) Mode. This operation of OFB mode is comparable to the operation of CFB mode. It eliminates the generation of plain text to cypher text in block-based adoption. Usage of internal feedback is common for both plain text and ciphertext [25, 26].

Counter (CTR) Mode. In this mode counter concept is used for different plain text. In the encryption process, XORing process is applied to plain text to produce ciphertext block without changing the original information. Simple in design, efficient and more secure are various advantages of this mode operation [27].

Our proposed method uses Cipher Block Chaining (CBC) Mode based Pipeline-AES algorithm to prove High-speed, secure communication.

4 Principle of Rijndael Cryptographic Algorithm

Rijndael cryptographic algorithm [1, 2] (AES) is one of the efficient block cypher, which is designed to encrypt and decrypt the given data. It includes an original key length variable size and collection of information length. For the general usage, The length of the AES group is perceived as 128-bit information blocks, and the length of the key is usually displayed at 128 bits, 192 bits and 256 bits.

Plain text and its necessary key are given in the original encryption phase. Due to the input information of 128 bytes, it is structured and manipulated as 4×4-byte matrices. Next stage is the usage of rounds and its transformations. Depends upon the different length of the key size it is considered the AES round function. For an instance, when the size of the key = 128 bit, then it needs to round for ten times. Thus for 192 bit, round count is 12, and for 256 bit, round count is 14 [28].

This paper considers 128 bit as key length size. In an initial round, the key addition concept is considered, and it is also known as AddRoundKey. Next, it is considered for nine times of round transformations and also known as Round. In the final stage round, it is known as FinalRound. Over all-around transformation is a combination of four main layers. From the layer one, byte substitution will take place (SubBytes). It is considered as S box along with the input size of 8 bit and output of 8 bit to form a state matrix. In the second layer, ShiftRows is considered, and in the third layer, MixColumns is considered. This will implement a 4×4 matrix process into line shift operation and mixed column operation. In the last layer key addition concept is used. This is also identified as AddRoundKey, meaning that every byte of key and also its respective state matrix byte performs XOR logical procedure. Figure 1 Illustrates the fundamental schematic diagram of a mechanism of AES encryption as well as decryption.

4.1   Pseudo-code: Encryption

Inputs and Outputs: Given Plaintext and Output ciphertext stored in one 128-bit register Re

```
128-bit round keys Rk 0, ... , Rk10
        Re = Re ⊕ Rk0;
        For i = 1 to 10 Do:
        {
        Re = Sub8(Re);
        Re = Per8(Re);
        If (i < 10) then Re = MixCol(Re);
        Re = Re ⊕ Rki;
        }
```

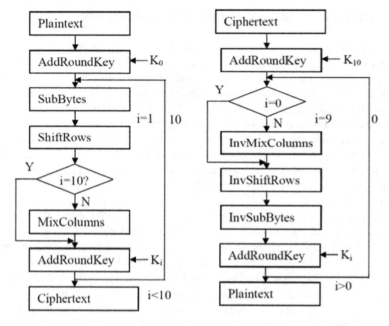

Fig. 1. AES encryption as well as decryption schematic diagram

4.2 Pseudo-code: Decryption

Inputs and Outputs: Given Ciphertext and Output Plaintext stored in one 128-bit register Re

```
128-bit round keys Rk0, ... , Rk10
        Re = Re ⊕ rk10;
        For i = 9 to 0 Do:
        {
        If (i >0) then Re = IMixCol(R);
        Re = IPer8(Re);
        Re = ISub8(Re);
        Re = Re ⊕ Rki;
        }
```

The above terms are given as,

```
Re-Registers
Rk1...Rk10 - Round keys
MixCol - MixColumn function
IMixCol – Inverse of mix column
Per8(Re) and IPer8(Re) - Byte of permutation and its
inverse
Sub8(Re) and ISub8(Re) - Byte of substitution and its
inverse
```

5 Proposed FPGA Rijndael Algorithm Implementation

The Pipeline-AES hardware implementation of Field Programmable Gate Array (FPGA) is suggested in this article. A hardware implementation design in each phase can obtain enhanced performance of this algorithm. The acceptance of a pipeline-based idea in the AES safety algorithm will create byte replacement, column mixing conversion and main expansion activities more effective. Pipeline-based AES encryption is presented in the following Fig. 2 and the decryption process is shown in the Fig. 3.

Pipeline encryption contains various stages like input multiplexing, storage, byte substitute and row and column mix along with key addition. In the pipeline, decryption is shown in the inverse encryption method processes like input multiplexing, storage, inverse column mix, inverse byte substitute and key addition.

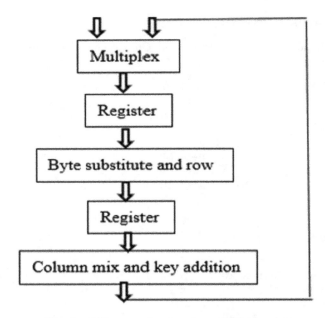

Fig. 2. AES encryption based on pipeline model

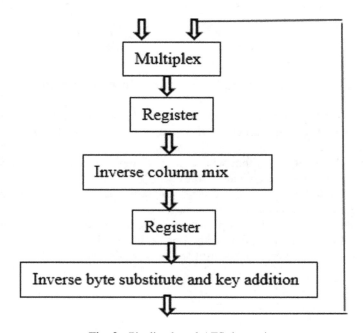

Fig. 3. Pipeline-based AES decryption

6 Experimental Results

Proposed pipeline-based AES algorithm is analyzed and verified with Modelsim hardware simulation tool. Cipher Block Chaining (CBC) Mode is considered in our proposed method with the data size of 128 bits. Two-stage pipeline concepts were considered in this proposed design. It is developed with VHDL [7] code and targeted to XILINX FPGA design. Our proposed pipeline-based approach is compared with the earlier existing method [2, 5, 20, 21], which was targeted to Spartan and Virtex based FPGA. Experimental results show the comparison of basic AES sequential core and Pipeline based AES core with earlier methods. Table 1 shows significant improvement in throughput value than other existing methods.

Table 1. Similarity of the implemented approach with an present approach

Method	FPGA device	Slice	BRAM	Throughput
T. Good method [2]	Spartan XC3S100E	124	2	–
G. Rouvroy method [5]	Spartan XC3S100E	146	3	358 Mbps
A. M. Borkar method [20]	Virtex XCV600	1853	–	0.352 Gbps
I. Algredo Badillo method [21]	Virtex4 XC4VLX40	1200	–	1.95 Gbps
Proposed Basic Sequential Core	Virtex4 XC4VLX100	1301	18	1.05 Gbps
Proposed Pipeline based Core	Virtex4 XC4VLX100	7873	45	10.74 Gbps

7 Conclusion

This paper focuses on a hardware-based high-speed, secure algorithm to fulfil the need of security-based IoT application. Pipeline-based AES crypto algorithm is proposed in this research article to improve the high-speed communications. XILINX FPGA based hardware is considered as the hardware target device. The hardware implementation indicates that the methodology presented is more effective than current methodologies that are in available. With sensible equipment excess, improved performance is accomplished. This paper also shows the tradeoff between different FPGA hardware and its throughput values. Thus, this method gives more comfort to high-speed IoT applications by providing improved high-speed AES in FPGA based hardware platform.

References

1. Daemen, J., Rijmen, V.: The Design of Rijndael: AES the Advanced Encryption Standard, 1st edn. Springer, Heidelberg (2002)
2. Good, T., Benaissa, M.: AES on FPGA from the fastest to the smallest. In: Cryptographic Hardware and Embedded Systems - CHES 2005, pp. 427–440 (2005)
3. Zhang, X., Parhi, K.: High-speed VLSI architectures for the AES algorithm. IEEE Trans. Very Large Scale Integr. (VLSI) Syst. **12**(9), 957–967 (2004)

4. Kaur, S., Vig, R.: Efficient implementation of AES algorithm in FPGA device. In: International Conference on Computational Intelligence & Multimedia Applications (2007)

5. Rouvroy, G., Standaert, F.X., Quisquater, J.J., Legat, J.D.: Compact and efficient encryption/decryption module for FPGA implementation of the AES Rijndael very well suited for small embedded applications. In: ITCC 2004, pp. 583–587 (2004)

6. FIPS 197, Advanced Encryption Standard (AES) (2001)

7. Saggese, G.P., Mazzeo, A., Mazocca, N., Strollo, A.G.M.: An FPGA based performance analysis of the unrolling, tiling and pipelining of the AES algorithm. In: FPL (2003)

8. Hodjat, A., Verbauwhede, I.: A 21.54 Gbits fully pipelined processor on FPGA. In: IEEE Symposium on Field-Programmable Custom Computing Machines, pp. 308–309 (2004)

9. Rashtchi, V., Mosavi, S.H.: Strengthened of AES encryption algorithm within the new logic topology. Mail. J. Electr. Eng. **12**(1), 87–94 (2018)

10. Carlet, C., Goubin, L., Prouff, E., Quisquater, M., Rivain, M.: Higher order masking schemes for S-boxes. LNCS, vol. 7549, pp. 366–384 (2012)

11. Ye, Y., Wu, N., Zhang, X., Dong, L., Zhou, F.: An optimized design for compact masked AES S-Box based on composite field and common subexpression elimination algorithm. J. Circ. Syst. Comput. **27**(11), 1850171 (2018)

12. Sakthivel, R., Vanitha, M., Kittur, H.M.: Low power high throughput reconfigurable stream cypher hardware VLSI architectures. Int. J. Inf. Comput. Secur. **6**(1), 1–11 (2014)

13. Abdellatif, K.M., Chotin-Avot, R., Mehrez, H.: Low-cost solutions for secure remote reconfiguration of FPGAs. Int. J. Embed. Syst. **6**(2/3), 257–265 (2014)

14. Gaspar, L., Drutarovsky, M., Fischer, V., Bochard, N.: Efficient AES S-boxes implementation for non-volatile FPGAS. IEEE Transaction Paper, pp. 649–653 (2009)

15. Wong, M.M., Wong, M., Zhang, C., Hijazin, I.: Circuit and system design for optimal lightweight AES encryption on FPGA. IAENG Int. J. Comput. Sci. **45**(1), 52–62 (2018)

16. Kshirsagar, R.V., Vyawahare, M.V.: FPGA implementation of high-speed VLSI architectures for AES algorithm. In: IEEE Conference on Emerging Trends in Engineering and Technology (2012)

17. Chakrabarti, S., Samanta, D.: Image steganography using priority-based neural network and pyramid. Emerg. Res. Comput. Inf. Commun. Appl., 163–172 (2016). https://doi.org/10.1007/978-981-10-0287-8_15

18. Ghosh, G., Samanta, D., Paul, M.: Approach of message communication based on twisty "Zig-Zag". In: 2016 International Conference on Emerging Technological Trends (ICETT) (2016). https://doi.org/10.1109/icett.2016.7873676

19. Hossain, M.A., Samanta, D., Sanyal, G.: Extraction of panic expression depending on lip detection. In: 2012 International Conference on Computing Sciences (2012). https://doi.org/10.1109/iccs.2012.35

20. Hossain, M.A., Samanta, D., Sanyal, G.: Statistical approach for extraction of panic expression. In: 2012 Fourth International Conference on Computational Intelligence and Communication Networks (2012). https://doi.org/10.1109/cicn.2012.189

21. Khadri, S.K.A., Samanta, D., Paul, M.: Approach of message communication using fibonacci series. In: Cryptology. Lecture Notes on Information Theory (2014). https://doi.org/10.12720/lnit.2.2.168-171

22. Choi, S.-K., Ko, J.-S., Kwak, J.: A study on IoT device authentication protocol for high speed and lightweight. In: Proceedings of the International Conference on Platform Technology and Service, PlatCon 2019 (2019). https://doi.org/10.1109/PlatCon.2019.8669418

23. Jindal, F., Mudgal, S., Choudhari, V., Churi, P.P.: Emerging trends in Internet of Things. Inf. Technol. Trends Emerg. Technol. Artif. Intell., 50–60 (2019). https://doi.org/10.1109/CTIT.2018.8649535

24. Borkar, A.M., Kshirsagar, R., Vyawahare, M.: FPGA implementation of AES algorithm. In: 3rd International Conference on Electronics Computer Technology (ICECT), vol. 3, pp. 401–405 (2011)
25. Algredo-Badillo, I., Feregrino-Uribe, C., Cumplido, R., Morales-Sandoval, M.: FPGA implementation and performance evaluation of AES-CCM cores for wireless networks. In: International Conference on Reconfigurable Computing and FPGAs, pp. 421–426 (2008)
26. Saravanan, S., Hailu, M., Gouse, G.M., Lavanya, M., Vijaysai, R.: Optimized secure scan flip flop to thwart side channel attack in crypto-chip. In: Zimale, F., Enku Nigussie, T., Fanta, S. (eds.) Advances of Science and Technology, ICAST 2018. Lecture Notes of the Institute for Computer Sciences, Social Informatics and Telecommunications Engineering, vol. 274. Springer, Cham (2019)
27. Saravanan, S., Hailu, M., Gouse, G.M., Lavanya, M., Vijaysai, R.: Design and analysis of low-transition address generator. In: Zimale, F., Enku Nigussie, T., Fanta, S. (eds.) Advances of Science and Technology, ICAST 2018. Lecture Notes of the Institute for Computer Sciences, Social Informatics and Telecommunications Engineering, vol. 274. Springer, Cham (2019)
28. Mohammed Gouse, G., Haji, C.M., Saravanan: Improved reconfigurable based lightweight crypto algorithms for IoT based applications. J. Adv. Res. Dyn. Control Syst. **10**(12), 186–193 (2018)

Credit Card Fraud Detection: A Systematic Review

C. Victoria Priscilla[1]([✉]) and D. Padma Prabha[2]

[1] Department of Computer Science, SDNB Vaishnav College for Women,
University of Madras, Chennai, India
aprofvictoria@gmail.com
[2] Department of Computer Applications, Madras Christian College,
University of Madras, Chennai, India
padmaj2000@gmail.com

Abstract. Due to the tremendous growth of technology, digitalization has become the key aspect in the banking sector. As online transaction increases, the fraud rate grows simultaneously. Even though many techniques are available to identify the fraudulent transaction, the fraudsters adapt their own paradigm. This review intends to present the research studies accomplished on Credit Card Fraud Detection (CCFD) by highlighting the challenge of class imbalance and the various Machine Learning techniques, it also extends the efficient evaluation metrics particularly for CCFD. As the dataset is more sensitive and less available we have outlined the web sources of available datasets and trending software tools used in the deployment of CCFD.

Keywords: Fraud detection · Class imbalance · Machine learning · Evaluation metrics

1 Introduction

Cashless transactions are the most popular method used to purchase products through e-commerce. Credit Cards are commonly used in online and physical stores. The Nilson survey projected the growth of purchase transactions worldwide for the six regions and reported "Among the global brand cards, those issued in the Asia-Pacific region generated 102.50 billion purchase transactions in 2017, accounting for 34.68% of the world wide purchase transaction total. This surpassed the United States for the first time. The Asia-Pacific is projected to reach 54.03% of the worldwide total by 2027" [1]. As the Card transaction increases the rate of fraudsters will also grow concurrently. Credit Card Fraud Detection (CCFD) is a challenging research undergone by the research community as the fraudsters change their behavioral pattern now and then which becomes an alarm for the banks to set a solution. The fraud damage is found after the fraudulent transaction is completed. Forecasting the fraud before it happens helps to prevent loss.

As the fraudsters increase, the technology for solving the problem plays a vital role. Any irregular pattern in the purchase transaction is considered as doubtful and further action is taken [2]. Machine Learning has various algorithms to overcome CCFD as it is

© Springer Nature Switzerland AG 2020
L. C. Jain et al. (Eds.): ICICCT 2019, LAIS 9, pp. 290–303, 2020.
https://doi.org/10.1007/978-3-030-38501-9_29

a binary classification problem, having the majority of normal transactions and the minority of fraudulent transactions. As the dataset is highly imbalanced when applying machine learning directly to the dataset, prediction inclines towards the majority class [3]. The important challenges faced by CCFD are the class imbalance - The nature of the data have less fraudulent behaviors than the legitimate [4], Concept drift - is due to the spending behavioral changes of the cardholders due to different circumstances that are classified into two; active and passive adaptation [5] finally misclassification of data – The legitimate transactions may be considered as fraudulent and inversely, therefore causes false alarm rate [6, 7] which create inconvenience to the cardholders.

The remainder of this paper is structured as follows. The following Sect. 2 addresses about the main challenge of class imbalance, Sect. 3 describes the CCFD techniques by different machine learning approach, Sect. 4 discusses the measures and metrics evaluated in CCFD of other research papers, Sect. 5 explains the Dataset and Tools used by other researchers, and Sect. 6 concludes with observations.

2 Class Imbalance

One of the challenging phases in CCFD is to train the classifier as the data have a minimum ratio of fraud, may lead to misclassification of data.

Preprocessing the data will balance and reduce the noise between the binary classes [8]. Therefore original dataset is balanced by performing oversampling to increase the fraudulent class or undersampling to remove the legitimate class [9], studies reveal that the performance of the classifier is better when oversampling technique SMOTE (synthetic minority oversampling technique) is used [3] to handle class imbalance. The two metrics; Balanced Classification Rate (BCR) and Matthews Correlation Coefficient (MCC) was applied and found FraudMiner performs well in handling class imbalance when compared with other classifiers. In [10] SMOTE increased the accuracy of the classifier by 2–4%. Though adaptive synthetic sampling (ADASYN), ranked minority sampling (RAMO) were recently suggested [11] but lead to a problem of class distribution when the number of iteration is high, hence the researchers suggest ensemble classifiers will have better results than a single classifier with unbalanced data.

Akila and Reddy [12] proposed an Overlapped-Majority Bagging method to handle class imbalance by selecting 60% data of minority class into a single bag with few majority data, without creating an internal imbalance in the bag. Awoyemi et al. [13] approached a hybrid sampling technique where the genuine class is undersampled and the fraud class is oversampled with the same number of iterations. Wang and Han [14] expressed an integrated clustering technique by taking random samples to improve the classification of imbalanced data where the SVM model made higher accuracy of minority and majority class [14]. Van Vlasselaer et al. [15] applied weighting method for neural networks and logistic regression and subsampling were used in random forests. The disadvantage of SMOTE is presented in [3] as it produces noise and chance of overlapping among classes and create overfitting and revealed undersampling is better than oversampling. Jurgovsky et al. [16] built an account based undersampling technique by randomly picking a legitimate account from the set of a compromised account by repeating the process to create training dataset. CLUE [4] employed undersampling for legitimate sessions by random skipping. The undersampling method effectively handle class imbalance problem [17–20]. Carneiro et al. [21] inferred that balanced dataset using undersampling majority class did not show a remarkable performance over unbalanced dataset when applied into a classifier, most of the classifiers do not classify the minority class in the dataset as the classifiers are biased by expressing the majority class.

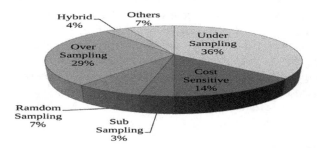

Fig. 1. Frequency of sampling techniques used in the literature.

Carneiro et al. [4] and Wang et al. [21] experimented random sampling to handle the data imbalance. Decision tree algorithm reduces the variance and overfitting of the dataset by adjusting the trees [22]. Clustering tree model [23] filters the majority class data based on the characteristic of the dataset considering class separable. Cost-sensitive methods solve the misclassification of data and made the classifier to consider the minority class [24–27].

3 Machine Learning Techniques of CCFD

The present techniques are based on supervised or unsupervised learning where attributes of the previous transaction are used in the training phase in supervised learning while unsupervised methods use clustering algorithms to group the customer behavior into fraud and legitimate patterns [28]. New algorithms are proposed by the researchers and some introduced hybrid techniques by combining the existing methods. An ensemble learning approach generated better performance when compared with other methods. The recent papers concentrate on Deep learning where the data should be huge in size.

3.1 Supervised and Unsupervised Learning

Supervised learning has labeled dataset to train the classifier. The samples of fraud and legal database are associated with their labels whereas unsupervised learning does not rely on labeled data to construct a model and does not need previous knowledge to train the model [6, 7]. Unsupervised learning extracts the significant features for data, [29] leveraging the accessibility of unlabeled data and adds dependency between data for training. Researchers proposed new methods and compared with the existing methods and found the classifier producing a good result. Few papers suggested a hybrid approach of the existing classifiers with a comparison report. The following are the various techniques proposed by the researchers using supervised and unsupervised learning algorithms.

Seeja and Zareapoor [30] proposed Fraud Miner using frequent itemset mining to identify legal and fraud transaction pattern of the customer in training stage by applying Apriori algorithm which returns the frequent itemset with more number of occurrences of legal and fraud patterns then proposed a matching algorithm during the testing phase to traverse the pattern database and match the new incoming transaction to detect fraud or legal. Hegazy et al. [31] created Frequent Pattern based on customer's previous transaction activity as Legal or Fraud transactions introducing using Rough Set and Decision Tree Technique clustering algorithm in Enhanced Fraud Miner algorithm which attained good improvement in finding the false alarm rate when compared to other models. de Sá et al. [24] proposed Fraud-BNC to generate Hyper-Heuristic Evolutionary Algorithm (HHEA) for dataset classification and it discovers the legitimate transaction when the class probability is less than 0.5 when compared with other methods the chance of missing a fraud class can be reduced by increasing the probability of fraud transaction. Van Vlasselaer et al. [15] suggested APATE (Anomaly Prevention using Advanced Transaction Exploration) used the RFM framework (Recency - Frequency – Monetary Value) based on the geographic location of the customer and achieved high AUC score and accuracy particularly for random forest method which exactly chooses the first transaction cycle of fraudulent.

Dai et al. [32] combined supervised and unsupervised approach by fusing various models to train and record the spending behavior for each cardholder based on their previous transactions and for every new transaction the fraud score is computed from the fraud pattern. Batani [33] designed a hybrid and adaptive method to detect fraud using cardholder's financial status, social status, and OTP by assigning weights using Artificial Neural Networks to produce card holder's social status and Hidden Markov Model (HMM) to extract financial profile from bank database. Behera and Panigrahi [34] combined fuzzy clustering and neural network where Fuzzy c-means clustering model groups the related dataset and neural network decrease the misclassification rate based on the parameters purchase amount, type and time. Jain et al. [35] presented a hybrid method to detect fraud by Rough Set and Decision Tree Technique where the mathematical tool Rough Set used to preprocess the data, a decision tree is built and applied to each row in the database and classification is acquired. Kamaruddin and Ravi [36] proposed a combined model of particle swarm optimization (PSO) and auto-associative neural network (AANN) to find an output on one-class classification (OCC) in the big data model using SPARK cluster and attained 89% true classification.

Santos and Ocampo [37] suggested Naïve Bayes method with clustering technique and produced accuracy by 81% using 3 clusters and an average accuracy of 79% with 2 clusters, recommended increasing the number of clusters to give the finest initialization of EM clustering model reporting the behavioral pattern of cardholders and fraudsters. The accuracy is reliant with the huge transaction of the fraudsters with the implication of high True Positive Rate (TPR) and overall accuracy. Hassan [38] conducted research by assigning different costs for false negative (FN) using the cost matrix and applied various cost-sensitive classifiers and investigated the performance by applying Bayes minimum risk. Yee et al. [39] tested the metrics for five Bayesian classifiers and reported that Bayesian classifiers are notably better after applying the Principle Component Analysis (PCA) method for the data to discard irrelevant attributes. Awoyemi et al. [13] compared Naïve Bayes, K-nearest neighbor, and Logistic regression models and concluded that K-NN performs well. Nur-E-Arefin and Islam [40] compared 16 classifiers to detect credit card fraud and reported that Multi-Class Classifier performs better.

Tran et al. [41] presented two real-time data-driven approaches, one class support vector machine (OCSVM) and T2 control chart which attained good accuracy and low false positives. Askari and Hussain [42] initiated Fuzzy-ID3 classifier to split the attributes possessing the highest information gain and the leaf nodes categorize the transactions as fraud or legitimate. Artikis et al. [43] introduced a prototype model SPEEDD for proactive event-driven decision-making for creating automated fraud pattern using the parameters country, merchant, amount and customer, applied OLED (Online Learning of Event Definitions) a statistical model to learn the pattern for a small set of data.

3.2   Ensemble Learning

Ensemble learning is supervised learning where many classifiers are combined and applied to make predictions. As there is an increase in computational power to train the classifier in an appropriate time frame can be a good method for CCFD [44].

Saia and Carta [9] proposed a novel method using Fourier transform and Wavelet transform to analyze and experiment the data by Random Forests method with k-fold cross-validation and compared with ten Data mining algorithms and concluded Random Forests as the best performing method. Dal Pozzolo et al. [8] compared Random Forests (RF) with Neural Network (NN) and Support Vector Machine (SVM) where the Random Forests performed well as expected and suggested the accuracy can be improved by increasing the training data size. Carneiro et al. [21] applied 10-fold cross-validation to Random Forests, SVM and Logistic regression and tested with balanced and unbalanced data then confirmed that Random Forests attained the best performance. Two techniques of Random-tree-based random forest and CART are applied [17] to train the dataset for legal and fraud pattern then compared their performance, voting mechanism inferred that both have equal weight. Noghani and Moattar [11] proposed an ensemble classification with cost-sensitive decision trees and increased the performance by 1.8% to 2.4%. Nami and Shajari [25] forwarded two stage novel approach using KNN and Dynamic Random Forest (DRF) by allotting weight to recent transactions can increase the performance of the system; Tree ensemble learns the pattern from both legal and fraud class based on their hierarchical structure which is more popular in solving problems having unbalanced data. Patil et al. [45] designed a framework for the fraud detection and prediction and concluded Random Forest Decision Tree model is best and the only difficulty is overfitting when data size expands. Wang et al. [3] proposed ensemble framework partitioning and hierarchical clustering (RFPH) with partitioning by Random Forest and clustering by C4.5. In [46] three classifiers are compared and pointed random forest performs well than the logistic regression and decision tree.

Zhang et al. [47] proposed stacking ensemble learning for each set of data to train different learner, the result produced by the first learner is considered as features to train meta-learner. Zareapoor et al. [22] compared bagging ensemble classifier with Naive Bayes, K-Nearest Neighbor, Support Vector Machines (SVM), and Ensemble classifier outperforms its competitors. Wang and Han [14] built an ensemble AdaBoost integrated learning algorithm having SVM as the base classifier and used cluster analysis for imbalanced dataset. Su et al. [48] implemented 10-fold cross-validation and suggested XGBoost as the best model, the output of several ensemble models revealed AUC over 0.9000 in MCC (Merchant Category Code) misuse. Kumari and Mishra [10] concluded that ensemble classifier performs well than any other classifier.

3.3 Deep Learning

Deep learning is one of the class in machine learning [49] which is based on artificial neural networks, In deep learning, the output of each level transformed to be an input data into a more abstract and composite representation.

Fu et al. [27] recommended Convolutional Neural Networks (CNN) framework to identify the fraud behavioral pattern and compared with SVM, NN, RF which exhibited CNN performs better than others. Jurgovsky et al. [16] proposed LSTM (Long Short-Term Memory) to aggregate previous purchase pattern of the cardholder and to improve the accuracy of the incoming transaction, compared sequence learner LSTM and static learner (Random Forest) where LSTM is prone to overfitting even with few nodes, hence suggested to increase the size of data. CLUE [4] was embedded by two deep learning methods Item2Vec and RNN including LSTM as default, the learning is based on the details of items browsed, and the time domain of browsing behavior. CLUE performs better than traditional methods. Roy et al. analyzed four deep learning method ANN, RNN, LSTM and Gated Recurrent Units (GRUs) using 10-fold cross-validation and the performance of GRU is considered as best with the accuracy of 0.916 and concluded ANN was the worst with accuracy score 0.889 [19]. Kim et al. [26] compared and confirmed Hybrid Ensemble with deep feed-forward Discriminative models to build multiple layers of abstraction, experimented with deep neural network which produced better performance. The 10-layer deep Variational Auto-Encoder (VAE) was applied in [51] and compared with Decision Tree, SVM and Ensemble Classifier (AdaBoost algorithm) where AdaBoost achieved high Precision and recall for VAE. Pumsirirat and Yan [29] presented Auto-Encoder and Restricted Boltzmann and confirmed supervised learning is appropriate for the historical transaction in credit card fraud detection. Rushin et al. [18] compared Deep Learning, Logistic Regression, and Gradient Boosted Tree, based on the AUC values of the feature set concluded deep learning is the best performer and LR (Logistic Regression) is the poor performer. Niimi [52] compared Deep learning with other supervised algorithms and confirmed the accuracy produced by Deep learning is close to Gaussian kernel SVM.

Table 1. Usage frequency of machine learning techniques in CCFD.

Machine learning	Methods	Usage frequency	References	Advantage	Disadvantage
Supervised Learning	Neural Network	10	[8, 9, 15, 17, 20, 27, 32–34, 38]	Highly accurate and reliable	Need to understand and label the input More computation time required for the training phase
	Support Vector Machine	16	[4, 8, 14, 17, 20–24, 27, 30, 38, 41, 48, 52, 53]		
	Bayesian Network Classifiers	15	[4, 9–11, 13, 17, 22, 24, 30, 32, 33, 37–40]		
	K-Nearest Neighbor	5	[9, 13, 22, 25, 30]		
	Logistic Regression	14	[4, 9, 13, 15, 20, 21, 23, 24, 26, 32, 38, 45, 46, 53]		
	Decision Tree	12	[3, 9, 11, 17, 20, 23, 26, 32, 38, 45, 46, 48]		
Unsupervised Learning	Expectation-Maximization	1	[37]	Easy to find unknown patterns and features of data	Computationally complex Less accurate due to unlabeled input
	K-Means	3	[22, 37]		
	Fuzzy C-Means	1	[34]		
	DBSCAN	1	[32]		
	Hidden Markov Model (HMM)	3	[2, 32, 33]		
	Self-Organizing Map (SOM)	1	[32]		
	LINGO	1	[31]		
Ensemble Learning	Random Forest	20	[3, 4, 8–10, 15–17, 20, 21, 23]–[25, 27, 30, 38, 45, 46, 48, 53]	Avoid the overfitting problem and gives better predictions when compared with a single model	Computation time is high Reduces model interpretability due to increased complexity
	Boosting	7	[9, 18, 20, 23, 48, 52, 53]		
	Bagging	3	[10, 12, 22]		
	Voting	1	[10]		
Deep Learning	Stochastic Gradient Descent	1	[9]	No need for feature extraction and labeling of data	A large amount of data is needed to find the pattern Create overfitting problem in the model
	Long short-term memory	3	[4, 16, 19]		
	Deep Feed Forward NN	3	[18, 26, 51]		
	Variational Autoencoder (VAE)	1	[51]		
	Auto Encoder (AE)	2	[29, 41]		
	Restricted Boltzmann Machines	1	[29]		
	Recurrent Neural Network	2	[4, 19]		
	Convolutional Neural Network	1	[27]		
	Generative Adversarial Network	1	[50]		

4   Classification Metrics for CCFD

Evaluation metrics identify the performance of the classifier when applied to the dataset. A 2×2 Confusion matrix [53] gives clarity by comparing the predicted and actual class. Based on the survey we found 20 evaluation metrics out of 41 papers. Then we summarized 12 metrics which are most commonly used by the researchers.

True Positive Rate (TPR). TPR is otherwise called as Sensitivity or Recall [54] is the number of actual positives which are predicted positive, in CCFD the actual positives are fraudulent transactions.

$$TPR = \frac{TP}{TP + FN} \tag{1}$$

True Negative Rate (TNR). Also called specificity, it is the proportion of actual negatives which are predicted negative, recognize how well the classifier identifies the legitimate transactions.

$$TNR = \frac{TN}{TN + FP} \tag{2}$$

False Positive Rate (FPR). The proportion of legitimate samples that were wrongly predicted as fraud [55, 56].

$$FPR = \frac{FP}{FP + TN} \tag{3}$$

Accuracy (Acc). The proximity measure of exactly classified samples [57] to the total number of instances.

$$Acc = \frac{TN + TP}{TP + FP + FN + TN} \tag{4}$$

Precision. Precision or Positive Predictive Value represents the proportion of positive samples that were correctly classified to the total number of positive predicted samples [56].

$$Precision = \frac{TP}{TP + FP} \tag{5}$$

F-measure. Otherwise referred as F1-score, and it represents the harmonic mean of precision and recall. The value ranges from 0 to 1, if the value is high then F-measure indicates high classification performance [56].

$$\text{F-measure} = 2 \times \frac{\text{Precision} \times \text{Recall}}{\text{Precision} + \text{Recall}} \tag{6}$$

Area Under the Curve (AUC). Represents the capability of distinguishing between classes. When AUC is higher the capability of prediction is better, the value bounds between 0 and 1.

Receiver Operating Characteristics (ROC). The receiver operating characteristics curve is a two-dimensional graph in which the TPR represents the y-axis and FPR is the x-axis at various threshold settings [56].

Balanced Classification Rate (BCR). The average of sensitivity and specificity [30]

$$\text{BCR} = \frac{1}{2}(\text{TPR} + \text{TNR}) \tag{7}$$

Matthews Correlation Coefficient (MCC). Represents the correlation between the observed and predicted classifications. MCC value is between $[-1, 1]$, if the coefficient is $+1$ indicates a perfect prediction, -1 represents prediction is exactly wrong [56].

$$\text{MCC} = \frac{\text{TP} \times \text{TN} - \text{FP} \times \text{FN}}{\sqrt{(\text{TP} + \text{FP})(\text{TP} + \text{FN})(\text{TN} + \text{FP})(\text{TN} + \text{FN})}} \tag{8}$$

Geometric Mean (GM). The Geometric Mean (GM) aggregates sensitivity and specificity [56] measured at one specific threshold.

$$\text{GM} = \sqrt{\text{TPR} \times \text{TNR}} \tag{9}$$

Cohen's Kappa Coefficient (K). Measures the chance of an agreement between the predicted and the actual classes [53], given by

$$K = \frac{P_0 - P_e}{1 - P_e} \tag{10}$$

where P_0 is the observed agreement, and P_e is the expected agreement.

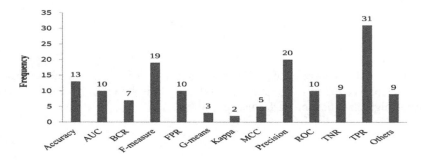

Fig. 2. Frequency of evaluation metrics used in the literature.

Precision is used to find the good classifier, [20] When the precision value is high it indicates the classifier is good. Accuracy [22, 30] being a common metric, it is not appropriate for CCFD due to data imbalance may lead to misclassification. The performance of the classifiers shows good results with preprocessed dataset than the raw dataset [46]. As Fig. 2 indicates that most of the papers have experimented Sensitivity (31), Precision (20), and F-measure (19) performance metrics to measure the efficiency of CCFD.

5   Dataset and Tools for CCFD

The publicly available dataset can be classified as Credit approval, Default payment, and Transactional dataset. The attribute class or default is the indicator of prediction with binary value 0(legitimate) and 1(fraud). Even in a very huge transactional dataset, the occurrence of fraud is very minimal which creates an imbalance problem. Most of the real data have a ratio of 99% of legal transaction and 1% of them are fraud [3]. Table 2 visualizes the experimental datasets used by the researchers in the field of CCFD. Few of the confidential real data is from the bank of China, Latin America, Singapore, South Korea, etc.

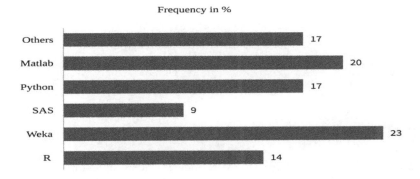

Fig. 3. Occurrence of software tools in the literature.

Figure 3 depicts Weka, Matlab, python are popularly used software tools to develop the CCFD system. 31% of the publication have not mentioned about the software used. When the research is based on big data analytics, there will be a huge volume of data which can be implemented in Apache Hadoop, Spark, etc. Tensorflow, H2O, Pytorch, Keras, etc. are the libraries imported in the application of deep learning.

Table 2. Dataset used for CCFD.

Dataset	Number of instances	Fraud instances	References	Source
UCSD-FICO DataMining Contest 2009	100,000	2293	[12, 22, 30, 38, 40]	https://www.cs.purdue.edu/commugrate/data/credit_card/
European – Credit card transaction	284, 807	492	[3, 9, 13, 23, 29, 41, 46, 50, 51]	https://www.kaggle.com/mlg-ulb/creditcardfraud
German– Credit	1000	300	[9, 10, 29, 35, 45]	UCI Machine Learning Repository
Taiwan– Default Credit Card	30000	6636	[20]	https://www.kaggle.com/uciml/default-of-credit-card-clients-dataset
Australian- Credit Approval	690	307	[10, 29, 52]	https://data.world/uci/credit-approval
cc Fraud	1048575	62739	[36]	https://packages.revolutionanalytics.com/datasets/
Confidential data from bank			[4, 8, 11, 12, 15, 17–19, 21, 23, 24–27, 42, 47, 48]	
Synthetic data			[33, 34, 43]	
Simulated data			[31, 32, 37, 39]	

6   Conclusion

The review concludes that Supervised Learning is widely used for CCFD than the unsupervised. We observed that the most commonly used fraud detection techniques are Support vector machine, Bayesian classifiers, Logistic Regression, and Decision Tree. It was also suggested that the performance of Hybrid methods are better than using a single classifier. Table 1 express that Ensemble learning showed good performance with Random Forest and the usage frequency is more when compared with other techniques. Although the recent researches are based on deep learning in predictive analytics, the performance when compared to supervised learning is still in the research process. Based on the study it was observed that Deep learning can perform well on image data than numeric. We observed that 55% of publications have used sampling method to handle the class imbalance and few have compared the performance using raw data and sampled data, where 45% have not reported about any sampling technique. From Fig. 1 it was noted that many of the research articles have used the undersampling technique. It was also spotted from Fig. 2 that Sensitivity, Precision, and F-measure have produced good results for the performance analysis in

CCFD and also suggested that the performance can be improved by increasing the size of the data.

In future cloud analytics becomes a popular analytical tool to extract information from massive data, implementing Machine learning with cloud or fog computing environment paves a way for better performance.

References

1. The Nilson Report 2019. https://nilsonreport.com/publication_newsletter_archive_issue. php?issue=1146. Accessed 03 June 2019
2. Prakash, A., Chandrasekar, C.: An optimized multiple semi-hidden Markov model for credit card fraud detection. Indian J. Sci. Technol. **8**(2), 176–182 (2015)
3. Wang, H., Zhu, P., Zou, X., Qin, S.: An ensemble learning framework for credit card fraud detection based on training set partitioning and clustering. In: IEEE SmartWorld, Ubiquitous Intelligence and Computing, Advanced and Trusted Computing, Scalable Computing and Communications, Cloud and Big Data Computing, Internet of People and Smart City Innovations, SmartWorld/UIC/ATC/ScalCom/CBDCom/IoP/SCI 2018, pp. 94–98 (2018)
4. Wang, S., Liu, C., Gao, X., Qu, H., Xu, W.: Session-based fraud detection in online e-commerce transactions using recurrent neural networks. In: Altun, Y., et al. (eds.) Machine Learning and Knowledge Discovery in Databases. Lecture Notes in Computer Science, pp. 241–252. Springer, Cham (2017)
5. Pozzolo, A.D., Boracchi, G., Caelen, O., Alippi, C., Bontempi, G.: Credit card fraud detection: a realistic modeling and a novel learning strategy. IEEE Trans. Neural Netw. Learn. Syst. **29**(8), 3784–3797 (2018)
6. Sourournejad, S., Zojaji, Z., Atani, R.E., Monadjemi, A.H.: A survey of credit card fraud detection techniques: data and technique oriented perspective (2016)
7. Abdallah, A., Maarof, M.A., Zainal, A.: Fraud detection system: a survey. J. Netw. Comput. Appl. **68**, 90–113 (2016)
8. Dal Pozzolo, A., Caelen, O., Le Borgne, Y.A., Waterschoot, S., Bontempi, G.: Learned lessons in credit card fraud detection from a practitioner perspective. Expert Syst. Appl. **41** (10), 4915–4928 (2014)
9. Saia, R., Carta, S.: Evaluating the benefits of using proactive transformed-domain-based techniques in fraud detection tasks. Future Gener. Comput. Syst. **93**, 18–32 (2019)
10. Kumari, P., Mishra, S.P.: Analysis of credit card fraud detection using fusion classifiers. In: Behera, H., Nayak, J., Naik, B., Abraham, A. (eds.) Computational Intelligence in Data Mining, vol. 711, pp. 111–122. Springer, Singapore (2019)
11. Noghani, F.F., Moattar, M.-H.: Ensemble classification and extended feature selection for credit card fraud detection. J. AI Data Min. **5**(2), 235–243 (2017)
12. Akila, S., Srinivasulu Reddy, U.: Cost-sensitive Risk Induced Bayesian Inference Bagging (RIBIB) for credit card fraud detection. J. Comput. Sci. **27**, 247–254 (2018)
13. Awoyemi, J.O., Adetunmbi, A.O., Oluwadare, S.A.: Credit card fraud detection using machine learning techniques: a comparative analysis. In: Proceedings IEEE International Conference Computing Networking Informatics, ICCNI 2017, January 2017, pp. 1–9 (2017)
14. Wang, C., Han, D.: Credit card fraud forecasting model based on clustering analysis and integrated support vector machine. Cluster Comput. **0123456789**, 1–6 (2018)
15. Van Vlasselaer, V., et al.: APATE: a novel approach for automated credit card transaction fraud detection using network-based extensions. Decis. Support Syst. **75**, 38–48 (2015)

16. Jurgovsky, J., et al.: Sequence classification for credit-card fraud detection. Expert Syst. Appl. **100**, 234–245 (2018)
17. Xuan, S., Liu, G., Li, Z., Zheng, L., Wang, S., Jiang, C.: Random forest for credit card fraud detection. In: ICNSC 2018 - 15th IEEE International Conference on Networking, Sensing and Control, pp. 1–6 (2018)
18. Rushin, G., Stancil, C., Sun, M., Adams, S., Beling, P.: Horse race analysis in credit card fraud—deep learning, logistic regression, and Gradient Boosted Tree. In: 2017 Systems and Information Engineering Design Symposium (SIEDS), pp. 117–121 (2017)
19. Roy, A., Sun, J., Mahoney, R., Alonzi, L., Adams, S., Beling, P.: Deep learning detecting fraud in credit card transactions. In: 2018 Systems and Information Engineering Design Symposium (SIEDS), pp. 129–134 (2018)
20. Akosa, J.: Predictive accuracy: a misleading performance measure for highly imbalanced data. In: Proceedings of the SAS Global Forum (2017)
21. Carneiro, N., Figueira, G., Costa, M.: A data mining based system for credit-card fraud detection in e-tail. Decis. Support Syst. **95**, 91–101 (2017)
22. Zareapoor, M., Shamsolmoali, P.: Application of credit card fraud detection: based on bagging ensemble classifier. Procedia Comput. Sci. **48**(C), 679–685 (2015)
23. Zhang, Y., Liu, G., Zheng, L., Yan, C., Jiang, C.: A novel method of processing class imbalance and its application in transaction fraud detection. In: 2018 IEEE/ACM 5th International Conference on Big Data Computing Applications and Technologies, vol. 1, pp. 152–159 (2018)
24. de Sá, A.G.C., Pereira, A.C.M., Pappa, G.L.: A customized classification algorithm for credit card fraud detection. Eng. Appl. Artif. Intell. **72**, 21–29 (2018)
25. Nami, S., Shajari, M.: Cost-sensitive payment card fraud detection based on dynamic random forest and k-nearest neighbors. Expert Syst. Appl. **110**, 381–392 (2018)
26. Kim, E., et al.: Champion-challenger analysis for credit card fraud detection: hybrid ensemble and deep learning. Expert Syst. Appl. **128**, 214–224 (2019)
27. Fu, K., Cheng, D., Tu, Y., Zhang, L.: Credit card fraud detection using convolutional neural networks. In: Hirose, A., Ozawa, S., Doya, K., Ikeda, K., Lee, M., Liu, D. (eds.) Neural Information Processing, pp. 483–490. Springer, Cham (2016)
28. Zareapoor, M., Yang, J.: A novel strategy for mining highly imbalanced data in credit card transactions. Intell. Autom. Soft Comput. 1–7 (2017)
29. Pumsirirat, A., Yan, L.: Credit card fraud detection using deep learning based on auto-encoder and restricted Boltzmann machine. Int. J. Adv. Comput. Sci. Appl. **9**(1), 18–25 (2018)
30. Seeja, K.R., Zareapoor, M.: FraudMiner: a novel credit card fraud detection model based on frequent itemset mining. Sci. World J. **2014**, 1–10 (2014)
31. Hegazy, M., Madian, A., Ragaie, M.: Enhanced fraud miner: credit card fraud detection using clustering data mining techniques. Egypt. Comput. Sci. **40**(03), 72–81 (2016)
32. Dai, Y., Yan, J., Tang, X., Zhao, H., Guo, M.: Online credit card fraud detection: a hybrid framework with big data technologies. In: Proceedings of 15th IEEE International Conference Trust Security and Privacy in Computing and Communication, 10th IEEE International Conference Big Data Science and Engineering, 14th IEEE International Symposium Parallel Distributed Processing, pp. 1644–1651 (2016)
33. Batani, J.: An adaptive and real-time fraud detection algorithm in online transactions. Int. J. Comput. Sci. Bus. Inform. **17**, 1–12 (2017)
34. Behera, T.K., Panigrahi, S.: Credit card fraud detection: a hybrid approach using fuzzy clustering & neural network. In: Proceedings of 2015 2nd IEEE International Conference on Advances in Computing and Communication Engineering, ICACCE 2015, pp. 494–499 (2015)

35. Jain, R., Gour, B., Dubey, S.: A hybrid approach for credit card fraud detection using rough set and decision tree technique. Int. J. Comput. Appl. **139**(10), 1–6 (2016)
36. Kamaruddin, S., Ravi, V.: Credit card fraud detection using big data analytics: use of PSOAANN based one-class classification. In: Proceedings of International Conference on Informatics Analytics – ICIA 2016, pp. 1–8 (2016)
37. Santos, L.J.S., Ocampo, S.R.: Bayesian method with clustering algorithm for credit card transaction fraud detection. Rom. Stat. Rev. **1**, 103–120 (2018)
38. Hassan, D.: The impact of false negative cost on the performance of cost sensitive learning based on Bayes minimum risk: a case study in detecting fraudulent transactions. Int. J. Intell. Syst. Appl. **9**(2), 18–24 (2017)
39. Yee, O.S., Sagadevan, S., Malim, N.: Credit card fraud detection using machine learning as data mining technique. J. Telecommun. Electron. Comput. Eng. **10**(1–4), 23–27 (2018)
40. Nur-E-Arefin, M., Islam, M.S.: Application of computational intelligence to identify credit card fraud. In: 2018 International Conference on Innovation in Engineering and Technology, ICIET 2018, pp. 1–6 (2018)
41. Tran, P.H., Tran, K.P., Huong, T.T., Heuchenne, C., HienTran, P., Le, T.M.H.: Real time data-driven approaches for credit card fraud detection, pp. 6–9 (2018)
42. Askari, S.M.S., Hussain, M.A.: Credit card fraud detection using fuzzy ID3. In: Proceeding - IEEE International Conference on Computing Communication and Automation ICCCA 2017, January 2017, pp. 446–452 (2017)
43. Artikis, A., et al.: A prototype for credit card fraud management: industry paper. In: Proceedings of the 11th ACM International Conference on Distributed and Event-Based Systems, pp. 249–260 (2017)
44. https://en.wikipedia.org/w/index.php?title=Ensemble_learning&oldid=896385411
45. Patil, S., Nemade, V., Soni, P.K.: Predictive modelling for credit card fraud detection using data analytics. Procedia Comput. Sci. **132**, 385–395 (2018)
46. Lakshmi, S., Kavila, S.D.: Machine learning for credit card fraud detection system. Int. J. Appl. Eng. Res. **13**(24), 16819–16824 (2018)
47. Zhang, Y., Liu, G., Luan, W., Yan, C., Jiang, C.: Application of SIRUS in credit card fraud detection. In: International Conference on Computational Social Networks, pp. 66–78 (2018)
48. Su, C.-H., et al.: A ensemble machine learning based system for merchant credit risk detection in merchant MCC misuse. J. Data Sci. **17**(1) (2019)
49. https://en.wikipedia.org/w/index.php?title=Special:CiteThisPage&page=Deep_learning&id=899278872#Wikipedia_talk_pages. Accessed 04 June 2019
50. Fiore, U., De Santis, A., Perla, F., Zanetti, P., Palmieri, F.: Using generative adversarial networks for improving classification effectiveness in credit card fraud detection. Inf. Sci. **479**, 448–455 (2017)
51. Raza, M., Qayyum, U.: Classical and deep learning classifiers for anomaly detection. In: Proceedings 2019 16th International Bhurban Conference on Applied Sciences and Technology, IBCAST 2019, pp. 614–618 (2019)
52. Niimi, A.: Deep learning for credit card data analysis. In: 2015 World Congress on Internet Security (WorldCIS), pp. 73–77 (2015)
53. Salo, F., Injadat, M., Nassif, A.B., Shami, A., Essex, A.: Data mining techniques in intrusion detection systems: a systematic literature review. IEEE Access **6**, 56046–56058 (2018)
54. https://en.wikipedia.org/wiki/Sensitivity_and_specificity. Accessed 05 June 2019
55. https://en.wikipedia.org/wiki/False_positive_rate. Accessed 05 June 2019
56. Tharwat, A.: Classification assessment methods. Appl. Comput. Inform. (2018, in press)
57. https://en.wikipedia.org/wiki/Accuracy_and_precision. Accessed 05 June 2019

Short Term Forecasting of Market Clearing Price (MCP) for an Electricity Bid Based on Various Algorithms in R

Niharika Singh[1], Gaurav Rajput[2(✉)], Aakash Malik[3], and Oshin Maini[4]

[1] School of CS University of Petroleum and Energy Studies, Dehradun, Utrakhand, India
niharika@ddn.upes.ac.in
[2] Department of CS&E, Accurate Group of Institutions, Greater Noida, Uttar Pradesh, India
gauravrajput3l@gmail.com
[3] Department of Electrical Engineering, Delhi Technological University, Delhi, India
malikakash22@gmail.com
[4] Department of Computer Science, Indira Gandhi Institute of Technology, Delhi, India
osh.maini@gmail.com

Abstract. The hydel merchandise foresee the market clearing price as one of the most crucial aspect. The ratio of supply to demand creates MCP as important because of the incredibly high money related stakes and the mammoth volume of the electrical influence. The forecasting of the MCP for an electricity bid has raised the tremendous demand due to the deregulation in the electric market. The working of the electrical power market can be just summed up to be founded on counter offering. The power makers, regardless of being a CGU (Central Generating Unit), SGU (State Generating Unit) or any IPP (Independent Power Producer) present their individual power and value offers available to be purchased, while the Consumers i.e. SDU (State Distribution Unit) or the DISCOs (Distribution Companies) present their power and value offers for acquiring purposes. There are various techniques for the MCP forecasting.

Keywords: Market clearing price · Genetics · Parallel algorithms · Thread

1 Introduction

In numerous nations around the world, power is currently exchanged on the basis of market rules utilizing spot and subsidiary agreements. In any case, power is an exceptionally uncommon product. It is monetary non-storable, and control framework soundness depends upon a steady harmony among generation and utilization. In the meantime, power request relies upon climate. In the meantime, power request relies upon climate (temperature, wind speed, rain, and so forth.) and the force of business and regular exercises (on-top versus off-top hours, weekdays versus ends of the week,

© Springer Nature Switzerland AG 2020
L. C. Jain et al. (Eds.): ICICCT 2019, LAIS 9, pp. 304–311, 2020.
https://doi.org/10.1007/978-3-030-38501-9_30

occasions and close occasions, and so forth.). From one perspective, these interesting and qualities prompt value elements not saw in some other market, showing regularity at the day by day, week by week and yearly levels, and unexpected, brief and for the most part unforeseen value spikes [1–3]. Then again, they have urged scientists to increase their endeavors in the improvement of better anticipating procedures. Flagrant value instability, which can majorly to two requests of extent higher than that of some other item or money related resource, has constrained market members to fence across volume chance as well as an against value developments. A generator, service organization or extensive modern purchaser who can gauge the unstable discount costs with a sensible level of precision can change its offering technique and its own creation or utilization plan for request to diminish the hazard or expand the benefits in day-ahead exchanging [4–6]. An assortment of techniques and thoughts have been striven for power value gauging (EPF), with fluctuating degrees of achievement. This audit arrangement means to clarify the many-sided quality of the accessible arrangements, with an exceptional accentuation on the qualities and shortcomings of the individual techniques [7, 8].

2 Backgrounds

ANNs are numerical models in view of the working of the human mind, ANNs are profoundly interconnected basic preparing units outlined in an approach to show how the human cerebrum plays out a specific task [1, 2]. Each unit have called neurons, frames a weighted total of its contributions, to which a steady term called predisposition is included. This aggregate is then gone over an exchange work: straight, sigmoid or hyperbolic digression. Figure 1 demonstrates the inner structure of a neuron; whose exchange work is sigmoid and b is predisposition for it [9–11].

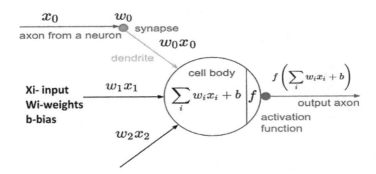

Fig. 1. Internal structure of a neuron

The neural network consists of two parts:

1. Feed forward
2. Back propagation

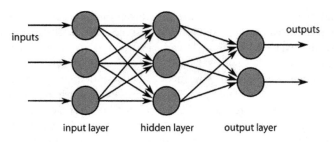

Fig. 2. Feed forward neural network

Back propagation, an abstract for proliferation of errors & quot is a general principal of training artificial neural networks operate in alliance with an extension method such as gradient descent. The technique figures the slope of a misfortune work as for every one of the weights in the system, with the goal that the inclination is nourished to the enhancement strategy which utilizes it to refresh the weights, trying to limit the misfortune work [12–14] (Figs. 2 and 3).

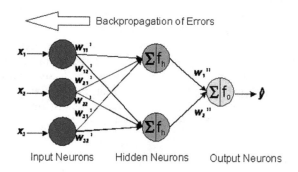

Fig. 3. Back propagation neural network

2.1 Supervised Training

In oversaw preparing, both the wellsprings of data and the yields are given. The system by then structures the wellsprings of information and looks at its following yields against the pined for yields. Blunders are then spread back through the structure, revealing the framework change the weights which control the system [15–18]. This technique occurs again and again as the weights are consistently changed. The blueprint of information which connects with the status is known as the "arranging set." During the arranging of a system a similar game-plan of information is dealt with ordinarily as the association weights are ever refined. The present business deal with progress packs offer instruments to screen how well a phony neural structure is joining on the capacity to predict the correct answer. While an administered characterization calculation figures out how to credit inputted names to pictures of creatures, its unsupervised partner will take a gander at innate likenesses between the pictures and separate them into bunches as needs be, appointing its own new name to each gathering. In a common-sense precedent, this sort of calculation is valuable for client division since it will return bunches considering parameters that a human may not consider due to prior predispositions about the organization's statistic.

2.2 Unsupervised Training

The other sort of getting ready is called unsupervised planning. In unsupervised setting up, the framework is outfitted with inputs yet not with needed yields. The structure itself should then pick what features it will use to accumulate the data. This is frequently implied as self- affiliation or adaption. At this moment, unsupervised learning isn't without a doubt knew [19]. This adaption to nature is the certification which would enable science fiction sorts of robots to continually learn alone as they encounter new conditions and new conditions.

2.3 K-Nearest Neighbor

K-Nearest Neighbor (KNN) is a most basic machine learning strategy. The decision of KNN is persuaded by its effortlessness and adaptability to consolidate diverse information composes. The fundamental thought of KNN is to construct estimation with respect to a settled number of perceptions, say k, which are nearest to the coveted yield. It is considered as a lethargic learning calculation since it doesn't assemble a model or capacity already, however yields the nearest k records of the preparation informational index that have the most astounding closeness to the test. KNN can be utilized both in discrete and nonstop basic leadership known as arrangement and relapse individually [18].

Finding the k-closest neighbors dependably and productively can be troublesome. Different measurements that the Euclidean can be utilized. The certain suspicion in utilizing any k-closest neighbors' system is that things with comparable ascribes tend to bunch together [20].

The system trains with the preparation set, and afterward predicts the estimation of the testing set [21, 22]. The outcomes are composed the anticipated esteem is contrasted and the first qualities, and the mean square blunder is figured.

3 Implementation

3.1 Artificial Neural Network

Here we are utilizing the Artificial Neural system calculation for foreseeing the Minimum Clearing Price (MCP) of the day ahead market for the Indian power advertise. We are utilizing the 'nnet' bundle of the R dialect for the same. The information comprises of different parameters, for example, the Selling Bid, Purchase Bid, Cleared Volume, The day no., The hour no, base cleared Volume (MCV) and the base Clearing Price (MCP). We are intrigued to conjecture the MCP. The dataset has 2979 lines, out of which: 75% of it has been chosen haphazardly as the preparation set. 25% of it has been chosen arbitrarily as the testing set. We are utilizing a 10-layer neural system and, taking MCP as the last yield, and every single other parameter as the contributions to the neural system. The system trains with the preparation set, and afterward predicts the estimation of the testing set. The outcomes are composed. The anticipated esteem is contrasted and the first qualities, and the mean square blunder is figured.

3.2 Support Vector Machine

The SVM calculation for anticipating the Minimum Clearing Price (MCP) of the day ahead market for the Indian power showcase. The 'e1071' bundle of the R dialect is utilized. The information comprises of different parameters, for example, the Selling Bid, Purchase Bid, Cleared Volume, The day no., The hour no, base cleared Volume (MCV) and the base Clearing Price (MCP). We are intrigued to foresee the MCP. The dataset demonstrates for foreseeing the best cost and gamma and, taking MCP as the last yield, and every single other parameter as the inputs to the machine. The model trains with the training set, and then predicts the estimation of the testing set. The anticipated esteem is contrasted and the first qualities, and the mean square error is calculated.

3.3 K-Nearest Neighbor

K-Nearest Neighbor (KNN) is a most basic machine learning strategy. The decision of KNN is persuaded by its effortlessness and adaptability to consolidate diverse information composes. The fundamental thought of KNN is to construct estimation with respect to a settled number of perceptions, say k, which are nearest to the coveted yield. It is considered as a lethargic learning calculation since it doesn't assemble a model or capacity already, however yields the nearest k records of the preparation informational index that have the most astounding closeness to the test. KNN can be utilized both in discrete and nonstop basic leadership known as arrangement and relapse individually.

Finding the k-closest neighbors dependably and productively can be troublesome. Different measurements that the Euclidean can be utilized. The certain suspicion in utilizing any k-closest neighbors system is that things with comparable ascribes tend to bunch together.

4 Results

Table 1. Comparison analysis between ANN, SVM, KNN

Model	Efficiency (%)	RMS error (%)
Artificial Neural Network	95.975	4.025
Support Vector Machine	96.551	3.449
K-Nearest Neighbor	88.313	12.637

The Table 1 above illustrates the two parameters on the efficiency and RMS for the comparative analysis of the three algorithms. The results validate that SVM is comparatively better as compared to KNN and ANN.

5 Plots

This section shows the graphical representation of the results (Figs. 4, 5 and 6).

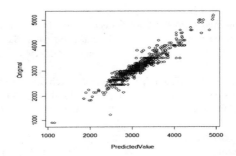

Fig. 4. Graphical representation of ANN

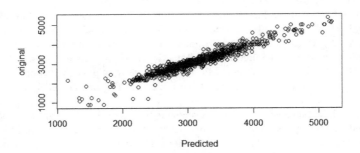

Fig. 5. Graphical representation of SVM

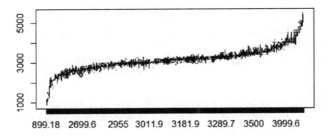

Fig. 6. Graphical representation of KNN

6 Conclusion

This paper gives the comparative results for three forecasting models i.e. ANN, SVM, KNN. In this study, experiment is conducted and compared the techniques according to efficiency and RMSE error. The experimental results proves that SVM provides better results compared to KNN and ANN.

References

1. Burges, C.J.C.: Simplified support vector decision rules. In: Proceedings 13th International Conference Machine Learning, pp. 71–77. Morgan Kaufmann, San Francisco (1996)
2. Vapnik, V.: Statistical Learning Theory. Wiley, New York (1998). ISBN 978-0-471-03003-4
3. Kushilevitz, E., Ostrovsky, R., Rabani, Y.: Efficient search for approximate nearest neighbor in high dimensional spaces. SIAM J. Comput. **30**, 457–474 (2000)
4. Cover, T.M., Hart, P.E.: Nearest neighbor pattern classification. IEEE Trans. Inf. Theory **13**, 21–27 (1967)
5. Breipohl, A.M.: Electricity price forecasting models. In: 2002 IEEE Power Engineering Society Winter Meeting, Conference Proceedings (Cat. No. 02CH37309), New York, NY, USA, vol. 2, pp. 963–966 (2002)
6. Ramsay, B., Wang, A.J.: A neural network based estimator for electricity spot-pricing with particular reference to weekend and public holidays. Neurocomputing **23**, 47–57 (1998)
7. Szkuta, B.R., Sanabria, L.A., Dillon, T.S.: Electricity price short-term forecasting using artificial neural networks. IEEE Trans. Power Syst. **14**, 851–857 (1999)
8. Hong, Y.-Y., Hsiao, C.-Y.: Locational marginal price forecasting in deregulated electricity markets using artificial intelligence. IEE Proc. Gener. Transm. Distrib. **149**, 621–626 (2002)
9. Zhang, L., Luh, P.B., Kasiviswanathan, K.: Energy clearing price prediction and confidence interval estimation with cascaded neural networks. IEEE Trans. Power Syst. **18**(1), 99–105 (2003)
10. Chakrabarti, S., Samanta, D.: Image steganography using priority-based neural network and pyramid. In: Shetty, N., Prasad, N., Nalini, N. (eds.) Emerging Research in Computing, Information, Communication and Applications, pp. 163–172. Springer, Singapore (2016). https://doi.org/10.1007/978-981-10-0287-8_15
11. Ghosh, G., Samanta, D., Paul, M.: Approach of message communication based on twisty "Zig-Zag". In: 2016 International Conference on Emerging Technological Trends (ICETT) (2016). https://doi.org/10.1109/icett.2016.7873676
12. Hossain, M.A., Samanta, D., Sanyal, G.: Extraction of panic expression depending on lip detection. In: 2012 International Conference on Computing Sciences (2012a). https://doi.org/10.1109/iccs.2012.35
13. Hossain, M.A., Samanta, D., Sanyal, G.: Statistical approach for extraction of panic expression. In: 2012 Fourth International Conference on Computational Intelligence and Communication Networks (2012b). https://doi.org/10.1109/cicn.2012.189
14. Khadri, S.K.A., Samanta, D., Paul, M.: Approach of message communication using Fibonacci series. In: Cryptology. Lecture Notes on Information Theory (2014). https://doi.org/10.12720/lnit.2.2.168-171

15. Kumar, H., Singh, A.K.: An optimal replenishment policy for non instantaneous deteriorating items with stock dependent, price decreasing demand and partial backlogging. **35**(8), 0974–6846 (2015). https://doi.org/10.17485/ijst/2015/v8i35/70140

16. Chan, C.J.S.: Development of a profit maximization unit commitment program. M.Sc. Dissertation, UMIST (2000)

17. Lim, B.I., Kim, S.-R., Kim, S.: The effect of the energy price increase on the energy poverty in Korea. Indian J. Sci. Technol. **8**, 790 (2015). https://doi.org/10.17485/ijst/2015/v8i8/69319

18. Jaggi, C.: An optimal replenishment policy for non-instantaneous deteriorating items with price dependent demand and time-varying holding cost. Int. Sci. J. Sci. Eng. Technol. **17**, 100–106 (2014)

19. Prabavathi, M., Gnanadass, R.: Energy bidding strategies for restructured electricity market. Int. J. Electr. Power Energy Syst. **64**, 956–966 (2015). https://doi.org/10.1016/j.ijepes.2014.08.018

20. Catalão, J.P.S., Mariano, S.J.P.S., Mendes, V.M.F., Ferreira, L.A.F.M.: Short-term electricity prices forecasting in a competitive market: a neural network approach. Electr. Power Syst. Res. **77**(10), 1297–1304 (2007)

21. Zhang, G., Patuwo, B.E., Hu, M.Y.: Forecasting with artificial neural networks: the state of the art. Int. J. Forecast. **14**(1), 35–62 (1998)

22. Yonaba, H., Anctil, F., Fortin, V.: Comparing sigmoid transfer functions for neural network multistep ahead streamflow forecasting. J. Hydrol. Eng. **15**(4), 275–283 (2010)

A Replica Structuring for Job Forecasting and Resource Stipulation in Big Data

G. Suhasini[(⊠)] and P. Niranjan

Mewar University, Chittorgarh, Rajasthan, India
suhasini.gadala@gmail.com

Abstract. Situation of Current days is in exponential growth of the data. When data is mounting up in large magnitude and increasing in size then analyzing data is extremely significant. This paper regards diverse possessions such as Big Data, Map reduce Framework, Performance Model, Job Scheduling, Profiling, the map reduce phase, platform model which illustrate the phase implementation as a function of data being processed and try-out results with respect to shuffle phase performance and accurac;y and efficiency of platform performance model.

Keywords: Hadoop · Big data · Job scheduling · Map reduce

1 Introduction

The organizations are confining the data constantly. Significance is given to data which effects in exponential growth of data for an instance, Social networking, World Wide Web and wireless sensor networks are generating data repeatedly. Once data is mounting up in enormous magnitudes and nurturing in size it presumes characteristics of big data. Evaluating such data is important for an organization to make important decisions. If data is not appropriately analyzed then organization will loose the prospects in business arena. In order to develop huge quantity of data a program representation by name Map Reduce has been introduced by Google. By this Map Reduce [1] has become a reliable model for routing big data. The motivation at the back of this is that Map reduce frame work is highly scalable; data parallel model and fault tolerant in nature. There are numerous usage of Map Reduce, for example, Hadoop, Dryad, Phoenix and Mars amidst them Hadoop is generally utilized system crosswise over globe. This is utilized for information concentrated applications that need to exploit parallel preparing intensity of appropriated programming systems. One of the good feature of Hadoop Map Reduce is it offer good support for cloud computing. Consequently organization can use this in pay per basis. Due to the financial limitations small and medium organization choose for this kind of model. Various service contributors like cloudera, Amazon with Elastic Map Reduce has capability to run Hadoop jobs. It is crucial to have job scheduling and resource provisioning mechanisms in cloudera. This cloud develops user contentment besides helping users to have most advantageous to make use of cloud resources. In order to estimate the time they applied various regression one among them is LWLR. In this case VM's are configured with large number of map Slots and reduce slot. In this research we focused on over provision

© Springer Nature Switzerland AG 2020
L. C. Jain et al. (Eds.): ICICCT 2019, LAIS 9, pp. 312–319, 2020.
https://doi.org/10.1007/978-3-030-38501-9_31

problem with dynamic over head of VMs in order to minimize resource over provisioning. This research also considers multiple hadoop jobs with dead line requirement for making a model for job scheduling and resource provisioning for efficient programming for big data processing.

2 Hadoop Map Reduce

Hadoop Map Reduce is consistent software to easily write application which takes care of bulk data and they work in parallel on large group of product hardware in imperfect module. A Map Reduce job usually divides the input data set into sovereign assessments and process them into independent actions by using map task. This map task choose the output which is given as input to the reduce task. The data that is given as input and the resultant output both are stored in the files that we provide in the file system. It will provide various advantages in data processing's like scalability, cost effective solution, flexible, fast and parallel processing We provide the map reduce function as a means to solve certain applications which provide necessary interfaces and classes to perform operations [2].

2.1 Map Reduce

This section provides a sensible amount of aspect on every user using a piece of Map Reduce in back ground. Let us first take the Mapper and Reducer boundaries. We are having many classes that are involved in running a client job.

2.1.1 Payload
Applications commonly actualize the Map and Reducer borders to give the map and reduce strategies.

2.1.2 Reduce
The job of this is to reduce by combining all the data in the shuffle phase into one tuple.
This can be done by a function call. Later these jobs can perform their cleanups [3, 4].
Reducer has three main phases

1. Shuffle
2. Sort
3. Reduce

Shuffle:
Here it returns array of values where all the values are appended but not aggregated.

Secondary Sort
If the analogous rule can be specified by class. This class is capable to assemble the intermediate key value pair.

Reduce Phase
In this phase the key value pair which are sorted are accumulated into a single value i.e. total count of occurrences of the key. The output of reducer is not determined prior [5].

2.1.3   Map Parameters
A record originated from the map will be chronologically accumulated in buffer and the Meta data will be stockpiled into the accounting buffers. But when the serialization buffer or the Meta data go beyond the threshold then the content of the buffer will be organized and written to the disk in background while the map go on with output records. If the buffer gets filled while the spill is in improvement the map thread will break apart. When the map is completed any permanent records are written to the disk and all on the disk subdivision are merged to a single file. Diminishing the number of slicks to disk can reduce map time but increase in the buffer also declines the memory available to the Mapper [6].

2.1.4   Shuffle/Reduce Parameter
The reducer gets the output to be paid to it by the partitioner via HTTP into memory and discontinuously merger these outputs to the disk. Map reduce is faster than traditional system, its written in java, the data processing involved is batch processing, it is complex and lengthy and does not support catching of data.

3   Job Scheduling

There are three significant arranging issues in Map Reduce, for example, region, harmonization and objectivity. Locality is basic issue troubling execution in an collective bunch situation, because of restricted system division data transmission. By synchronization of these two phases Map and reduce improved the overall execution of the Map reduce Model. A Map Reduce occupation of with an overwhelming remaining task at hand may oversee use of shared group, so some mutual computational employments might not have wanted reaction time. So reasonable outstanding task at hand to each activity conveyance bunch ought to be considered. To take care of these issues numerous calculations have been anticipated in past decades.

3.1   FIFO Scheduler

FIFO is a default Hadoop scheduler. The primary goal of the FIFO is to timetable employments dependent on their needs in First started things out premise. The FIFO scheduler works utilizing a FIFO line. Occupation is isolated into autonomous errand and after that they are put into the line and afterward allocated to Task Tracker hubs dependent on the free openings accessible. Occupation Tracker pulls the old employment first from the activity line and it doesn't think about the need and the size of the activity. FIFO has some confinement like poor reaction time for short employments contrasted with enormous occupations. Another confinement is it gives low execution to run various sort employments and great execution when run single kind of occupation.

3.2   Fair Scheduler

This was created at face book to deal with the hadoop group the goal of this scheduler is to appoint each utilization a decent amount of the cluster limit over a period. Client may dole out employment to pools, and each pool is apportioned a base number of Map reduce slots. Free spaces in wasteful pools might be allotted to the new pools. This is a primitive system which implies that the scheduler kills the undertaking in the pool running with over limit and gives the opening to the pool running under limit. Priority is even doled out to the pools. Here assignment are booked in interleaved way dependent on need (Fig. 1).

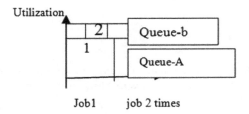

Fig. 1. Capacity scheduling

3.3   Longest Approximate Time to End (Late)

It isn't unprecedented for a specific task to keep on progress gradually. This might be because of a few reasons like high CPU load on the hub, moderate background process and so on. All task need to be done unblemished of the activity. LATE scheduling algorithm attempts to improve the Hadoop by calculating the time remaining to run the task.

LATE depends on three standards:

1. To give priority to the jobs
2. Selecting the node to run first job
3. Protects the task from destroying

The benefit of LATE methodology is power to node heterogeneity, since just slowest task are restarted. This strategy does not break the synchronization stage between map and reduce phases reduce stages, yet just makes action on moderate tasks [7].

3.4   Delay Scheduling

This technique performs well in Hadoop on the grounds that the undertaking in Hadoop is short with respect to employments and there are different areas where an errand can rushed to get to the information square. At the point when a hub demand for a specific undertaking and in the event that it can't be doled out to a neighbourhood task scheduler skirts that errand and begins searching for the following employment. In any case if work has been skipped for quite a while we begin it to dispatch non-nearby errands to keep away from starvation. This technique improves the issue of area by approaching the occupations to sit tight for planning opportunity on hub with nearby information. This paper introduces a way to deal with another model for occupation planning and asset provisioning to limit over provisioning of assets and considerably think about single and different jobs in single and numerous wave models. The Map Reduce frame work and distributed computing standards don't give resource provisioning. The client are mindful to settle on choices on resource need and frequently the clients are unconscious of the asset necessities so in this paper the approach has arrangement for assessing the execution time of given occupation which has a due date prerequisite. In a half breed cloud condition the estimation of time is made. This aides in mechanizing asset provisioning and occupation planning to guarantee that the activity is executed with in the given due date [8].

4 Literature Survey

In this paper we present a re-evaluate of literature on Job Scheduling and resource provisioning in cloud computing with respect to Map reduce program paradigm.

4.1 Scheduling of Jobs Using Map Reduce

The jobs that are executed by server depend on Map Reduce or open source usage Hadoop. Such applications which are executed consume more resources. Experiments analyses have been directed to investigate the presentation of these calculations. By using the scheduling we can minimise the usage of resources and even efficiently manage the resources for the jobs which are vital and leate for the for jobs which are not that important [9].

4.2 Optimal Resource Provisioning

Given assortment of resources accessible in public cloud it is hard to decide the number and sort of resource to apportion for a given movement. So as to settle this they initially characterized the calculations as far as Integer Linear Programming and we utilize the solver to provide the choice in few seconds [10].

4.3 Methodology

So our methodology depends on the two significant measurements job execution time with dead line. We evaluate our methodology with two classes Map reduce application and simulation. A significant preferred position of our methodology is that, it is demonstrated optimal by solver [11].

5   Performance Modelling Frame Work

We presented an exhibition demonstrating frame work that expects to appraise the map reduce application fruition time as a component of the allotted resource the input informational index and job settings [12]. The instinct of our work originates from two sections:

(1) In map reduce jobs keep running on new information for example facebook, Yahoo and so on.

We can extricate an agent employment profile that mirrors the presentation attributes of the modified map reduce functions and utilize the activity profiles to anticipate the future execution of same application when executed on various arrangement of info information.

The demonstrating frame work comprises of three execution models that predict the application finish time at various frameworks and application level. The accompanying Fig. 2 demonstrates an outfit of execution model intended for assessing the application culmination time [13].

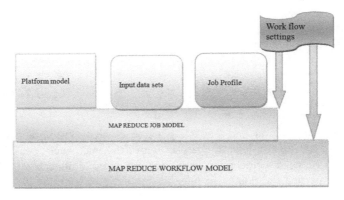

Fig. 2. Ensemble of model.

The stage model intends to predict the fruition time of various map Reduce stages as a component of prepared information. This methodology incorporates all inclusive statement and scalability.

5.1   Profiling Map Reduce Phases

For the map task we profile the following generic phases:

Read Phase: Here we concentrate on what and how much date is being read.

Collect Phase: This phase describes the execution of map with client characterized guide work. We measure the time taken to support the map phase throughput into memory and the measure of produced intermediate information [18, 19].

Merge Phase: This phase describes the information about We measure the ideal opportunity for combining distinctive spill documents to solitary spill record for each task in reduce phase [14].

Shuffle Phase: We measure the time taken to move transitional information from map task to reduce task and combine and sort them.

Write Phase: This phase describes the data regarding the execution of the reduce phase that executes the custom reduce task. We measure the time taken to compose the reduce throughput to HDFS.

Map Phase: We measure the span of whole map and number of prepared records [20].

For exact performance displaying it is attractive to limit the overheads presented by extra observing and profiling procedure. There are two distinct methodologies for executing stage profiling [16, 17].

Counter based profiling: The present Hadoop usage incorporates a few counters to record data, for example, number of bytes read and composed. We changed the Hadoop code by including the counters that measure the span of six conventional stages to the current counter reporting component.

For structure the stage execution mode we execute a lot of small scale bench marks and measure the term of six nonexclusive stages for processing diverse measure of information: read, gather, spill and union periods of map task execution and shuffle and compose stages in reduce task preparing and finally feel that this methodology will try to improve the resource utilization and scheduling in Big data [15].

6   Conclusions

This paper presented various job scheduling and resource provisioning techniques and a methodology for running jobs by the user with dead line, since user is unaware of what resources he need we need to prove the resources and allocate it to them. New algorithms are needed for cloud scheduler for further improving the job execution.

References

1. Abadi, D.J.: Data management in the cloud: limitations and opportunities. IEEE Data Eng. Bull. **32**(1), 3–12 (2009)
2. Amazon redshift. http://aws.amazon.com/redshift/
3. Apache S4: distributed stream computing platform. http://incubator.apache.org/s4/
4. Assunção, M.D., et al.: J. Parallel Distrib. Comput. **79–80**, 3–15 (2015)
5. Announcing Suro: Backbone of Netflix's Data Pipeline. http://techblogy.netflix.com/2013/12/announcing-suro-backbone-of-netflixs.html
6. Andrienko, G., Andrienko, N., Wrobel, S.: Visual analytics tools for analysis of movement data. SIGKDD Explor. Newsl. **9**(2), 38–46 (2007)
7. Ananthanarayanan, R., Gupta, K., Pandey, P., Pucha, H., Sarkar, P., Shah, M., Tewari, R.: Cloud analytics: do we really need to reinvent the storage stack? In: Proceedings of the

Conference on Hot Topics in Cloud Computing (HotCloud 2009). USENIX Association, Berkeley, USA (2009)

8. Announcing Suro: Backbone of Netflix's Data Pipeline. http://techblog.netflix.com/2013/12/announcing-suro-backbone-of-netflixs.html

9. Armbrust, M., Fox, A., Griffith, R., Joseph, A.D., Katz, R.H., Konwinski, A., Lee, G., Patterson, D.A., Rabkin, A., Stoica, I., Zaharia, M.: Above the clouds: a Berkeley view of cloud computing. Technical report UCB/EECS-2009-28, Electrical Engineering and Computer Sciences, University of California at Berkeley, USA, February 2009

10. Attention, shoppers: Store is tracking your cell, New York Times. http://www.nytimes.com/2013/07/15/business/attention-shopper-storesare-tracking-your-cell.html

11. Balmin, A., Beyer, K., Ercegovac, V., Ozcan, J.M.F., Pirahesh, H., Shekita, E., Sismanis, Y., Tata, S., Tian, Y.: A platform for eXtreme analytics. IBM J. Res. Dev. **57**(3–4), 4:1–4:11 (2013)

12. Cloud9 Analytics. http://www.cloud9analytics.com

13. Kumar, A., Niu, F., Ré, C.: Hazy: making it easier to build and maintain big-data analytics. Commun. ACM **56**(3), 40–49 (2013)

14. Chohan, N., Gupta, A., Bunch, C., Prakasam, K.: Hybrid cloud support for large scale analytics and web processing. In: Proceedings of the 3rd USENIX Conference on Web Application Development (WebApps 2012), Boston, USA (2012)

15. Chen, Q., Hsu, M., Zeller, H.: Experience in continuous analytics as a service (CaaaS). In: Proceedings of the 14th International Conference on Extending Database Technology, pp. 509–514. ACM, New York, USA (2011)

16. Chakrabarti, S., Samanta, D.: Image steganography using priority-based neural network and pyramid. In: Shetty, N., Prasad, N., Nalini, N. (eds.) Emerging Research in Computing, Information, Communication and Applications, pp. 163–172. Springer, Singapore (2016). https://doi.org/10.1007/978-981-10-0287-8_15

17. Ghosh, G., Samanta, D., Paul, M.: Approach of message communication based on twisty "Zig-Zag". In: 2016 International Conference on Emerging Technological Trends (ICETT) (2016). https://doi.org/10.1109/icett.2016.7873676

18. Hossain, M.A., Samanta, D., Sanyal, G.: Extraction of panic expression depending on lip detection. In: 2012 International Conference on Computing Sciences (2012a). https://doi.org/10.1109/iccs.2012.35

19. Hossain, M.A., Samanta, D., Sanyal, G.: Statistical approach for extraction of panic expression. In: 2012 Fourth International Conference on Computational Intelligence and Communication Networks (2012b). https://doi.org/10.1109/cicn.2012.189

20. Khadri, S.K.A., Samanta, D., Paul, M.: Approach of message communication using Fibonacci series. In: Cryptology. Lecture Notes on Information Theory (2014). https://doi.org/10.12720/lnit.2.2.168-171

Using the Information Technology in Studying Anthropometry Indices of the Vietnamese People and Orients of Health Education for Students Now

Mai Van Hung[1(✉)] and Nguyen Ngoc Linh[2]

[1] VNU University of Education, Hanoi, Vietnam
hungmv@vnu.edu.vn
[2] National College for Education, Hanoi, Vietnam

Abstract. This study is to identify the reality of basic anthropometry indices and health status of Vietnamese people from these anthropometry indices and health status of people and build methods for health education skills. Using the method to measure anthropometry indices. After anthropometry indices, the information technology was used to assess anthropometry indices of people. The results show that; the increase of mean anthropometry indices of Vietnamese people follow a rule of body growth. The mean anthropometry indices and health status of people of Vietnam are weaker the mean anthropometry of the many people of other country in the world. So that health education in university including: swimming, gymnastics, running, yoga practicing following specific lessons, plays an important role in increasing anthropometry indices of people.

Keywords: Anthropometry · Weight · Height · Information · Technology

1 Introduction

The anthropometry indices are results from the measurements of living human individuals for the understanding human morphological variation. The goals of this study is to identify the reality of basic anthropometry indices and health status of Vietnamese people, find out the marked differences between anthropometry indices and health status of people. From there, it can be build solutions to enhance human quality for Vietnamese people for "the strategy of Vietnamese development on period 2011–2020".

1.1 Methodology

About 1800 people aged from 19 to 22. They were enrolled in 6 universities in Hanoi, Vietnam.
University of Education: 300 students (150 male and 150 female).
University of Natural Sciences: 300 students (150 male and 150 female).
University of Economic: 300 students (150 male and 150 female).
University of Technology: 300 students (150 male and 150 female).
University of Law: 300 students (150 male and 150 female).
University of Foreign language: 300 students (150 male and 150 female).

© Springer Nature Switzerland AG 2020
L. C. Jain et al. (Eds.): ICICCT 2019, LAIS 9, pp. 320–330, 2020.
https://doi.org/10.1007/978-3-030-38501-9_32

Time period of the research: from January, 2018 to July, 2018. Anthropometry parameters were measured, including weight, height, head, neck, mid-upper arm, chest, abdomen, and hip. After anthropometry indices, IT software (WHO AnthroPlus) was used to assess anthropometry indices of people.

1.2 Results and Discussion

After measure anthropometry indices, the obtained results are as follows:

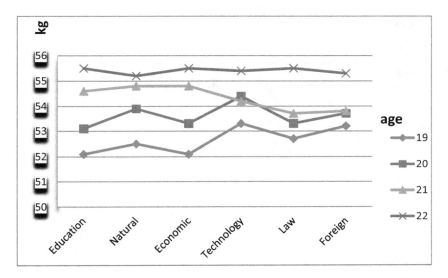

Fig. 1. Weight for age of males

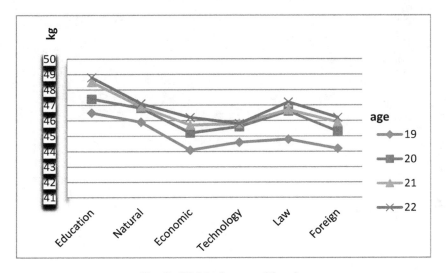

Fig. 2. Weight for age of females

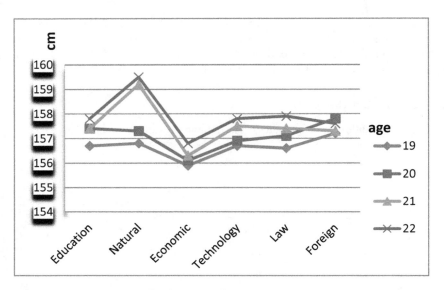

Fig. 3. Height for age of males

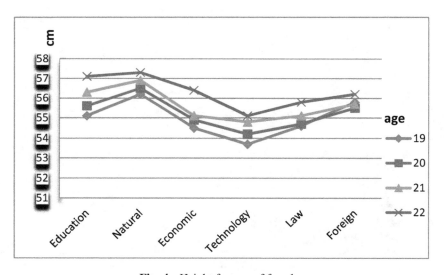

Fig. 4. Height for age of females

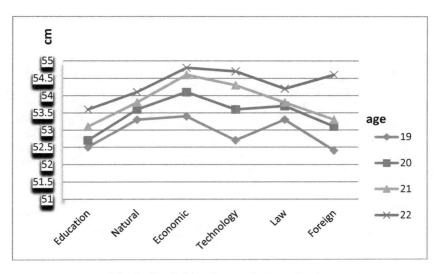

Fig. 5. Head circumference for age of males

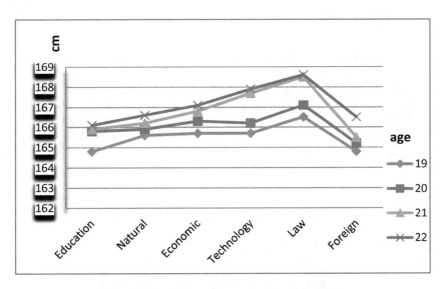

Fig. 6. Head circumference for age of females

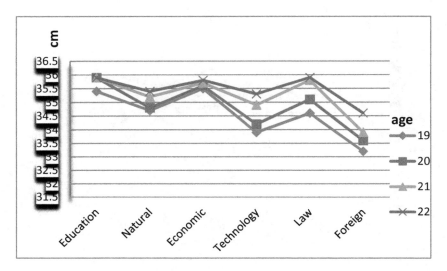

Fig. 7. Neck circumference for age of males

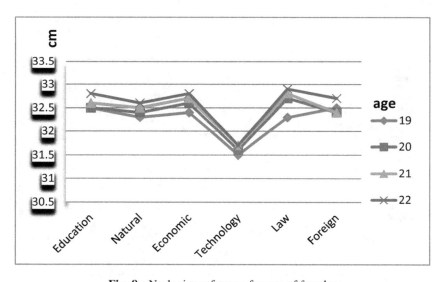

Fig. 8. Neck circumference for age of females

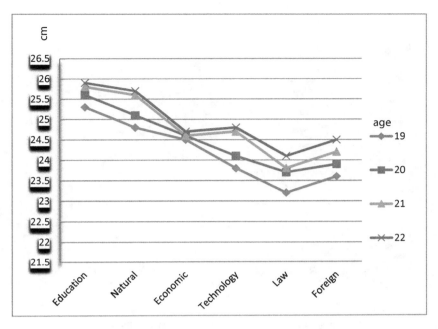

Fig. 9. Mid upper arm circumference for age of males

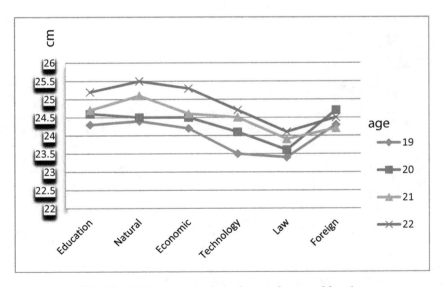

Fig. 10. Mid upper arm circumference for age of females

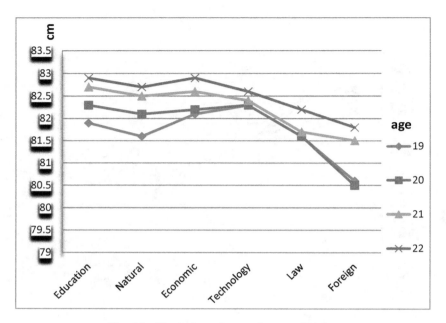

Fig. 11. Chest circumference for age of males

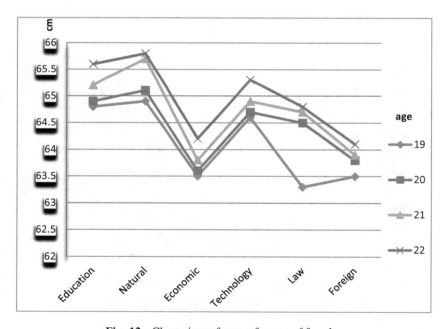

Fig. 12. Chest circumference for age of females

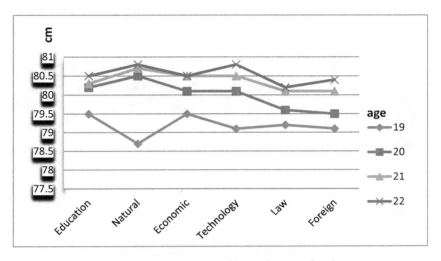

Fig. 13. Abdomen circumference for age of males

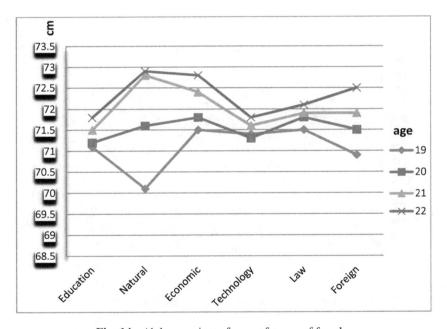

Fig. 14. Abdomen circumference for age of females

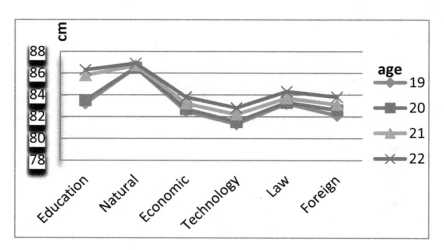

Fig. 15. Hip circumference for age of males

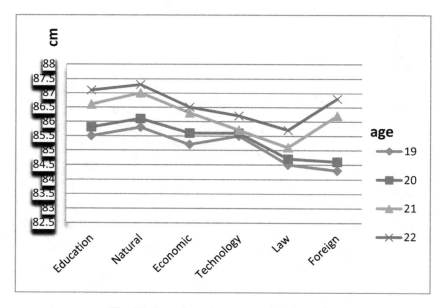

Fig. 16. Hip circumference for age of females

Results showed that the anthropometry indices of people aged from 19 to 22 include: Weight-for-age (Figs. 1 and 2), Height-for-age (Figs. 3 and 4), Head circumference-for-age (Figs. 5 and 6), Neck circumference-for-age (Figs. 7 and 8), Mid upper arm circumference for age (Figs. 9 and 10), Chest circumference-for-age (Figs. 11 and 12), Abdomen circumference-for-age (Figs. 13 and 14) and Hip circumference-for-age (Figs. 15 and 16) followed a rule of body growth.

To contribute to the success of this strategy by the Government, this research is urgently needed about the theoretical basis and practical basic as apply the information on the status of the anthropometry characteristics of Vietnamese people in other ecology regions, which help to built strategic planning in the future, at the same time find out the causes affect on the anthropometry index. From this research results build solutions to improve the quality of Vietnamese people next time. On the other hand research results provide the latest general information to the anthropology of Vietnamese people as the basis for the local implementation of the project strategies in each university in Vietnam.

What makes it complex is that there are too many factors affect the human height such as: genetic factor, nutrition, absorbent ability, psychology, environment, hormone level,... Investigations have been reported recently show that the average anthropometry indices of Vietnamese has been increasing one centimeter per every 10 years while it is two centimeters for Thai and Chinese. Italian and American scientists have discovered two genes showing effect on increasing the anthropometry indices. They are HMGA2 gene and GDF5 gene. Vietnamese now is one of five ethnic groups which have the lowest anthropometry indices in the world.

Orients of health education in university.

This research reveals that there might be some factors affect their anthropometry indices so that, the exercise habit including swimming, gymnastics, running, yoga practicing following specific lessons, plays an important role in increasing anthropometry indices of students.

2   Conclusion

The main anthropometry indices of people in some universities of Vietnam are weaker the main anthropometry indices standard of the people in the world now (aged 19–22). So that Orients of health education in university including: swimming, gymnastics, running, yoga practicing following specific lessons, plays an important role in increasing anthropometry indices of people.

WHO AnthroPlus is useful software to assess growth of students. So that, the WHO AnthroPlus software could be extensively used in health organizations, universities to evaluate and monitor health education for students.

References

1. Ministry of Health: Bological Indices of Vietnameses in 1990s. Medical Publishing House, Hanoi (2003)
2. Cho, S.C.: Anthropometry studies have shown that height at maturity. Master's thesis, Seoul National University, College of Social, Korea (2006)
3. Fernandez, J.E., et al.: Anthropometry of Korean female industrial workers. Pub Med, indexed for Medicine (2010)
4. Montagu, M.F.A.: A Handbook of Anthropometry. University of Minnesota, USA (1960)

5. Spencer, J.: Atlas for Anthropology. University of Minnesota, WM.C. Brown Company Publishers, Minneapolis, Dubuque (1969). Second Printing
6. Stanley, M., Garn, M.D: Anthropometry in clinical appraisal of nutritional status. In: Yearbook of Physical Anthropology, vol. 10 (1962)
7. Hermanussen, M., Scheffler, C., Groth, D., Aßmann, C.: Height and skeletal morphology in relation to modern life style. J. Physiol. Anthropol. **34**, 41 (2015)
8. Van Hung, M.: Morphological and Physical Indexes of Vietnamese People. LAP LAMBERT Academic Publishing, Saarbrücken (2015)
9. Van Hung, M., Thu, M.V.T.: Some anthropometry indices of Vietnamese people of Kinh, Mong, Dao, and Thai ethnics in the North West of Vietnam. In: RVP International Conference "Being Human in Multicultural Traditions", 29th–31st March, India (2016)

Skew Handling Technique for Scheduling Huge Data Mapper with High End Reducers in MapReduce Programming Model

B. Arputhamary[(✉)]

Department of Computer Applications, Bishop Heber College, Trichy, India
arputhambaskaran@rediffmail.com

Abstract. In recent years, to handle huge amount of data, MapReduce Programming Model has been used in parallel which ensures distributed processing. However, data skew is an important issue in MapReduce model which occur when the data are unevenly distributed between the mapper and the reducer which increase the waiting time of the reducer with small sized partition. Several Skew Handling techniques were proposed which tune this process by repartitioning the data into equal sized chunks and distribute them to the reducers. Repartitioning will be suitable for only independent datasets and will degrade the accuracy in dependent datasets. Time Series datasets such as Online Sales, weather and sensor data are dependent data where the prediction of future depends on the historical data. In such case, repartitioning will reduce the prediction accuracy and makes the decision process tedious. A skew handling technique is proposed, where the mapper with huge sized data will be assigned with high-end reducers dynamically.

Keywords: Big data · Data skew · MapReduce

1 Introduction

In Big Data Analytics (BDA) a huge amount of data is analyzed that is generated from heterogeneous sources in a high speed manner. Predictive Analytics (PA) is the part of Big Data Analytics where the future values are predicted by analyzing the historical pattern of the past. Today data are generated in an unprecedented manner and they are very important for the business people in order to take right decisions at right time. There are many traditional data processing and analytical algorithms available with inefficiency in handling the Big Data. At present MapReduce Programming models are widely used to analyze large and diverse datasets in parallel. In this paper, over the many advantages of MapReduce model, computational imbalance between the mapper and the reducer is focused. This unbalanced situation is called as data skew which degrades the performance of MapReduce model by keeping some of the machines idle for a long time while the other nodes are in processing with huge data. Hence, this paper propose a skew handling technique which could effectively mitigate data skew in MapReduce jobs and improves computational efficiency (Fig. 1).

© Springer Nature Switzerland AG 2020
L. C. Jain et al. (Eds.): ICICCT 2019, LAIS 9, pp. 331–339, 2020.
https://doi.org/10.1007/978-3-030-38501-9_33

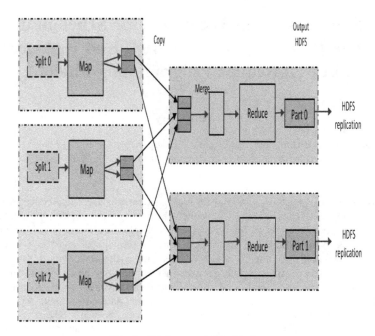

Fig. 1. Hadoop and MapReduce

And another kind of partitioning is range where the input data are split into multiple chunks based on range values. For example, in sales value prediction input data may be decomposed on quarterly basis with the help of range values. In most of the cases, range partitioning leads to uneven split of input data which may cause data skew in partitioning [6–10].

The following Table 1 gives the study report on various partitioning methods and their significance with respect to complexity, scalability, consistency and distributed transactions.

Table 1. Analysis on partitioning techniques

Partition/Data	Complexity	Scalability	Consistency	Distributed transactions
Range	1	1	4	3
Hash	1	1	2	3
Graph	3	1	1	3
Schema	2	2	1	1
Vertical	1	2	1	1
Workload Driven	2	3	3	1

From the study on complexity, scalability, consistency and distributed transactions parameters of different partitioning techniques it is found that Range partitioning is used where ease of use and strong consistency is required. Vertical partition is used to reduce the number of distributed transactions [29, 30]. Workload based partitioning technique is the suitable choice where scalability, consistency and less number of distributed transactions are main considerations [11].

2 Skew Handling Technique with MapReduce

To improve the performance of the MapReduce Programming Model a Skew Handling mechanism is introduced to shrink data skew by dynamically assigning the huge data sets to the node with lower delay. Figure 2 shows the MapReduce with Skew Handler.

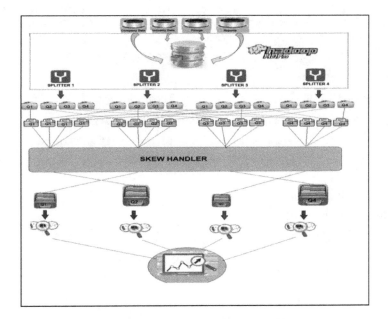

Fig. 2. MapReduce with Skew Handler

In the MapReduce model data skew is the main issue which degrades the performance of heterogeneous multi-core environments by keeping some of the cores idle while the other completes its job. Therefore, the occurrence of data skew will definitely degrade the performance [12–15]. From the background study, Range partitioning of data caused data skew which is considered in this proposed work [28]. The problem of data skew is usually solved by using hash and dynamic partitioning or by using samplings. Traditional approaches cannot be used in Time series data as these data are time dependent and making further partitions will degrade the accuracy in prediction. But if data skew problem is not taken for consideration, it will increase the idling time of reducers which are assigned a lesser size data [16–20]. Therefore it is necessary to assign the mapper to the appropriate reducer in an effective manner. In this work, skew handler(a new decision making approach) is proposed for the appropriate assignment of mapper to the reducer [26].

2.1 Proposed Work

```
Step  1:  Read  input  data  and  store  into  Hadoop
Distributed File System(HDFS)

Step 2: Mapper Stage

Step 3: Shuffling and Sorting

Step 4: Skew Handler
```

Let R be the set of reducers with different computing capability and M be the set of all data chunks which has different sizes.

$M \leftarrow \{ M_1, M_2, M_3, \ldots\ldots, M_n\}$ and $R \leftarrow \{ R_1, R_2, R_{3,\ldots\ldots}, R_n\}$

```
for each R_i in R do

t_i ← estDelay(R_i) // t data transfer delay

e_i ← estComputeTime(R_i) // e computation time

// The estimation is made for fixed unit of data
```

Assign the mapper with huge data to the reducer with lower delay

```
Step 5: Reducer Stage

Reduce<list(key),Value>

// Apply Prediction using Holt Winter's Prediction
Model

Step 6: Visualize
```

3 Experimental Summary

Let R_i be the number of resources with different computing capability and M_i be the set of all data chunks which has different sizes.

Let $m = r$ (i.e.) m are number of mappers and n are number of reducers.

The delay for processing unit of data on R_i be d_i and assume that each mapper can be assigned to one reducer and the delay for executing data on i^{th} mapper with j^{th} reducer is denoted as d_{ij}.

On the whole this model deals with the assignment of m mapper to n reducer in such a way that to minimize the time for overall completion of the tasks (Table 2).

Table 2. Number of resources R_i and Data chunks M_i

	M_1	M_2	M_3	M_n
R_1	d_{11}	d_{12}	d_{13}......	d_{1n}
R_2	d_{21}	d_{22}	d_{23}.........	d_{2n}
R_3	d_{31}	d_{32}	d_{33}.........	d_{3n}
.				
.				
R_n	d_{41}	d_{42}	d_{43}........	d_{4n}

Let d_{ij} be the delay for processing the ith mapper by j^{th} reducer. Then,

$$X_{ij} = \begin{cases} 1 \text{ if the } i^{th} \text{ mapper is assigned to } j^{th} \text{ reducer} \\ 0 \text{ if the } i^{th} \text{ mapper is not assigned to } j^{th} \text{ reducer} \end{cases}$$

Here the objective is to reduce the overall time required to complete all the mapper tasks. Clearly the first set of constraints indicates that each mapper must be assigned to exactly one reducer and the second set of constraints assures that each mapper can be performed by exactly one reducer.

Hence by using assignment problem large scale data chunks are assigned to a node with large bucket size (i.e.) low delay.

Illustration (1). Assume that there are 4 mappers with different size of data.

$$M_1 = 10000, \ M_2 = 30000, \ M_3 = 50000 \text{ and } M_4 = 15000.$$

Let R_1, R_2, R_3 and R_4 are the set of resources with different capability.

Consider that the delay for processing unit of data (i.e.) 1000 records on each machine is,

$$R_1 = 3, \ R_2 = 4, \ R_3 = 1 \text{ and } R_4 = 5$$

Then the delay for processing data units are calculated based on the delay of processing unit of data on each machine (Table 3).

Table 3. Delay for processing data units

R_i/M_i	M_1 10000	M_2 30000	M_3 50000	M_4 15000
$R_1(3)$	30	90	150	45
$R_2(4)$	40	120	200	60
$R_3(1)$	10	30	50	15
$R_4(5)$	50	150	250	75

By solving the above assignment problem.

M_2 is assigned to R_1, M_4 is assigned with R_2, M_3 is with R_3 and M_1 with R_4. Hence the mapper with huge amount of data is assigned to the reducer with low delay (i.e.) R_4.

3.1   Findings on Real Time Environment

The quintessence of the proposed work is that - It identifies the most efficient node in terms of minimal execution time as the high end node. The data imbalance is handled effectively by using the skewness formula and assigning the distributed data to the high end node identified. The principles of data skew handling by avoiding straggler or idle times by using skew handling technique with MapReduce algorithm effectively operates in the heterogeneous environment.

Figure 3 depicts that the nodes are identified in their order of execution time from the least to the highest.

The proposed approach is tested with the online sales data (dependent data) and the results prove that the partitioning of time series data with MapReduce model and the skew handling increases the performance with respect to prediction accuracy and also in execution time. It is evident that the proposed method gives the highest accuracy rate i.e. least execution time. So in terms of accuracy the proposed method overrides the existing technique. Figure 4 represents the execution time of the proposed system is relatively less than the existing method.

Fig. 3. Representation of high end machine with least execution time in their order

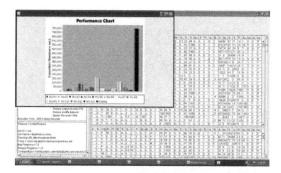

Fig. 4. Calculating accuracy of existing and proposed techniques in data skew handling

The execution time taken after skew handling is found to be better than before in Table 1.

Table 4. Comparison analysis

Months	Reducers	Execution time (Nano Secs) before skew handling	Execution time (Nano Secs) after skew handling
M1	R0	18750	17297
M2	R1	9891	9110
M3	R2	5672	5203
M4	R3	4672	4453
M5	R4	3172	3062
M6	R5	4796	4703
M7	R6	4406	4297
M8	R7	2875	2812
M9	R8	2937	2875
M10	R9	2719	2641
M11	R10	2703	2594
M12	R11	2938	2859

The monthly prediction value and threshold value is achieved in Fig. 5 by using the online sales data, in which, the actual data is marked in red and the predicted data is marked in blue. It is clear that the predicted data is higher than the actual data (Table 4).

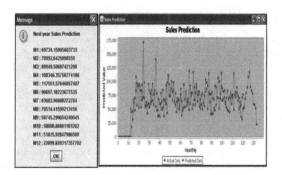

Fig. 5. Representation of prediction using Holt Winters Method

4 Conclusion

Today data are generated in an unprecedented manner and prediction analytics plays important role in managing this situation. In this a skew handling technique is proposed to solve the problem of data skew in MapReduce programming model which is the wide choice for Big Data Analytics. This proposed approach improved the performance of MapReduce Model by reducing the idle time of the reducers in heterogeneous multi core environment. The working principle of this approach is mathematically proved. The proposed work also be analyzed in real time Hadoop and MapReduce environment with different number of nodes.

References

1. Chen, Q., Yao, J., Xiao, Z.: LIBRA: light weight data skew mitigation in MapReduce. IEEE Trans. Parallel Distrib. Syst. **26**(9), 2520–2533 (2014)
2. Kwon, Y., Balazinska, M., Howe, B., Rolia, J.: SkewTune: mitigating skew in MapReduce applications. In: Proceedings of the ACM SIGMOD International Conference on Management of Data (2012)
3. Chen, Q., Liu, C., Xiao, Z.: Improving MapReduce performance using smart speculative execution strategy. IEEE Trans. Comput. (TC) **63**(4), 954–967 (2014)
4. Fan, Y., Wu, W., Xu, Y., Chen, H.: Improving MapReduce performance by balancing skewed loads. China Commun. **11**(8), 85–108 (2014)
5. DeWitt, D., Naughton, J., Schneider, D., Seshadri, S.: Practical skew handling in parallel joins. In: VLDB (1992)
6. Slagter, K., Chung, Y., Zhang, D., Hsu, C.H.: An improved mechanism for optimizing massive data analysis using MapReduce. J. Supercomput. **66**(1), 539–555 (2013)
7. Wajid, N., Satish, S., Manjunath, T.N.: A survey on HadoopMapReduce framework and the data skew issues. Int. J. Sci. Res. Eng. Technol. **4**(4) (2015). ISSN 2278-0882
8. Kwon, Y., Balazinska, M., Howe, B., Rolia, J.: A study of skew in MapReduce applications. In: The 5th International Open Cirrus Summit (2011)
9. Kwon, Y., Balazinska, M., Howe, B., Rolia, J.: SkewTune: mitigating skew in MapReduce applications. In: Proceedings of the ACM SIGMOD International Conference on Management Data, pp. 25–36 (2012)
10. Gufler, B., Kemper, A., Reiser, A., Augsten, N.: Handling data skew in MapReduce. In: International Conference on Cloud Computing and Services Science, CLOSER (2011)
11. Vinutha, J., Chandramma, R.: Skew types mitigating techniques to increase the performance of MapReduce applications. Int. J. Emerg. Technol. Adv. Eng. **5**(2) (2015). ISSN 2250-2459 (Online)
12. Nawale, V.A., Deshpande, P.: Survey on load balancing and data skew mitigation in MapReduce application. Int. J. Comput. Eng. Technol. **6**(1), 32–41 (2015). ISSN 0976-6367 (Print), ISSN 0976-6375 (Online)
13. Sridhar, S., Choudary, S., Raigur, N., Kumar, D., Kumar, S.: Dynamic mitigation of data skew in MapReduce. SS Int. J. Multidisciplinary Res. **2**(4) (2016). ISSN 2395-7964 (Online)
14. Kwon, Y., Balazinska, M., Howe, B., Rolia, J.: Skew-resistant parallel processing of feature-extracting scientific user defined functions. In: Proceedings of the First SOCC Conference, pp. 75–86 (2010)
15. Gufler, B., Kemper, A., Reiser, A., Augsten, N.: Load balancing in MapReduce based on scalable cardinality estimates. In: Proceedings of the 28th ICDE Conference, pp. 522–533 (2012)
16. Kumar, A., Yadav, J.S.: A review on partitioning techniques in database. Int. J. Comput. Sci. Mob. Comput. (IJCSMC) **03**(05), 342–347 (2014)
17. Praveen, B., Umarani, N., Anand, T., Samanta, D.: Cardinal digital image data fortification expending steganography. Int. J. Recent Technol. Eng. **8**(3) (2019). ISSN 2277-3878

18. Dhanush, V., Mahendra, A.R., Kumudavalli, M.V., Samanta, D.: Application of deep learning technique for automatic data exchange with Air-Gapped Systems and its Security Concerns. In: Proceedings of IEEE International Conference on Computing Methodologies and Communication, Erode, 18–19 July 2017 (2017)
19. Kumar, R., Rishabh, K., Samanta, D., Paul, M., Vijaya Kumar, C.M.: A combining approach using DFT and FIR filter to enhance Impulse response. In: Proceedings of IEEE International Conference on Computing Methodologies and Communication, Erode, 18–19 July 2017 (2017)
20. Ghosh, G., Samanta, D., Paul, M., Kumar Janghel, N.: Hiding based message communication techniques depends on Divide and Conquer Approach. In: Proceedings of IEEE International Conference on Computing Methodologies and Communication, Erode, 18–19 July 2017 (2017)
21. Singh, R.K., Begum, T., Borah, L., Samanta, D.: Text encryption: character jumbling. In: Proceedings of IEEE International Conference on Inventive Systems and Control @IEEE, Coimbatore, 19–20 January 2017 (2017)
22. Chakrabarti, S., Samanta, D.: Image steganography using priority-based neural network and pyramid. In: Emerging Research in Computing, Information, Communication and Applications, pp. 163–172 (2016). https://doi.org/10.1007/978-981-10-0287-8_15
23. Ghosh, G., Samanta, D., Paul, M.: Approach of message communication based on twisty "Zig-Zag". In: 2016 International Conference on Emerging Technological Trends (ICETT) (2016). https://doi.org/10.1109/icett.2016.7873676
24. Hossain, M.A., Samanta, D., Sanyal, G.: Extraction of panic expression depending on lip detection. In: 2012 International Conference on Computing Sciences (2012). https://doi.org/10.1109/iccs.2012.35
25. Hossain, M.A., Samanta, D., Sanyal, G.: Statistical approach for extraction of panic expression. In: 2012 Fourth International Conference on Computational Intelligence and Communication Networks (2012). https://doi.org/10.1109/cicn.2012.189
26. Khadri, S.K.A., Samanta, D., Paul, M.: Approach of Message Communication Using Fibonacci Series: In Cryptology. Lecture Notes on Information Theory (2014). https://doi.org/10.12720/lnit.2.2.168-171
27. Abouzeid, A., Bajda-Pawlikowski, K., Abadi, D., Rasin, A., Silberschatz, A.: HadoopDB in action: building real world applications. In: Proceedings of the 36th ACM SIGMOD International Conference on Management of Data, pp. 1–3 (2010). ISBN 978-1-4503-0032-2
28. Aditya, B., Patel, M.B., Nair, U.: Addressing big data problem using Hadoop and MapReduce. In: Nirma University International Conference on Engineering (NUICONE), pp. 114–121 (2012)
29. Dean, J., Ghemawat, S.: MapReduce: simplified data processing on large clusters. ACM Digit. Libr. **51**(01), 107–113 (2014)
30. Kaur, K., Laxmi, V.: Partitioning techniques in cloud data storage: review paper. Int. J. Adv. Res. Comput. Sci. (IJARCS) **08**(05), 219–221 (2017). ISSN 976-5697

Protecting Medical Research Data Using Next Gen Steganography Approach

B. Praveen[1(✉)], Debabrata Samanta[2], G. Prasad[3],
Ch. Ranjith Kumar[4], and M. L. M. Prasad[5]

[1] University of Mysore, Mysore, India
praveen071205@gmail.com
[2] Department of CS, Christ (Deemed to be University), Bangalore, India
debabrata.samanta369@gmail.com
[3] VBIT, Hyderabad, India
[4] BVCITS, Amalapuram, India
[5] JBREC, Hyderabad, India

Abstract. In this paper our main aim is to protect medical research information, when data either images or information shared via internet or stored on hard drive 3rd person can't access without authentication. As needs be, there has been an expanded enthusiasm for ongoing years to upgrade the secrecy of patients' data. For this we combined different techniques to provide more security. Our approach is a combination of cryptography, Steganography & digital watermarking we named this technique as Next Gen. We used cryptography for encrypting the patient's information even if they find image is stegonized and digital watermarking for authenticity and for Steganography we used most popular least significant bit algorithm (LSB). The experimental outcomes with various inputs show that the proposed technique gives a decent tradeoff between security, implanting limit and visual nature of the stego pictures.

Keywords: Medical image security · Digital watermarking · Steganography

1 Introduction

Nowadays, the transmission of data is a day by day common and it is important to locate an effective method to transmit them over the net for sensitive data like research information, patient details. The wide utilizations of different types of systems and IoT the huge development in cloud computing are the surprising test to the image steganography [1–4]. As of late, different government and private medicinal associations are constantly relocating into the cloud and portable situations with the expanding utilization of systems administration [5]. These systems require enormous scale medicinal server farms for restorative information stockpiling. So in order to provide more security to this type very sensitive and high priced research data we decrypt the patient information and with the help of the digital watermarking we kept it in to image by using a secret key then after enter image is converted into another form of the image using steganography [6, 8]. A Secure sharing key shares between the receiver and sender by using this secure key only they can access the details of the medical image [7].

© Springer Nature Switzerland AG 2020
L. C. Jain et al. (Eds.): ICICCT 2019, LAIS 9, pp. 340–348, 2020.
https://doi.org/10.1007/978-3-030-38501-9_34

2 Literature Survey

When all is said in done, security signifies "the quality or condition of being secure to be free from threat". Depending upon the sort of substance security can be performed at various layers of substance. Information security gives the confirmation of data security which is stored, transmitted via utilization of various protocols, systems & hardware. Information security can be categorized as measures received to antedate the unapproved use/alteration of utilization of data or capacities [25].

2.1 Security Attacks

The information is we are transmitted from source system to destination system which is referred to as its typical flow as appeared in the figure. While transmitting the information from source system to destination system intruder/hacker/programmer whatever name you say he may see information or access or change the actual content of the originally shared data. These kinds of attacks are officially known as security assaults/attacks [9–12]. A hacker can troubled this typical stream by executing the various kinds of systems/programming codes over the information and system in following ways [26].

2.2 Watermarking

Watermarking is an implanted picture or pattern in paper, can see it by reflected light. Watermarking is regularly utilized as security highlights of banknotes, travel papers, postage stamps and different reports. Additionally, computerized watermarking is some implanting in development in an advanced sign [13]. It is utilized to confirm the computerized signal's legitimacy or the character of its proprietors. Normal medium on advanced watermarking is sound, picture, or video [27]. A few distinctive advanced watermarks can be installing in one sign simultaneously, and in the event that the sign is replicated, at that point the information on it will likewise be duplicated and conveyed in the duplicate [14, 15, 30]. Presently days, innovation is growing increasingly quick, it is assuming a significant job in people's life and work [20–23]. With the quick improvement of system and advanced innovation, we generally utilize the Internet and computerized sign to transmit information [16, 17, 28]. Computerized watermarking is utilized for a wide scope of uses, for example, source following, copyright insurance, and communicate checking, secretive correspondence, charges security. Now a day Digital watermarking has-been using in a wide range of applications, such as: protection of copyright, source tracking, monitoring of broadcasting services, communication conversion, security for billing and authenticity identification [18, 19, 29].

2.3 Steganography

Steganography is a technique to hide or conceal the information in to same object without changing the visual appeal. For example, a picture or a sound clasp and sent

[24–27]. The presence of the concealed message isn't known aside from by the sender and recipient. In Fig. 1 explains how the information is stored in images and resultant image after the steganography.

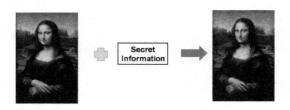

Fig. 1. Flow diagram for transmission of data

3   Proposed System

In our proposed system sensitive data is converted into the encrypted form and then applied to digital watermarking algorithm. The out of the system is digitally watermarked information along with the medical image. Now this digital watermarked image is given input to the Steganography module here information is added at LSB of each pixel. Now the resultant image is stegonized with encrypted patient data this output is given as input for Encrypted algorithm then final output secured medical image.

3.1   Digital Watermarking

As we discussed earlier digital watermarking provides the authenticity for the information. In our approach digital watermarking provides the authenticity for patient information that means particular reports is for the particular patient. Algorithm steps for digital watermarking:

Input: Medical research image
Output: Watermarked image
Step 1: Patient image as input
Step 2: Get Watermark text
Step 3: Apply Given image and text to wavelet decomposition function
Step 4: In this step exploit wavelet coefficients into separate blocks
Step 5: Perform image into binary form
Step 6: Insert watermark individual blocks into the chosen wavelet coefficients
Step 7: above step result is given as input for PNN Training
Step 8: Apply reconstruction function for watermarked image.

4   Sample Code

```
//choose medical image
File dimage = new File("patientimage.jpg");
// read buffered image
Image patientsrc = ImageIO.read(dimage);
//get width
intpwidth = patientsrc.getWidth(null);
intpheight = patientsrc.getHeight(null);
System.out.print("image width= "+pwidth+" and image height = "+pheight);
//buffering image
BufferedImagepimage = new BufferedImage(pwidth, pheight,
BufferedImage.TYPE_INT_RGB);
Graphics gp = pimage.createGraphics();
gp.drawImage(patientsrc, 0, 0, pwidth, pheight, null);
gp.setColor(Color.red);
gp.setFont(new Font("timesnewroman", Font.PLAIN, 24));
Font f=new Font("timenewroman", Font.BOLD, 20);
gp.drawString("praveen", 5, pheight - f.getSize() / 2 - 5);
gp.dispose();
FileOutputStreamoutp = new FileOutputStream("patinedt.jpg");
JPEGImageEncoder encoder = JPEGCodec.createJPEGEncoder(outp);
encoder.encode(pimage);
```

5   Steganography

In this module digital watermarked image is given as input for the steganography LSB algorithm. Here patient information is embedded into each pixel LSB part of the image. Algorithm steps for steganography:

Input: Digital watermarking image & patient information
Output: Steganography image
Step 1: Take Input as watermarked image
Step 2: Read Patient information
Step 3: Convert Step 2 input into encrypted form by using secret key
Step 4: Converted patient information into binary form.
Step 5: Convert image into binary form
Step 6: Insert each bit at LSB positions of the image
Step 7: Repeat step 6 until information is inserted at LSB positions.

5.1   Sample Code

```
byteOutPatient= new ByteArrayOutputStream();
byteArrayPatIn= new byte[inputpFileSize];
DataInputStreaminp= new DataInputStream(new FileInputStream(masterFile));
inp.read(byteArrayPatIn, 0, inputpFileSize);
inp.close();
bytetempByte[]= new byte[4];
for(i=24, j=0; i>=0; i-=8, j++)          {
temppInt= inputpFileSize;
temppInt>>= i;
temppInt&= 0x000000FF;
temppByte[j]= (byte) temppInt;          }
embedpBytes(temppByte);
private static void embedpBytes(byte[] bytesp)   {
intsizep= bytesp.length;
for(int i=0; i<sizep; i++)
     { bytep1= bytesp[i];
for(int j=6; j>=0; j-=2)
{ bytep2= bytep1;
          bytep2>>= j;
          bytep2&= 0x03;
          bytep3= byteArrayIn[inputOutputMarkerp];
          bytep3&= 0xFC;
          bytep3|= bytep2;
byteOut.write(bytep3);
inputOutputMarkerp++;       }}}
```

6   Encryption

In this module given stegonized image as input output will be the encrypted form of the stegonized image. Here secret key is shared in between the sender (source) and receiver (destination). By using this secret key only information is visible to the authenticated persons.

6.1   Sample Code

```
import sha1packge;
SecureRandomsrp= SecureRandom.getInstance("SHA1PRNG");
srp.setSeed(sbp.toString().getBytes());
BufferedImageFSImgp=ImageIO.read(new File(sPath));
for (intpwidth = 0; pwidth<FSImgp.getWidth(); pwidth++) {
for (intpheight = 0; pheight<FSImgp.getHeight();pheight++) {
Color colorp=new Color(FSImgp.getRGB(pwidth, pheight));
Color newColor=new
```

Color(colorp.getRed()^srp.nextInt(255),
colorp.getGreen()^srp.nextInt(255),
colorp.getBlue()^srp.nextInt(255));
FSImgp.setRGB(width, height, newColor.getRGB());}}
System.out.println("Completed!");

7 Project Results

Medical image is converted into the form of the binary this is shown in Fig. 6 (Figs. 2, 3, 4, 5, 7 and 8).

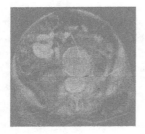

Fig. 2. Binary representation of image

Fig. 3. Histogram, of the given image

Fig. 4. Encryption & decryption section screen

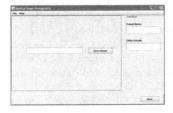

Fig. 5. Patient information embedding screen

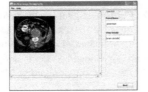

Fig. 6. Secret key to encrypt

Fig. 7. Processing   **Fig. 8.** Final Encrypted Image with hidden information

8 Conclusion

In this paper we proposed a secure transmission of medical images over the network but there is a limitation for media such as it is developed for images only, in extension of this project we can it for audio and video formats along with more secure feature using ancient techniques like tap code but not exactly.

References

1. Mahalakshmi, V., Satheeshkumar, S., Sivakumar, S.: Performance of steganographic methods in medical imaging. IJCAM **12**(1), (2017). ISSN 1819-4966
2. Ashwin, S., Ramesh, J., Kumar, S.A., Gunavathi, K.: Novel and secure encoding and hiding techniques using image steganography: a survey. In: Emerging Trends in ICETEEEM (2012)
3. Hemalatha, S., Acharya, U.D., Renuka, A.: Audio data hiding technique using integer wavelet transform. Int. J. Electron. Sec. Digit. Forensics **8**, 131–147 (2016)
4. Samanta, D., Podder, S.K.: Level of green computing based management practices for digital revolution and New India. Int. J. Eng. Adv. Technol. (IJEAT) **8**(3S), (2019). ISSN 2249-8958
5. Sivakumar, P., Nagaraju, R., Samanta, D., Sivaram, M., Hindialraj, N., Amiri, S.: A novel free space communication system using nonlinear InGaAsP microsystem resonators for enabling power-control toward smart cities. Wirel. Netw. J. Mob. Commun. Comput. Inf. (2019). ISSN 1022-0038
6. Mahua, B., Podder, S.K., Shalini, R., Samanta, D.: Factors that influence sustainable education with respect to innovation and statistical science. Int. J. Recent Technol. Eng. **8**(3), (2019). ISSN 2277-3878
7. Praveen, B., Umarani, N., Anand, T., Samanta, D.: Cardinal digital image data fortification expending steganography. Int. J. Recent Technol. Eng. **8**(3), (2019). ISSN 2277-3878
8. Dhanush, V., Mahendra, A.R., Kumudavalli, M.V., Samanta, D.: Application of deep learning technique for automatic data exchange with Air-Gapped Systems and its Security Concerns. In: Proceedings of IEEE International Conference on Computing Methodologies and Communication, Erode, 18–19 July 2017 (2017)

9. Kumar, R., Rishabh, K., Samanta, D., Paul, M., Vijaya Kumar, C.M.: A combining approach using DFT and FIR filter to enhance impulse response. In: Proceedings of IEEE International Conference on Computing Methodologies and Communication, Erode, 18–19 July 2017 (2017)

10. Ghosh, G., Samanta, D., Paul, M., Kumar Janghel, N.: Hiding based message communication techniques depends on divide and conquer approach. In: Proceedings of IEEE International Conference on Computing Methodologies and Communication, Erode, 18–19 July 2017 (2017)

11. Singh, R.K., Begum, T., Borah, L., Samanta, D.: Text encryption: character jumbling. In: Proceedings of IEEE International Conference on Inventive Systems and Control, Coimbatore, 19–20 January 2017. IEEE (2017)

12. Chugh, G.: Image steganography techniques (2013)

13. Jain, M., Lenka, S.K.: Diagonal queue medical image steganography with Rabin cryptosystem. Brain Inform. 3, 39 (2016)

14. Hossain, Md.A., Samanta, D., Sanyal, G.: A novel approach to extract panic facial expression based on mutation. In: Proceedings of International Conference on Advanced Communication Control and Computing Technologies, ICACCCT 2012, Tamil Nadu, India, 23–25 August 2012. IEEE (2012)

15. Hossain, Md.A., Samanta, D., Sanyal, G.: Statistical approach for extraction of panic expression. In: Proceedings of International Conference on Computational Intelligence and Communication Networks (CICN 2012), Mathura, India, 3–5 November 2012, pp. 420–424. IEEE (2012)

16. Hossain, Md.A., Samanta, D., Sanyal, G.: Automated smiley face extraction based on genetic algorithm. In: Proceedings of the International Workshop on Signal Image Processing and Multimedia (SIPM-2012), India, 13-15 July 2012, pp. 31–37 (2012)

17. Hossain, Md.A., Samanta, D., Sanyal, G.: Extraction of panic expression depending on lip detection. In: Proceedings of the International Conference on Computing Sciences (ICCS), 14–15 September 2012, pp. 473–477. IEEE (2012)

18. Hossain, Md.A., Samanta, D., Sanyal, G.: A novel approach to extract region from facial expression based on mutation. In: Proceedings of the Second International Conference on Advances in Computing, Control and Communication (CCN-2012), Delhi NCR, India, 16–17 June 2012, pp 15–18. https://doi.org/10.3850/978-981-07-2579-2 CCN-390. ISBN 978-981-07-2579-2

19. Hossain, Md.A., Samanta, D., Sanyal, G.: Extraction of panic expression from human face based on histogram approach. In: Venugopal, K.R., Patnaik, L.M. (eds.) Proceedings of Sixth International Conference on Information Processing, ICIP 2012, CCIS 292, 10–12 August 2012, pp. 411–418 (2012). Springer, Heidelberg (2012)

20. Samanta, D., Ghosh, A.: Automatic obstacle detection based on Gaussian function in RoboCar. In: Proceedings of International Conference on Competitiveness and Innovativeness in Engineering, Management and Information Technology (ICCIEMI-2012), 29 January 2012, Haryana, India (2012)

21. Samanta, D., Sanyal, G.: A novel approach of SAR image classification using color space clustering and watersheds. In: Proceedings of International Conference on Computational Intelligence and Communication Networks (CICN 2012), Mathura, India, 3–5 November 2012, pp. 237–240. IEEE (2012)

22. Samanta, D., Sanyal, G.: Novel Shannon's entropy based segmentation technique for SAR images. In: Venugopal, K.R., Patnaik, L.M. (eds.) Proceedings of Sixth International Conference on Information Processing, ICIP 2012, CCIS 292, 10–12 August 2012, pp. 193–199. Springer, Heidelberg (2012)

23. Samanta, D., Sanyal, G.: Segmentation technique of SAR imagery based on fuzzy C-means clustering. In: Proceedings of the IEEE International Conference on Advances in Engineering, Science and Management (IEEE-ICAESM 2012), Nagapattinam, India, 30–31 March 2012, pp. 610–612. IEEE (2012). ISBN 978-81-909042-2-3

24. Samanta, D., Sanyal, G.: SAR image segmentation based on higher order moments. In: Proceedings of International Conference on Systemics, Cybernetics and Informatics (ICSCI - 2012), Andhra Pradesh, India, 15–18 February 2012, pp. 323–326 (2012)

25. Samanta, D., Sanyal, G.: A novel approach of SAR image processing based on Hue, Saturation and Brightness (HSB). In: Proceedings of the 2nd International Conference on Computer, Communication, Control and Information Technology (C3IT-2012), West Bengal, India, 25–26 February 2012. Procedia Technology (2012). ISSN 2212-0173

26. Chakrabarti, S., Samanta, D.: Image steganography using priority-based neural network and pyramid. In: Emerging Research in Computing, Information, Communication and Applications, pp. 163–172 (2016). https://doi.org/10.1007/978-981-10-0287-8_15

27. Ghosh, G., Samanta, D., Paul, M.: Approach of message communication based on twisty "Zig-Zag". In: 2016 International Conference on Emerging Technological Trends (ICETT) (2016). https://doi.org/10.1109/icett.2016.7873676

28. Hossain, M.A., Samanta, D., Sanyal, G.: Extraction of panic expression depending on lip detection. In: 2012 International Conference on Computing Sciences (2012). https://doi.org/10.1109/iccs.2012.35

29. Hossain, M.A., Samanta, D., Sanyal, G.: Statistical approach for extraction of panic expression. In: 2012 Fourth International Conference on Computational Intelligence and Communication Networks (2012). https://doi.org/10.1109/cicn.2012.189

30. Khadri, S.K.A., Samanta, D., Paul, M.: Approach of message communication using Fibonacci series: in cryptology. In: Lecture Notes on Information Theory (2014). https://doi.org/10.12720/lnit.2.2.168-171

Pedestrian Detection - A Survey

C. Victoria Priscilla[(⊠)] and S. P. Agnes Sheila

Department of Computer Science, S.D.N.B. Vaishnav College for Women
University of Madras, Chennai, India
aprofvictoria@gmail.com, agnessheilaisrael@gmail.com

Abstract. Pedestrian detection is one of the important fields of computer vision. Computer vision is getting more information from the digital images and videos and distinguishes between objects and classifies the object. Pedestrian detection is being applied in a wide range of applications such as video surveillance, automated driving, etc. This pedestrian detection has been a growing research area and then so many techniques have been used for detection. This paper focuses on various techniques applied to pedestrian detection and discusses the outcome of every technique and its accuracy and miss rate.

Keywords: Pedestrian detection · Video surveillance · Automated driving · Computer vision

1 Introduction

Detecting the objects in an image or video frames is the main task in object detection research. Machine learning is an application of artificial intelligence which is used to automatically learn and improve from experience. For the nature of the learning ability of machine learning techniques, object detection adopts this technology for improving the speed and accuracy. The detection of the pedestrian has done in two steps. First, it extracts the features from the still images or video frames. Second, it classifies the features of whether the person is present or not in the frame. Figure 1 describes the overall structure of pedestrian detection. The detector contains the two stages one is the training stage and another one is the testing stage. In the training stage, sample input images are given and compute feature vectors and train the detector. Sample images are including pedestrian and non pedestrian images. In the testing stage, the test image is given as input and classifies whether the person is present in the image not. Pedestrian detection having many challenges such as illumination changes, occlusion, scaling. This paper reviews the related work in terms of the challenges.

2 Pedestrian Detection Challenges

Though so many algorithms are used for detecting the pedestrian detection still there are many challenges such as illumination changes such as day and night time and bad weather condition such as rainy and fog, various style of clothing in appearance, object scaling, occluding accessories and occlusion between pedestrian. The posture of the

L. C. Jain et al. (Eds.): ICICCT 2019, LAIS 9, pp. 349–358, 2020.
https://doi.org/10.1007/978-3-030-38501-9_35

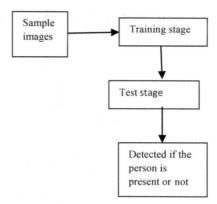

Fig. 1. Structure of pedestrian detector

person also comes into a challenge for detection. Camera and persons are moving at the same time in advance driver assistance system. In that scenario, detection becomes very tough in terms of speed and accuracy. And above mentioned challenges also involved in Advance Driver Assistance System (ADAS).

2.1 Pedestrian Detection Methods Under the Daytime

The most powerful and widely used method for detection is the Histogram of Oriented Gradients (HOG). Dalal *et al.* [1] proposed the approach, which was based on shape based detector. First, it computes the Histogram of Oriented Gradients (HOG) descriptor from the image and that feature descriptor fed into the support vector machine classifier. It detects the outline of a person. And it is used for extracting strong features from images. You *et al.* [2] investigate filtered channel features. This work combines a Histogram of Oriented Gradients (HOG) and LUV (a color space composing of Luminance (L) and components of two Chrominance (UV)) method for improving low level featured and having more convolution neural network. This method has been implemented in three different datasets and achieved the results. But this method suffers from the dimension explosion. Li *et al.* [3] deal with fast pedestrian detection and dynamic tracing within a vehicle to vehicle (v2v) environment. it was implemented in real-time. For fast pedestrian detection, using cascade classifiers with selected haar-like features. This work does not deal with the effect of bad weather. Billal *et al.* [12] using orientation histogram of locally significant gradient and Lookup Table (LUT) based Histogram Intersection Kernel (HIK) Support Vector Machine (SVM) classifier. Convolution neural network has been recently used for detection in pedestrians. Yoshihashi *et al.* [15] stated that two convolution network has been trained separately for spatial and temporal streams and take the difference in frames. This paper also stated that transfer learning was used to transfer the still images and motion features from the dataset.

　　Cao *et al.* [6] using different approaches for pedestrian detection, they used the non-neighbouring features, that is, side inner difference features (SIDF) and symmetrical similarity features (SSF). SIDF is the difference between the background image and pedestrian and the difference between the bounding box of pedestrian and inner parts. SSF is used to obtain the similarity of the shape of the pedestrian. Miss rate of these techniques was 23.06%. Dollar *et al.* [26] performed a survey of the pedestrian detector. For that sixteen detectors had been analysed and compared its performance. Features are very important in the detection, so features like gradient, grayscale, color, texture, self-similarity, motion and classifiers, detection details, implementation details were given in that paper. Li *et al.* [27] used a Fully Convolutional Network (FCN) to extract the features which are pixel-dense output during the semantic segmentation. Combine the output of various layers of FCN and tuning the result. So the features are determined whether the bounding box contains the pedestrian. Advanced Driver Assistance System (ADAS) used in cars for avoiding the collision with the pedestrian.

Gerónimo *et al.* [28] investigated various techniques used in ADAS and examine the cameras used in that system. The area is also playing an important role because of the detection of a person in an urban area and market street more complex than the highway. The urban and market area contains so similar things related to the person. Any vertical object like lamppost and tree can be falsely identified as a person. So it should be avoided. And in ADAS, both the camera and person are moving whereas the camera is stable in other detection which is very difficult to track the person.

Ling *et al.* [29] exploit the Mean local binary pattern method to extract features from gray images. Much information has been lost during the conversion of a colour image to gray image. To avoid that colour based on mean of the local binary pattern was calculated. This texture-based method was compared with other methods and found that it gave better results. Rajnoha *et al.* [30] suggests a new method and created a new dataset that was shot through a public camera. This detector contains pre-trained Visual Geometry Group (VGG16) (16 refers to a number of the layers) in the first layer and Convolutional Neural Network (CNN) in remaining layers. So this method predicts the person from low resolution image. Cao et al. [5] proposed the technique which was called as Multilayer Channel Features (MCF). It consists of three steps, first, it constructs the multilayer channel and feature extraction is the second step. In the last step, the cascade AdaBoost classifier was used to learn the features. Miss rate of this method on Caltech dataset is 7.98%. It improves the speed of detection by 1.43. A framework was proposed by Li et al. in [9] which handles the object detection, shadow removal, tracking, and occlusion. For pedestrian detection Gaussian mixture model was used. Varga *et al.* [11] exploit the multi-scale center symmetric local binary pattern for detecting the pedestrian in real-time video surveillance. And this paper stated that it was also good at handling the occlusion.

Dow *et al.* [34] exploit the technique of motion based HOG for detecting the person. To avoid the accident in the road intersection, the system detects the person intends to cross near the zebra cross using the motion HOG technique. Al-Bsool *et al.* [37] explore the HOG in deeply and the canny edge was used to detect the outer shape of the object. Here Feature extraction was done using weighted HOG at a different scale. Finally, combine all the features and recognize the object. A new system was suggested by Lahmyed et al. in [40]. Hypothesis Generation (HG) was used to find the Region of interest (ROI). Hypothesis Verification used to ensure the presence of the pedestrian. ROI has been computed from a visible image and thermal image of the same place. A random forest classifier was used in HV. Feature extraction can be done by color based HOG and Histogram of oriented optical flow (HOOF). Choudhury *et al.* [10] proposed a framework to detect person by extracting moving objects from frame. It also handled the occlusion and produce better results.

2.2   Pedestrian Detection Under Nighttimes

Most of the researches focused the pedestrian detection in the daytime because of good lighting. Adverse illumination such as bad weather, night time and rain and fog time, that time detection of pedestrian from the image or video are very difficult. Because the picture is not clearly visible and extracts the features from that image is also a very complex one. So detection becomes very crucial. Thermal cameras are used for taking the pictures in the night time because it is less dependent on the surrounding light. However, the image missed the details of pedestrian compared to daytime images. Chen *et al.* [8] proposed a new fusion technique for integrating two convolution layer. This paper stated that three layers are used for extracting the features. It also detects the pedestrian at various scaling.

Tsimhoni *et al.* [17] experimented with the instrumented vehicle. An instrumented vehicle consists of the NIR system (near infrared) which captures the light reflected from the object and FIR system (far infrared) which captures the light emitted from the object and one more camera mounted on the dashboard. This experiment mainly focused on the distance between the vehicle and pedestrians. A particular distance, it detects the pedestrian and a rectangle box appeared on the pedestrian in a movie clip. It drew the attention of the driver and drive accordingly. Pawlowski *et al.* [18] Analysed that the low resolution images from NIR and FIR were passed through different image processing algorithms for optimization. And also [18] stated that a Histogram of Oriented Gradients (HOG) was used for pedestrian detection and SVM was used as a classifier. Night-time Pedestrian Dataset, Laboratorio de Sistemas Inteligentes/ Intelligent System Lab Far Infrared Pedestrian Dataset, Thermal pedestrian datasets were used for this experiment.

Ge *et al.* [19] explores very detail for detecting the pedestrian in the night time. All night vision system uses the thermal camera for taking the images. This paper stated that the Region of Interest (ROI) was achieved by the dual segmentation algorithm. Pedestrian detection strongly depends upon the features, so HOG and Haar-like feature descriptors are used for detecting whether the image contains the pedestrian or other objects. Then the AdaBoost Classifier was used. Kalman algorithm was used for tracking. A new method proposed by Biswas *et al.* [32] detection in thermal infrared images. This method exploits Local steering Kernal (LSK) with the maximum a margin framework. LSK works well in noisy images. Experiments conducted on the datasets OSU-T and OSU-CT and it produced the results with a lower miss rate. Lee *et al.* [35] investigated the Nighttime Pedestrian Detection (NPD) method using part based detection. Head, head-shoulder and Leg parts are identified individually and group the features for detecting the person. In night time some kind of clothing absorbs the light and visibility becomes very low to detect the parts. This method achieved better results in identifying the person at night time. Liu *et al.* [36] exploits the temperature matrix combined with HOG intensity for obtaining the person from the infrared images.

Region of interest can be identified through the temperature matrix and then built the histogram of the intensity of every cell. From this method, the person has been identified. Sun et al. [39] suggested the Multilayer and multi-size Superpixel Segmentation (MMSS). This approach finding the difference between the object and background based on the center surrounded superpixel. This method detects different size objects. The scheme is tested on the Infrared Interested Object Image Dataset (IIOID) which was created by Sun et al.

2.3 Pedestrian Detection Under Scaling

Scaling is important in pedestrian detection because usually, the object near to the camera is visible as the big one. When the object is far from the camera, it seems to be very small. Here the distance between the camera and object becomes a significant point. Scaling is also not covered by many researchers; they concentrate on the pedestrian within particular distance so the images are visible and having enough details of the pedestrian. Zhang et al. [4] employ the bounding box proposal using multilayer ResNet. From the initial bounding box proposal, localization policy is applied for determining the pedestrian in the image at different scaling. Scale aware Fast R-CNN is proposed in [13]. It uses two sub networks for detecting the pedestrian with large and small sizes respectively.

Sang et al. [23] proposed the new method for detecting the persons without the extraction of proposals. It uses two layers; one layer is used for detecting a person in a different scale and other layers extract the contextual information. Zhang et al. [24] designed to obtain the unique characteristics of features and modified Recurrent Convolutional Layer (RCL) is used to obtain the other information surrounding the person without loss of resolution. Crowd counting is also a hot field of computer vision, Ma et al. [25] analysed the crowd counting with a different scale. Here head counting is taken into account, as the whole body, the head also appears small when the person is far from the camera. So Ma et al. [25] used the foreground separation using gradient difference and split the images into different scale by overlapping patches. Patches were classified into different groups according to the gradient distribution and density map was calculated and summed with a perspective map.

Miss rate is used for finding the performance of the detector by calculating the average miss rate. Table 1 shows the miss rate comparison on two popular benchmarks such as Caltech and ETH. Two techniques proposed in [4] and [13] were tested on the datasets. The result shows that scale aware localization policy gave better detection under the scaling. Pedestrian detection in the traffic scenes also including abnormal behavior. That kind of activity also to be detected and give an alarm to the officials to avoid the bad situation. Wang et al. [33] researched in that context and gave the solution. Detecting the pedestrian using various feature extraction methods and tracking the pedestrian. From that, motion information about the pedestrians were collected. This method was tested on real-time video surveillance in the area of Shanghai Highway. It achieved good results with less miss rate and false alarm.

Table 1. Miss rate comparison

Technique	Miss rate	
	Caltech benchmark	ETH benchmark
SA Localization policy	1.88%	17.63%
SA Fast RCNN	9.32%	34.64%

2.4 Pedestrian Detection Under Occlusion

The occluded pedestrian is pedestrian whose body parts are not visible in the image or hiding by some other objects like car, clothes, and other humans. When in the crowded scene many humans are walking parallel, then they occluded by others. To detect the occluded pedestrian is more complex because the full detail of the pedestrian is not visible. Ouyang et al. [16] investigate that problem and developed a framework using a deep learning model. This model is used to find the relationship among the visibility part at different layers. A new CUHK dataset has been developed by Ouyang *et al.* Because existing datasets for pedestrian detection was not handling occlusion evaluation. Xu et al. [14] exploit Deep omega model. This detects the omega shape human that is head shoulder part of human instead of a full part of human. It uses Region based Fully ConvNet for detecting the occluded pedestrian, especially for top view camera scenes. Crowded people cause heavy occlusion in the images, Sidla *et al.* [20] investigate that HOG was trained with a full-body, head-left, head-right, the right and left side of people images. And polynomial SVM classifier used to detect the person. INRIA dataset was used and observed that the low false positive rate was achieved.

Wang *et al.* [21] explore a new method for detecting the occluded pedestrian in the crowd scene. It stated that the part based detector can be used in occluded person via scale prior method. The scale prior is calculated based on the information provided the neighbouring detection. Patil *et al.* [31] uses two stage detectors. The first stage was used for non occluded detection. For that trilinear interpolation was applied in the HOG feature vector calculation. Part based detection was achieved in the second stage. A new framework was proposed by Wang *et al.* in [22]. In which a deformable part based model was used for detecting the person. In that, head detection was focused because head was less occluded and other body part detection is suppressed. Computation speed is also relatively fast in this approach. Zhu *et al.* [7] investigated the multi-task algorithm. It learned the common features of different levels of occlusion. And boosted detector was used to detect the pedestrian from the background. Ghaneizad *et al.* [38] exploits the novel method for handling the occlusion. In this approach, 3D images have been constructed from 2D images. This has been achieved through placing the camera in multiple viewpoints. So an array of images has been taken through camera from different angles. Using the depth estimation occluded target object has been reconstructed.

3   Dataset

Datasets are most important in the research area and some datasets are available free in online for the research community and some researchers are developing their dataset for the research. From Table 2, it has been observed that Caltech, INRIA and KITTI datasets are used in most of the paper. Caltech dataset contains 10 h video. The KITTI dataset contains the video which was taken by high resolution camera. INRIA dataset contains pedestrian videos in different scenarios. Datasets were clearly explained in [26] and scale statistics, occlusion statistics, and position statistics were provided in that. Dataset comparison was given in that paper. Tsimhoni *et al.* [17] and Ge *et al.* [19] exploits real-time video rather existing datasets. Real time video was taken in night time using NIR and FIR sensor and camera fitted in the car.

Table 2. Pedestrian detection techniques

Paper no	Year	Datasets	Techniques/algorithm
[1]	2005	MIT pedestrian database, INRIA	HOG and SVM classifier
[2]	2018	Caltech, INRIA, and KITTI	Extended filtered channel framework (ExtFCF)
[3]	2017	Real-Time	Haar-like based cascade classifiers, For tracking using CamShift algorithm together with extended Kalman filtering (EKF)
[4]	2018	Caltech, Imagenet	Multiplelayer neuronal representations RSNET
[5]	2017	Caltech, INRIA, ETH data set, TUD-Brussels, KITTI	Convolutional neural network (CNN) and traditional handcrafted features (HOG+LUV)
[6]	2016	INRIA, KITTI, Caltech	Side-inner difference features (SIDF) and symmetrical similarity features (SSFs)
[7]	2015	INRIA, Caltech, TUD-Brussels	Multi-task extension of LDCF (*Locally Decorrelated Channel Features*)
[8]	2018	KAIST, UTokyo multispectral datasets, OSU color–thermal dataset	Multi-layer fused convolution neural network (MLF-CNN)
[9]	2014	Real traffic videos	Gaussian mixture model for extracting, Segmented by mean-shift technique
[10]	2018	INRIA, CVC-01 dataset Daimler 2006 dataset	Silhouette orientation histogram, Golden ratio principles for part based detection

(*continued*)

Table 2. (*continued*)

Paper no	Year	Datasets	Techniques/algorithm
[11]	2016	CAVIAR	Multi-scale Center-symmetric Local Binary Pattern operator
[12]	2016	INRIA, ETH and Caltech pedestrian datasets	Orientation histograms of locally significant gradients, LUT based HIK SVM classification
[13]	2017	Caltech, INRIA, ETH, and KITTI	Extracting common features and joining the outputs of the two sub-network by a scale-aware weighing mechanism
[14]	2018	Caltech dataset, DukeMTMC dataset, Bronze dataset	Deep Omega-shape feature learning, and multipath detection
[15]	2018	Caltech Pedestrian Detection Benchmark, Daimler Mono Pedestrian Detection Benchmark	A two-stream network architecture for activity recognition and convolutional channel features for pedestrian detection

4   Conclusion

This paper has discussed various techniques used for pedestrian detection. Every technique has given a good result within its limitations. In pedestrian detection speed and accuracy play a major role. The algorithms have tried to satisfy both speed and accuracy. It was observed that the miss rate in every detection techniques. Research may take into the direction to reduce the miss rate in detection. Pedestrian detection challenges exist which is troublesome in real-time applications and future research will be taken place in that area to overcome the problem.

References

1. Dalal, N., Triggs, B.: Histogram of oriented gradients for human detection. In: 2005 IEEE International Conference on Computer Vision and Pattern Recognition, CVPR 2005, pp. 886–893 (2005). Print ISSN 1063-6919
2. You, M., Zhang, Y., Shen, C., Zhang, X.: An extended filtered channel framework for pedestrian detection. IEEE Trans. Intell. Transp. Syst. **19**(5), 1640–1651 (2018)
3. Li, F., Zhang, R., You, F.: Fast pedestrian detection and dynamic tracking for intelligent vehicles within V2V cooperative environment. IET Image Process. **11**(10), 833–840 (2017)
4. Zhang, X., Cheng, L., Li, B., Hu, H.M.: Too far to see? Not really:- pedestrian detection with scale-aware localization policy. IEEE Trans. Image Process. **27**(8), 3703–3715 (2018)
5. Cao, J., Pang, Y., Li, X.: Learning multilayer channel features for pedestrian detection. IEEE Trans. Image Process. **26**(7), 3210–3220 (2017)
6. Cao, J., Pang, Y., Li, X.: Pedestrian detection inspired by appearance constancy and shape symmetry. IEEE Trans. Image Process. **25**(12), 5538–5551 (2016)

7. Zhu, C., Peng, Y.: A boosted multi-task model for pedestrian detection with occlusion handling. IEEE Trans. Image Process. **24**(12), 5619–5629 (2015)

8. Chen, Y., Xie, H., Shin, H.: Multi-layer fusion techniques using a CNN for multispectral pedestrian detection. IET Comput. Vis. **12**(8), 1179–1187 (2018)

9. Li, Q., Shao, C.F., Zhao, Y.: A robust system for real-time pedestrian detection and tracking. J. Cent. South Univ. **21**(4), 1643–1653 (2014)

10. Choudhury, S.K., Sa, P.K., Prasad Padhy, R., Sharma, S., Bakshi, S.: Improved pedestrian detection using motion segmentation and silhouette orientation. Multimed. Tools Appl. **77** (11), 13075–13114 (2018)

11. Varga, D., Szirányi, T.: Robust real-time pedestrian detection in surveillance videos. J. Ambient Intell. Humaniz. Comput. **8**(1), 79–85 (2017)

12. Bilal, M., Khan, A., Khan, M.U.K., Kyung, C.M.: A low-complexity pedestrian detection framework for smart video surveillance systems. IEEE Trans. Circuits Syst. Video Technol. **27**(10), 2260–2273 (2017)

13. Li, J., Liang, X., Shen, S., Xu, T., Feng, J., Yan, S.: Scale-aware fast R-CNN for pedestrian detection. IEEE Trans. Multimed. **20**(4), 985–996 (2018)

14. Xu, Y., Zhou, X., Liu, P., Xu, H.: Rapid pedestrian detection based on deep omega-shape features with partial occlusion handing. Neural Process. Lett. **49**, 1–15 (2018)

15. Yoshihashi, R., Trinh, T.T., Kawakami, R., You, S., Iida, M., Naemura, T.: Pedestrian detection with motion features via two-stream ConvNets. IPSJ Trans. Comput. Vis. Appl. **10** (1), 12 (2018)

16. Ouyang, W., Zeng, X., Wang, X.: Partial occlusion handling in pedestrian detection with a deep model. IEEE Trans. Circuits Syst. Video Technol. **26**(11), 2123–2137 (2016)

17. Tsimhoni, O., Flannagan, M.: Pedestrian detection with night vision systems enhanced by automatic warnings. In: Proceedings of the Human Factors and Ergonomics Society Annual Meeting, vol. 50, no. 22, pp. 2443–2447 (2012)

18. Pawlowski, P., Piniarski, K., Dabrowski, A.: Pedestrian detection in low resolution night vision images. In: Signal Processing: Algorithms, Architectures, Arrangements, and Applications (SPA), vol. 2015-Decem, pp. 185–190 (2015)

19. Ge, J., Luo, Y., Tei, G.: Real-time pedestrian detection and tracking at nighttime for driver-assistance systems. IEEE Trans. Intell. Transp. Syst. **10**(2), 283–298 (2009)

20. Sidla, O., Rosner, M.: HOG pedestrian detection applied to scenes with heavy occlusion. In: Intelligent Robots and Computer Vision XXV: Algorithms, Techniques, and Active Vision, vol. 6764, pp. 676408 (2007)

21. Wang, L., Xu, L., Yang, M.H.: Pedestrian detection in crowded scenes via scale and occlusion analysis. In: Proceedings of the International Conference on Image Processing, ICIP, vol. 2016-Augus, pp. 1210–1214 (2016)

22. Wang, B., Chan, K.L., Wang, G., Zhang, H.: Pedestrian detection in highly crowded scenes using 'online' dictionary learning for occlusion handling. In: 2014 IEEE International Conference on Image Processing, ICIP 2014, no. October 2014, pp. 2418–2422 (2014)

23. Direct Multi-Scale Dual-Stream Network For Pedestrian Detection Sang-Il Jung and Ki-Sang Hong POSTECH Department of Electrical Engineering Pohang, Korea, pp. 156–160 (2017)

24. Zhang, C., Kim, J.: Multi-scale pedestrian detection using skip pooling and recurrent convolution. Multimed. Tools Appl. **78**(2), 1719–1736 (2019)

25. Ma, T., Ji, Q., Li, N.: Scene invariant crowd counting using multi-scales head detection in video surveillance. IET Image Process. **12**(12), 2258–2263 (2018)

26. Dollár, P., Wojek, C., Schiele, B., Perona, P.: Pedestrian detection: an evaluation of the state of the art. IEEE Trans. Pattern Anal. Mach. Intell. **34**(4), 743–761 (2012)

27. Li, C., Wang, X., Liu, W.: Neural features for pedestrian detection. Neurocomputing **238**, 420–432 (2017)
28. Gerónimo, D., López, A.M., Sappa, A.D., Graf, T.: Survey of pedestrian detection for advanced driver assistance systems. IEEE Trans. Pattern Anal. Mach. Intell. **32**(7), 1239–1258 (2010)
29. Ling, X., Yongjun, Z., Qian, W., Yuewei, L.: Improved local texture features for pedestrian detection. In: 2018 3rd IEEE International Conference on Image, Vision and Computing Comput, ICIVC 2018, pp. 60–65 (2018)
30. Rajnoha, M., Povoda, L., Masek, J., Burget, R., Dutta, M.K.: Pedestrian detection from low resolution public cameras in the wild. In: 2018 5th International Conference on Signal Processing and Integrated Networks, SPIN 2018, pp. 291–295 (2018)
31. Patil, S.S., Palanisamy, P.: Pedestrian classification in partial occlusion. In: 2017 4th International Conference on Signal Processing, Communication and Networking, ICSCN 2017, vol. 2, no. 2, pp. 2–7 (2017). https://doi.org/10.1109/icscn.2017.8085642
32. Biswas, S.K., Milanfar, P.: Linear support tensor machine with LSK channels: pedestrian detection in thermal infrared images. IEEE Trans. Image Process. **26**(9), 4229–4242 (2017)
33. Wang, X., Song, H., Cui, H.: Pedestrian abnormal event detection based on multi-feature fusion in traffic video. Optik (Stuttg) **154**, 22–32 (2018)
34. Dow, C., Lee, L., Huy, N.H., Wang, K.: A human recognition system for pedestrian crosswalk. In: HCI International 2018 - Posters' Extended Abstracts, vol. 852. Springer (2018)
35. Lee, Y.S., Chan, Y.M., Fu, L.C., Hsiao, P.Y.: Near-infrared-based nighttime pedestrian detection using grouped part models. IEEE Trans. Intell. Transp. Syst. **16**(4), 1929–1940 (2015)
36. Liu, L., Bao, H., Pan, W., Xu, C.: Night-time pedestrian detection based on temperature and HOGI feature in infra-red images. Int. J. Simul. Syst. Sci. Technol. **17**(28), 14.1–14.7 (2016)
37. Al-Bsool, M.H.: Pedestrian recognition based on multi-scale weighted HOG. J. Comput. Sci. **14**(11), 1431–1439 (2018)
38. Ghaneizad, M., Kavehvash, Z., Aghajan, H.: Human detection in occluded scenes through optically inspired multi-camera image fusion. J. Opt. Soc. Am. A **34**(6), 856 (2017)
39. Sun, N., Jiang, F., Yan, H., Liu, J., Han, G.: Proposal generation method for object detection in infrared image. Infrared Phys. Technol. **81**, 117–127 (2017)
40. Lahmyed, R., El Ansari, M., Ellahyani, A.: A new thermal infrared and visible spectrum images-based pedestrian detection system. Multimed. Tools Appl. **78**(12), 15861–15885 (2019)

Overview of an Ovarian Classification and Detection PCOS in Ultrasound Image: A Study

N. Priya[1,2(✉)] and S. Jeevitha[1,2]

[1] Department of Computer Science, S.D.N.B. Vaishnav College for Women,
University of Madras, Chennai, India
drnpriya2015@gmail.com, jeevisivanandham@gmail.com
[2] Department of Computer Application, S.D.N.B. Vaishnav College for Women,
University of Madras, Chennai 600 044, India

Abstract. PCOS (polycystic ovary syndrome) is the main cause of forming multiple follicles in ovaries which leads to infertility, obesity, diabetes, heart disease for women. Hormone level imbalance cause to get irregular periodic cyclic because of PCOS. According to the different disorder stages of Ovary, Cyst classified into three categories such as Normal Cyst, Ovarian Cyst and Poly Cyst Ovary. Currently Adult female is affected frequently by polycystic ovaries disease (PCOD). Doctors and Medical Analysts conforming the scan reports based on the number of cysts present in ovary. This may leads to inconsistency of finding proper count of follicles, prolonged time consumption and vulnerable error. To avoid this issue, computer based techniques helps the doctors for the easy detection of follicles and PCOS with different hormone issues. Based on study, Image processing and machine learning techniques play a vital role for PCOS detection and Ovary classification. In this paper, several techniques implemented by many researchers were discussed along with the features and metrics of Ovarian Classification and the detection of PCOS.

1 Introduction

X-rays, CT, MRI and Ultrasound images all used to examine the internal organ of the body and 3d scanning is used to scan external body of the organ [1]. Along with this technology, doctors can easily solve the problem in reducing the risk level in diagnosis of ovary. More than a decade ultrasound image is used to determine the gynaecology issue without any side effect like radiations. Some of the waste masses in the body can affect the organs which leads to form ovarian cysts, PCOS, Fibroids other related problems.

1.1 Classification of Ovary

Finding a tumour in right time is more important in diagnosis and it is based on the size, diameter of the tumour and classification of ovary like normal ovary, ovarian cyst and poly cyst ovary explained clearly. Polycyst ovary identified by the number of presence of follicles around 12 or more and size 2–9 mm. PCOS can be find out various

L. C. Jain et al. (Eds.): ICICCT 2019, LAIS 9, pp. 359–365, 2020.
https://doi.org/10.1007/978-3-030-38501-9_36

measures like blood test, hormone level test (both male and female), thyroid issues and sugar level [5]. To estimate the hormonal measurement, steroid level, oestradiol and progesterone along with urine sample also included to predict ELG and PDG in the hormone test. By this they observed the growth of the follicles in ovary, stage by stage in the periods cyclic of women. Aarti [21] used Soft Computing methods to find the classification of ovary with fuzzy logic and genetic algorithm. In fuzzy logic author examined three steps such as Fuzzification, Inference and Defuzzification.

Figures 1, 2 and 3. [3] Ultra Sound Image of Normal Ovary with dominant Follicles, cyst ovary and polycystic.

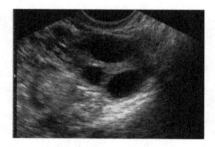

Fig. 1. [3] Normal ovary with dominant follicles

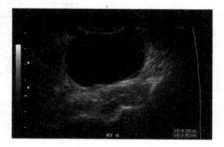

Fig. 2. [3] Cyst ovary image

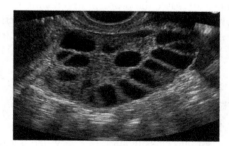

Fig. 3. [3] Image of ovary with polycystic structure

Table 1, Explained the different sizes and count of follicles in the ovary to detect the various cyst type and identity the PCOS easily from the ultrasound image.

Table 1. Size and Count of Follicle in ovary

Type of Cysts	Count of Follicle	Size
Normal Cyst	8 to 10 in.	2 mm–28 mm
Ovarian Cyst	3 to 4 in.	Pea size
PolyCysticOvary	12 in.	9 mm

1.2 Data Set

Most of the images referred from the radiology info and ovary research dataset for the diagnosis of ovary [6]. Usha [7] used more than 50 Transvaginal images of ovary can be taken from GE-LOGIQ dataset. To conclude the conformation of PCOS and Ovarian Classification by the average error percentage of major and minor axis length of the ovary. In medilab diagnostics, image processing techniques are applied the ovary image and detect the follicles based on size and count.

2 PCOS Detection Using Image Processing Methods

In this section, method and approaches are used to detect follicle and PCOS in stage by stage process. Image Processing at different stages like Image acquisition, Image Enhancement such as Histogram-based techniques, Spatial Filters, Gaussian low-pass Filters etc., Image Segmentation such as Edge detection, Corner Detection, Thresholding, Region-growing, Active Contour Models etc., and Image Features Extraction such as Boundary Representation, Texture features, Wavelet Transformation etc., MSE, PSNR, FSIM, Fuzzy classification, CNN, NNL. AI, region of interest etc. are the metrics and classification methods used to conclude the conformation of PCOS and Ovarian Classification with the number and size of follicles.

2.1 Segmentation Techniques for PCOS

Many researcher executed different kinds of segmentation process to detecting PCOS. Nasrul [2] Determined two types of method threshold and edge detection. Threshold method used to detect boundaries and sobel edge detection which helps to identify the smooth edges in noise reduction of image. To modify the gradient of image, (3) approached Watershed segmentation and edge based method and canny operator used to measure de-noise of image. [4, 5, 7] proposed morphology operation like opening, closing, dilation and erosion are used to adding and removing pixel of the image boundaries. [7] explained segmentation based on global enhancement along with binary thresholding to execute the final binary image contain ovary and other fake region. Anthony [10] considered a watershed method to determine the ROI (Region Of Interest) automatically to the inner border which is used to search outer border of follicle and the inner border is identified and lead to isolation of antrum. This can be done in instinctive way because the antrum feature is dark in intensity. Detecting inner follicle border can be done in four more steps like pre-processing, watershed segmentation of gray scale image, watershed of binary image and post-pre-processing.

Setiawati [11] developed a new method to finding solution for optimization problem using PSO algorithm (particle swarm optimization). In this algorithm, clustering process is used to examine the follicle segmentation. PSO is an evolutionary computational method generally modelled on social stimulus of birds group. Sandy [17], Ranjitha [14] and Bozidar [12] implemented Region growing method for automated analysis of ovary. The Region Growing Method make the homogenous region to grow till overlap with the boundaries and examine the follicles position with ovary, after post processing method. Bedy [13] used canny method for image segmentation to detect follicle edges and cropped separately. Each image of follicles consider as new image for next processing like extraction. In this segmentation process author differentiate the background image from the fixed object. Prasanna [16] developed segmentation of polycyst ovary using Active Contour without edge detection method to determine small follicles and fast segmentation of image. Bozidar Potocnik [19] approached Optimal threshold selection method to divide the images as subimage to minimize the problem of follicle detection. From the subimage the follicle is identified by the dark region using threshold selection method. Palak Mehrotra [22] proposed recognition of follicles for PCOS in dark image. Finding object in dark image is rare, so author utilized the horizontal and vertical scanline threshold method. Finally, Image Merging is implemented to find the follicles regions. Ashika [15] introduced Tophat and bottomhat filtering techniques to rectify the denoise and add contrast in the image. Similar opening and closing methods in morphology is used for enhanced the image. Figure (a) and (b) shows the enhanced image using above methods.

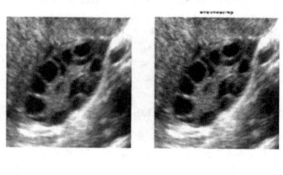

Fig(a)   Fig(b)

2.2   Image Enhancement in PCOS

Different image enhancement method has been implemented by many researchers to improve the standard of the image to further processing level. Some of the enhancement methods like histogram, spatial filtering, box filtering, Gaussian filters, butterworth low-pass filters. [2] proposed histogram equalization and linear grey scale level to find about the image negatives. Padmapriya [5, 26] Author, discussed about PCOS classification used image enhance technique as histogram for spread intensity range uniformly in the image. Sandy [17] approach to extract the image feature using Geometrical and Textural feature method, which show the exact measure size and position of the ovary based on some parameters such as area, major and minor length axis etc., this can be help to detect the major difference between the follicles and cyst in the ovary. Padmapriya [24] explained Histogram method is used to enhance the global contrast of image to show the equal intensity distribution and further author applied filtering method to reject the noise from the image, An Average Filter for smoothing is used to decrease the noises in the image rather than using edge detection method.

2.3   Features Extraction and Metrics in PCOS

[24] distinguished the difference between the original image to compression quality by using MSE and to find the peak error using PSNR for Gaussian and speckle noise by winer filter and mean filter used for overall performance. [4] implemented Fuzzy C-Mean Clustering algorithm to detect the follicles. In this data, clusters are formed with similar characteristics. By apply this method they reduced the dataset to get refined output [4]. Hiremath [29] used Mean Square Error to measure the performance of segmentation. If the value of MSE is lower than the executed segmentation value is effective one and this way they used the MSE technique to detect the cyst in the ovary. Hiremath [29] utilized contourlet transformation to remove the speckle noise from the image and applied the histogram equalization method to enhance the image. Instead of edge detection method author used active contourlet to detect the follicle. Prasanna [6] proposed improved total variation algorithm (ITV) to reduce the noise specking in the image ITV uses some of the filtering parameter like MSE. PSNR, FSIM, SSIM used to clear the speckle noise, by the example of three ovary they examined the speed variation of removing noise from the ovary image. In that ITV Method obtained very less average time as 11.87 value. Kiruthika [23] proposed to find PCO and classification of ovary using Discrete Wavelet Transform and also texture and intensity based ovarian method is used to find the recognition of follicles. Table 2 represent the different authors of detection of follicles present in the ovary with different metrics.

Table 2. Represent the follicle recognition and false acceptance rate.

Author	Metrics	Recognition of follicles	False acceptance rate (FAR)
[28]	MSE	3.684	2.1222
[20]	KNN and SVM	92.86%	31.1%
[18]	BVF	61%	1%
[27]	Active contour	92.3%	12.6%
[9]	Contourlet transform	75.2%	24.1%
[30]	AI	89.4%	7.45%
[27]	Fuzzy logic	97.61% and 98.18%	9.05% and 4.52%
[31]	SMV	99.67%	2.00%

3 Conclusion

This paper overview the ovarian classification and detection of PCOS in different methods using Image Processing and Machine Learning methods. From the study it shows that the factor and level of the factor identifying the PCOS in ultrasound images are identified and the result is compared with different metrics. Even though so many implementation done, using different techniques with proper valued outcome and many tasks are intensify for future work.

References

1. Haleem, A., Javaid, M.: 3D scanning applications in medical field: a literature-based review. Clin. Epidemiol. Glob. Health **7**(2), 199–210 (2019)
2. Mahmood, N.H., Ahmmad, S.N.Z., Naqiah, H., Rani, A.: Ovary ultrasound image edge detection analysis: a tutorial using MATLAB. Int. J. Eng. Res. Appl. (IJERA), **2**(3), 2248–9622 (2012)
3. Vasavi, G., Jyothi, S.: Classification and detection of ovarian cysts in ultrasound Images. In: International Conference on Trends in Electronics and Informatics ICEI. IEEE (2017). 978-1-5090-4257-9
4. Raj, A.: Detection of cysts in ultrasonic image of ovary. Int. J. Sci. Res. (USR) **2**(8), 2319–7064 (2013)
5. Padmapriya, B., Kesavamurthy, T.: Detecting of follicles in poly cystic ovarian syndrome in ultrasound images using morophological operations. J. Med. Imaging Health Inform. **6**, 240–243 (2016)
6. Prasanna Kumar, H., Srinivasan, S.: Despeckling of polycystic ovary ultrasound images by improved total variation method. Int. J. Eng. Technol. (UET) **6**(4) (2014). 0975–4024
7. Usha, B.S., Sandyas, S.: Measurement of ovarian size and shape parameters. In: Annual IEEE Indian Conference (INDICON) (2013). 978-4799-2275-8
8. Hiremath, P.S., Tegnoor, J.R.: Follicle detection and ovarian classification in digital ultrasound images of ovaries (2013). https://doi.org/10.5772/56518
9. Hiremath, P.S., Tegnoor, J.R.: Automatic detection of follicles in ultrasound images of ovaries using edge based method. Int. J. Comput. Appl. Spec. Issue Recognit. **8**(3), 120–125 (2010)
10. Krivanek, A., Sonka, M.: Ovarian ultrasound image analysis; follicle segmentation. IEEE Trans. Med. Imaging **17**(6) (1998). 0278-0062/98
11. Setiawati, E., Tjokorda, A.: Particle swarm optimization on follicles segmentation to support PCOS detection. In: 3rd International Conference on Information and Communication Technology (ICoICT) (2015). 978-1-4799-7752-9/15
12. Potocnik, B., Cigale, B., Zazula, D.: The XUltra project-automated analysis of ovarian ultrasound images. In: Proceedings of the 15th IEEE Symposium on Computer-Based Medical Systems (CBMS 2002). IEEE (2002). 1063-7125/2
13. Purnama, B., Wisest, U.N., Nhita, A.F., Gayatri, A., Mutiah, T.: A classification of polycystic ovary syndrome based on follicle detection of ultrasound images. In: 3rd International Conference on Information and Communication Technology (ICoICT). IEEE (2015). 987-1-4799-7752

14. Aitheswaran, R., Malarkhodi, S.: An effective automated system in follicle identification for polycystic ovary syndrome using ultrasound images. In: International Conference on Electronics and Communication System (ICECS) (2014)

15. Raj, A.: Ovarian follicle detection for polycystic ovary syndrome using Fuzzy C-means clustering. Int. J. Comput. Trends Technol. (IJCTT) 4(7), 2231–2803 (2013)

16. Prasanna Kumar, H., Srinivasan, S.: Segmentation of polycystic ovary in ultrasound images. In: 2nd International Conference on Current Trends in Engineering and Technology (ICCCTE), pp. 237–240. IEEE (2014)

17. Rihana, S., Moussallem, H., Skaf, C., Yaacoub, C.: Automated algorithm for ovarian cysts detection in ultra sonogram. In: 2nd International Conference on Advances in Biomedical Engineering. IEEE (2013). 978-1-4799-0251-4/13

18. Deng, Y., Wang, Y., Senoor, Chen, P.: Automated detection of polycystic ovary syndrome from ultrasound images. In: 30th Annual International IEEE EMBS Conference, Vancouver, British Columbia, Canada. IEEE (2008). 978-1-4244-1815-2/08

19. Potocnik, B., Zazula, D., Korze, D.: Automated computer-assisted detection of follicle in ultrasound images of ovary. J. Med. Syst. 21(6), 445–457 (1997)

20. Lawrence, M.J., Eramian, M.G., Pierson, R.A., Eramian, M.G., Neufeld, E.: Computer assisted detection polycystic ovary morphology in ultrasound images. In: Fourth Canadian Conference on Computer and Robot Vision (CRV 2007). IEEE (2007). 7695-8/07

21. Parekh, A.M., Shah, N.B.: Classification of ovarian cyst using soft computing technique. In: 8th ICCCNT. IEEE (2017). 40222

22. Mehrotra, P., Chakraborty, C., Ghoshdastidar, B., Ghoshdastidar, S.: Automated ovarian follicle recognition for polycystic ovary syndrome. In: International Conference on Image Information Processing (ICIIP). IEEE (2011). 978-1-61284-861-7

23. Kiruthika, V., Sathiya, S., Ramya, M.M.: Automatic texture and intensity based ovarian Classification. J. Med. Eng. Technol. 0309–1902 (2019)

24. Padmapriya, B., Kesavamurthy, T.: Diagnostic tool for PCOS classification. In: 7th WACBE World Congress on Bioengineering, pp. 182–185 (2015)

25. Vigil, P., Cortés, M.E., Zúñiga, A., Riquelme, J., Ceric, F.: Scanning electron and light microscopy study of the cervical mucus in women with polycystic ovary syndrome. J. Electron Microsc. 58(1) (2009)

26. Padmapriya, B., Kesavamurthy, T.: Diagnostic tool for PCOS classification. Springer (2015). https://doi.org/10.1007/978-3-319-19452-3_48

27. Hiremath, P.S., Tegnoor, J.R.: Automated ovarian classification in digital ultrasound images. Int. J. Biomed. Eng. Technol. 11(1), 46–65 (2013)

28. Mehrotra, P., Chakraborty, C., Ghoshdastidar, B., Ghoshdastidar, S.: Automated ovarian follicle recognition for polycystic ovary syndrome. In: 2011 International Conference on Image Information Processing (ICIIP). IEEE (2011). 978-1-61284-861-7/11

29. Hiremath, P.S., Tegnoor, J.R.: Automatic detection of follicles in ultrasound image of ovaries using active contours method. In: Proceedings of IEEE International Conference on Computational Intelligence and Computing Research (ICCIC) (2010). Health Informatics, 19(4). IEEE (2015)

30. Deng, Y., Wanga, Y., Shenb, Y.: An automated diagnostic system of polycystic ovary syndrome based on object growing. Artif. Intell. Med. 51, 199–209 (2011)

31. Tegnoor, J.R.: Automatic ovaries classification in digital ultrasound image using SVM. Int. J. Eng. Res. Technol. (IJERT) 1(6), 1–17 (2012)

Real-Time Application of Document Classification Based on Machine Learning

Abhijit Guha[1] and Debabrata Samanta[2(✉)]

[1] Eagle Labs, First American India Private LTD., Bangalore, India
abhijitguha.research@gmail.com
[2] Department of Computer Science, Christ University (Deemed to be),
Bangalore, India
debabrata.samanta369@gmail.com

Abstract. This research has been performed, keeping a real-time application of document (multi-page, varying length, scanned image-based) classification in mind. History of property title is captured in various documents, recorded against the said property in all the countries across the world. Information of the property, starting from ownership to the conveyance, mortgage, refinance etc. are buried under these documents. This is by far a human driven process to manage these digitized documents. Categorization of the documents is the primary step to automate the management of these documents and intelligent retrieval of information without or minimal human intervention. In this research, we have examined a popular, supervised machine learning technique called, SVM (support vector machine) with a heterogeneous data set of six categories of documents related to property. The model obtained an accuracy of 88.06% in classifying over 988 test documents.

Keywords: Document classification · SVM · tf-idf · t-SNE · Text analytics · Balanced accuracy · OCR

1 Introduction

In this era of digitization, humongous amount of electronic data is generated, stored and ready to be consumed by computer systems, for faster, intelligent and automated processing [1–5]. One such area which has tremendous potential for application of AI is property title examination which mostly deals with scanned documents of various types, for example: Deeds, Mortgages, Agreements, Taxes and so on, recorded in the registrar office against the property. Classification of these documents is the primary task for automated management and processing of the information buried inside them. A well-managed property data has direct influence on the economic well-being of a country. What makes the task of classification difficult is that the documents are of various kind, varying length, unstructured format and are closely related in terms of the feature content [6–10]. An advanced methodology of unstructured data mining, a systematic exploration and engineering of the features become very important to model a multi-class classifier to obtain a reasonable accuracy in this task. In the proposed method, we systematically explored property related documents of six categories and

© Springer Nature Switzerland AG 2020
L. C. Jain et al. (Eds.): ICICCT 2019, LAIS 9, pp. 366–379, 2020.
https://doi.org/10.1007/978-3-030-38501-9_37

generated a high dimensional feature space using $tf-idf$ technique of word embedding, a state-of-the-art mechanism of converting unstructured text data into n-dimensional feature space. We applied SVM (support vector machine) with linear kernel for this research [11–13].

Below sections of this paper systematically depicts the methodology, data set, feature engineering technique, through exploratory analysis to approach classification in Machine Learning followed by the result of the experiment, conclusion and the future scope.

2   Research Methodology

The methodology comprises of three stages. In the first stage, we have performed an exploratory analysis using advanced text analytic and data mining techniques to understand the feasibility of classification using "Machine Learning" approach. The second stage, we have presented the "feature engineering" technique followed by actual classification and validation of result of the experiment. The proposed architecture of the system consists of two modules, namely Training and Testing. In training module, the system learns from the existing labelled data, extracted and engineered through pre-processing and the testing module validates the accuracy of the model. OCR [14–16] (optical character recognition) becomes an important step in this research as we are dealing with scanned, image-based documents [22–28]. Extraction of the texts from the images is an already established field of research. Once extracted, cleaning of the text which is called the text pre-processing is done followed by state-of-the-art word embedding technique known as $tf-idf$ vectorization.

In this research, we have used the data set of six categories of property documents related to Deed, Purchase Agreements (PSA), Taxes, Insurance Commitment (TC), Property Detail Report (PDR), Voluntary Lien Report (VLR). The distribution of each type of document is as shown in the below plot, Fig. 1. In the next sections, we have examined these documents in greater detail using advanced Text Analytic and Data Mining techniques and applied pre-processing techniques to extract low level features out of the text corpus [29, 30].

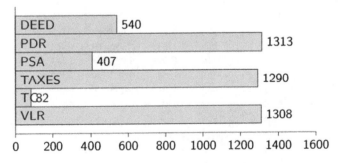

Fig. 1.   Category wise document count

3   Exploratory Analysis

We have performed a thorough exploratory analysis on the data set to understand the features through a visual representation of the same [17–21]. This exploration helps us model the approach of classification better. There are three steps involved in the exploration: 1. pre-processing of the texts and examination of the uni-grams with associated probability of the tokens being part of a class, 2. feature extraction and building the feature vector space and 3. Visualize the feature space into a low dimensional space.

3.1   Pre-processing

Text pre-processing steps are performed on the OCR extracted texts for 4940 documents of six categories. In this step, we have removed the word tokens and characters having less or no contribution in defining the class of the document. Standard text pre-processing techniques like *punctuation removal, non-ASCII characters removal, stop-words removal, numbers removal, lower-casing, rare tokens removal* etc. are applied to the data.

After applying the above pre-processing on the corpus, we have got 12392 unique tokens (uni-gram) in the corpus. Below is the list of top 20 tokens (uni-gram and bi-gram) in terms of their frequency of presence in documents.

Table 1: Captures top 20 uni-grams along with their term frequency, document frequency and respective inverse document frequency present in the whole text corpus.

Table 1.   Top 20 uni-grams in the corpus

Term	TermFreq	DocFreq	idf
Sale	94941	3596	0.31754321459045
Date	86264	4542	0.08399788751646
Type	73775	4015	0.20732798380110
Mortgage	68727	1707	1.06262788738753
Property	67951	4750	0.03922071315328
Document	65783	3203	0.43327746057281
Seller	63122	1007	1.59038971746341
Title	61293	3589	0.31949171903436
Recording	57450	3285	0.40799867926325
Buyer	52828	354	2.63582369704819
Agreement	49716	1769	1.02695091602218
Transfer	45521	1866	0.97356822877467
Amount	35969	4301	0.13851777739823
Finance	30098	1380	1.27528183203072
Closing	27656	456	2.38262780066758
Orig	26324	1237	1.38467623778948
Loan	26215	1642	1.10145032016959
Rate	25848	2026	0.89130192537334
Company	25376	2300	0.76445620826472
History	23043	2651	0.62242840389294

Table 2: Captures top 20 bi-grams along with their term frequency, document frequency and respective inverse document frequency present in the whole text corpus.

Table 2. Top 20 bi-grams in the corpus

Term	TermFreq	DocFreq	idf
recording_date	43815	2646	0.62431626550706
mortgage_rate	23709	1304	1.33192886769537
document_type	20398	1305	1.33116229042517
history_record	20363	1304	1.33192886769537
data_trace	16688	1237	1.38467623778948
title_company	16092	16092	1.06497392951928
date_sale	14955	3151	0.44964546842469
sale_date	14837	3166	0.44489636974101
sale_price	14348	2644	0.62507240921039
type_sale	13405	1275	1.35441915258944
sale_type	12951	2615	0.63610123368294
type_mortgage	12334	1293	1.34040023141011
mortgage_loan	12153	1350	1.29726073874949
loan_amount	12118	1337	1.30693703308002
transfer_type	12079	1304	1.33192886769537
loan_type	12069	1307	1.32963089655775
finance_title	12068	1305	1.33116229042517
rate_type	12065	1304	1.33192886769537
mortgage_recording	12047	1306	1.33039630034559
date_mortgage	12035	1306	1.33039630034559

The philosophy of classification of documents in $tf-idf$ feature space is that the documents are constructed of varying types and frequency of tokens. Through exploration, we have tried to get an insight of the assumption and have taken a decision on the feasibility of a machine learning based classification approach. In the below section, (Tables 3, 4, 5, 6, 7 and 8) we have inspected top 20 tokens from each category of documents and the conditional probability of the tokens being present within that class. Below results show that the words and their associated probabilities are pretty much an indicator of a good classification model. The word boundaries seem to be well separated for the categories, represented in the Venn diagram below (Fig. 2).

(Tables 3, 4, 5, 6, 7 and 8) Captures below the top 20 uni-gram tokens per document category with the associated term frequency, document frequency, inverse document frequency and associated conditional probability of each token being part of the said category.

We see that there are intersections among the groups in terms of the top 20 frequent tokens, maximum between the group 'DEED' and 'TC'.

Table 3. Top 20 uni-grams in DEED documents with conditional probability

Term	TermFreq	DocFreq	idf	Prob
County	3665	540	0	0.028330035248283
State	1967	536	0.007434978	0.015204687403376
Instrument	1549	505	0.067010710	0.011973594706575
Deed	1753	539	0.001853568	0.013550491620803
Recorded	1461	467	0.145239881	0.011293364665144
Grantor	1682	376	0.361979996	0.013001669655556
Notary	1086	482	0.113625025	0.008394657102227
Public	1110	484	0.109484232	0.008580174386249
Grantee	1216	333	0.483426649	0.009399542390698
Commission	819	442	0.200259257	0.006330777317423
Page	1045	376	0.361979996	0.008077731742004
Acknowledged	801	513	0.051293294	0.006191639354393
Expires	750	440	0.204794412	0.005797415125842
Warranty	1003	462	0.156004248	0.007753076494960
Oregon	708	67	2.086876520	0.005472759878795
Plat	738	364	0.394415271	0.005704656483829
Seal	651	365	0.391671785	0.005032156329231
Feet	813	110	1.591088773	0.006284397996413
Consideration	694	454	0.173471941	0.005364541463112
Official	631	338	0.468523244	0.004877558592542

Table 4. Top 20 uni-grams in PDR documents with conditional probability

Term	TermFreq	DocFreq	idf	Prob
Property	7867	1313	0	0.06786986791818
Sale	6545	1309	0.0030511083	0.05646476236487
Number	3927	1309	0.0030511083	0.03387885741892
Prior	3927	1309	0.0030511083	0.03387885741892
Information	3927	1309	0.0030511083	0.03387885741892
Date	5236	1309	0.0030511083	0.04517180989190
Reference	2618	1309	0.0030511083	0.02258590494595
Assessors	2618	1309	0.0030511083	0.02258590494595
Address	2618	1309	0.0030511083	0.02258590494595
Parcel	2692	1309	0.0030511083	0.02322431478781
Legal	2622	1313	0	0.02262041358605
Price	2618	1309	0.0030511083	0.02258590494595
Water	1467	1292	0.0161231899	0.01265604375695
Medium	1319	1313	0	0.01137922407322
Amounttype	1309	1309	0.0030511083	0.01129295247297
Alternate	1309	1309	0.0030511083	0.01129295247297
Neighborhood	1311	1309	0.0030511083	0.01131020679302
Characteristics	1309	1309	0.0030511083	0.01129295247297
Municipality	1309	1309	0.0030511083	0.0112929524729
Carrier	1306	1306	0.0053455644	0.01126707099289

Table 5. Top 20 uni-grams in PSA documents with conditional probability

Term	TermFreq	DocFreq	idf	Prob
Seller	62471	406	0.00246002584	0.0270118349463642
Buyer	52795	339	0.1828130780	0.0228280294215444
Agreement	45737	406	0.00246002584	0.0197762208855606
Closing	27564	402	0.0123610968	0.0119183976318865
Property	49485	406	0.00246002584	0.0213968185609456
Purchaser	18643	214	0.6428371704	0.0080610465480793
Escrow	18806	401	0.01485175813	0.0081315261161390
Section	16545	394	0.03246227614	0.0071538923530533
Sellers	14345	401	0.01485175813	0.0062026343792414
Purchase	13055	406	0.00246002584	0.0056448512945972
Party	14427	404	0.00739830748	0.0062380903582653
Buyers	11587	313	0.26260999490	0.0050101027920718
Notice	12977	401	0.01485175818	0.0056111248755257
Parties	10867	403	0.00987662341	0.0046987820006425
Agent	10105	385	0.05556985118	0.0043693008297131
Lease	9569	321	0.237372062312	0.0041375397960935
Days	9010	402	0.01236109682	0.0038958337927477
Provided	9496	402	0.01236109682	0.0041059753269625
Period	8268	394	0.03246227614	0.0035750004215802
Obligations	7935	395	0.029927420541	0.0034310145555441

Table 6. Top 20 uni-grams in TAX documents with conditional probability

Term	TermFreq	DocFreq	idf	Prob
Data	18309	1244	0.0363102240	0.04639288890690
Trace	16787	1236	0.0427618593	0.04253631689771
Customer	12015	1244	0.0363102240	0.03044462069017
Order	12165	1287	0.0023282897	0.03082470334548
Taxes	8540	1256	0.0267101503	0.02163937250887
Report	7291	1244	0.0363102240	0.01847455093234
Search	6090	1254	0.0283037761	0.01543135580550
Information	7130	1257	0.0259142887	0.01806659554897
Warranty	5933	1236	0.0427618593	0.01503353595962
Certificatetax	4843	1206	0.0673331200	0.01227160199771
Total	4927	1257	0.0259142887	0.01248444828468
Land	4898	1258	0.0251190600	0.01241096563799
Service	3800	1285	0.0038835000	0.00962876060113
Assessments	3684	1244	0.0363102240	0.00933483001436
Angeles	3648	746	0.5476718971	0.00924361017709
Paid	3796	1255	0.0275066457	0.00961862506366
Results	2527	1248	0.0330999484	0.00640312579975
Cover	2454	1236	0.0427185933	0.00621815224084
Taxing	2457	1235	0	0.00622575389394
Park	2509	745	0.5490132789	0.0063575158811

Table 7. Top 20 uni-grams in TC documents with conditional probability

Term	TermFreq	DocFreq	idf	Prob
Commitment	2589	74	0.1026541540	0.01728327481007
Information	2681	71	0.1440393702	0.01789743521275
Policy	2070	81	0.0122700925	0.01381860906020
Company	2680	82	0	0.01789075955620
Insured	1506	81	0.0122700925	0.01005353876553
Recorded	1586	81	0.0122700925	0.01058759128960
Schedule	1453	80	0.0246926125	0.00969972896834
Insurance	1494	82	0	0.00997343088692
American	1364	81	0.0122700925	0.00910559553532
Land	1541	82	0	0.01028718674481
Page	1384	81	0.0122700925	0.00923910866633
Records	1206	81	0.0122700925	0.00805084180029
Public	1085	73	0.1162598061	0.00724308735764
Feet	1174	53	0.4364273337	0.00783722079066
Shown	876	79	0.0372713947	0.00584787513851
Proposed	755	75	0.0892311337	0.00504012069587
Book	735	66	0.2170645052	0.00490660756485
Estate	828	82	0	0.00552744362408
Interest	924	82	0	0.00616830665295
East	728	54	0.4177352006	0.00485987796899

Table 8. Top 20 uni-grams in VLR documents with conditional probability

Term	TermFreq	DocFreq	idf	Prob
Sale	79588	1304	0.003062789530545	0.08815948195171
Type	68271	1304	0.003062789530545	0.07562366176214
Mortgage	67106	1304	0.003062789530545	0.07433319339413
Document	61562	1304	0.003062789530545	0.06819211473980
Recording	52853	1304	0.003062789530545	0.05854517137752
Date	52829	1304	0.003062789530545	0.05851858662143
Transfer	42125	1304	0.003062789530545	0.04666178541006
Finance	29993	1304	0.003062789530545	0.03322319121196
Orig	26172	1133	0.143630270989138	0.02899067650450
Amount	27345	1304	0.003062789530545	0.03029000645788
Title	32676	1304	0.003062789530545	0.03619514540200
Loan	24397	1304	0.003062789530545	0.02702451225280
Rate	24163	1304	0.003062789530545	0.02676531088102
History	21668	1304	0.003062789530545	0.02400160394691
Record	20364	1304	0.003062789530545	0.02255716553330
Bookpage	17446	1133	0.143630270989138	0.01932490227332
Lender	12094	1304	0.003062789530545	0.01339650166763
Stamp	11638	1304	0.003062789530545	0.01289139130213
Owner	13091	1304	0.003062789530545	0.01450087674310
Term	12077	1304	0.003062789530545	0.01337767079875

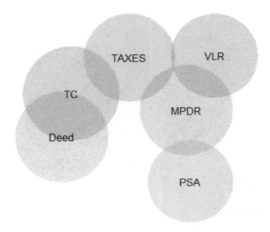

Fig. 2. Venn diagram of top 20 uni-grams from six categories of documents.

3.2 Feature Extraction

In this research, we have used *tf*−*idf* statistics for feature embedding. The *tf*−*idf* measure is the product of two metrics, term frequency $f(t,d)$ where t represents the terms and d represents the document of the corpus, and inverse document frequency. Term frequency $tf(t,d)$, gives the count of number of times that term t occurs in document d. If, the count is represented by $f_{t,d}$,

$$tf(t,d) = f_{t,d} \tag{1}$$

The *inverse document frequency* is a measure which provides the information about the commonality or the rarity of a term. We get measure of document frequency for a term by the ratio of the total number of documents represented by N, present in the corpus and the number of documents containing the term t is represented by D. *Inverse document frequency* is the inverted logarithm of the document frequency.

$$idf(t,D) = \log \frac{N}{|\{d \in D : t \in d\}|} \tag{2}$$

tf−*idf* is calculated as

$$tf{-}idf(t,d,D) = tf(t,d) * idf(t,D) \tag{3}$$

Applying the above-mentioned technique of feature embedding, we get a matrix of dimension 4940×12393 where the row number 4940 represents the number of documents and the column number 12393 represents the feature dimension length. As, it is impossible to visualize the data in such a high dimensional space we use *t-SNE* (t-distributed Stochastic Neighbor Embedding), a state-of-the-art technique of high dimensional data visualization. 2D representation of the data is shown below in Fig. 3.

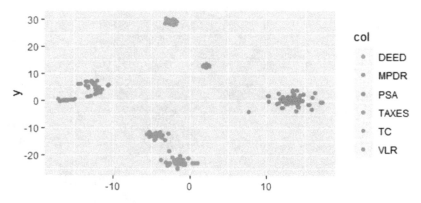

Fig. 3. 2D visualization of n-dimensional features

t-SNE [17] (*t-distributed Stochastic Neighbor Embedding*) is state-of-the-art statistical technique to reduce high dimensional data into lower dimensions using non-linear method of dimensionality reduction. By creating probability distribution over pair of high dimensional data points, this methodology groups the nearby data points with high probability and distant data points with lower probability.

If $\zeta \in R^d$ where ζ denotes the data points and d is the dimensionality of the space. If ζ_i and ζ_j are two points in the space, the Euclidean distance between ζ_i and ζ_j is denoted by $|\zeta_i - \zeta_j|$. The conditional similarity between the two points are denoted by,

$$\zeta_{j|i} = \frac{exp\left(-||\zeta_i - \zeta_j||^2 / 2\psi_i^2\right)}{\sum_{k!=i} exp\left(-||\zeta_i - \zeta_k||^2 / 2\psi_i^2\right)} \tag{4}$$

This statistic calculates the probability of the distance of ζ_i *is from* ζ_j considering a normal distribution around ζ_i with a variance ψ^2. The variance depends on the density of the data points around the point in question. Similarity of the two points are defined by the measure below.

$$\zeta_{i,j} = \frac{\zeta_{j|i} + \zeta_{i|j}}{2N} \tag{5}$$

3.3   Visualization in 2D Space

The 2D visualization of the high dimensional feature vectors using *t-SNE* [17], confirms that the similar documents are clustered together and are well separated in the two-dimensional space. We conclude from the results of EDA, that a supervised classification algorithm is a good choice for learning the patterns hidden in the low-level features within the documents and high accuracy of classification is expected.

4 Classification

If there are two non-empty sets D and C where domain of D is the set of all documents and domain of C is the set of the categories of the documents. Mathematically f : $D \to C\{0 \, to \, 1\}$ is a surjection logistic function where $y = f(x), x \in D$ and $y \in C$. The classifier approximates the true function f with another function $f' : D \to C\{0 \, to \, 1\}$ where the approximation results to error as minimal as possible. A classifier function is built by training a subset of D with labelled classes from C. The challenge in scanned document classification lies in the presence of huge number of word features (high dimensional) and understanding the complex semantics of the terms within the documents. Support vector machine has been used as the classifier in this research.

4.1 Support Vector Machine

SVM, since its inception has been one of the most popular supervised machine learning algorithms in practice. In this methodology the data points in high dimensional plane are separated by a hyper-plane with maximum possible margin. This algorithm can perform non-linear classification using different kernel tricks which internally maps the data points into a higher dimensional plane for a better separation of the points in that space.

Given a set of training data points as pairs of data and class $(x_i, y_i), i = 1, \ldots, l$ where $x_i \in R^n$ and $y \in 1, -1$, the optimization which SVM tries to perform is,

$$\frac{min}{w, b, \tau} \frac{1}{2} w^T w + C \sum_{i=1}^{l} \tau_i \tag{6}$$

$$subject \, to \quad y_i \left(w^T \varphi(x_i) + b \right) \geq 1 - \tau_i, \tau_i \geq 0 \tag{7}$$

The training data points (n dimensional) x_i are projected on a higher dimensional space using the function φ. During the training, the algorithm tries to find out the equation of the hyper-plane which maximizes the separation of the classes. C is the regularization parameter. The kernel function is,

$$K(x_i, x_j) = \varphi(x_i)^T \varphi(x_j) \tag{8}$$

There are four basic types of kernels used in SVM.

$$Linear \, kernel : K(x_i, x_j) = \sum_{k=0}^{n} x_i * x_j \tag{9}$$

$$Polinomial \, kernel : K(x_i, x_j) = 1 + \left(\sum_{k=0}^{n} x_i * x_j \right)^d \tag{10}$$

$$Radial\ kernel : K\left(x_i, x_j\right) = exp^{\gamma *\left(\sum_{k=0}^{n} x\_i * x\_j\right)^d} \tag{11}$$

Where $\gamma, r,\ and\ d$ represent the kernel parameters.

An 80 : 20 random split has been done to generate the training and the testing set. Training set is used to train the SVM classification model and testing set which is unseen to the model, is used for testing the accuracy of the prediction. Below section shows the result of the model. Test set consists of 988 documents and the distribution of the documents in the test set is shown below in Fig. 4.

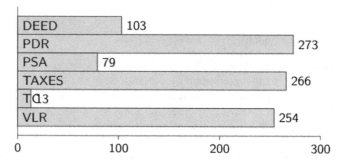

Fig. 4. Category wise document count for test set.

4.2   Result

Here, we have examined below three metrics for accessing the accuracy of the model. As we are dealing with highly class imbalanced data, overall accuracy may not be the best metric to assess the model performance. Sensitivity and specificity are also taken into consideration for calculating the balanced accuracy.

Overall Accuracy. It is the representation of the ratio of number of accurate predictions made by the classifier to the total number of test samples.

$$Accuracy = \frac{Number\ of\ correct\ predictions}{Total\ predictions\ made} \tag{12}$$

Sensitivity (True Positive Rate). It is the representation of the ratio true positives and the sum of false negatives and true positives. Positive data points which are correctly classified as positive are the true positive set whereas, the positive data points which are categorized as negative are the false negative set.

$$Sensitivity = \frac{TP}{FN + TP} \tag{13}$$

Specificity (False Positive Rate). Specificity is measured with the ratio of false positives and the sum of false positives and true negatives. Negative data points that are mistakenly considered as positive by the classifier are the false positives and the negative data points which are truly identified as negatives by the classifier represents the true negative set.

$$Specificity = \frac{FP}{FP + TN} \tag{14}$$

Below tables capture different accuracy metrics for the research (Tables 9 and 10).

Table 9. Confusion matrix

	Deed	PDR	PSA	TAXES	TC	VLR
DEED	45	7	0	7	0	8
PDR	0	228	0	0	0	0
PSA	72	8	82	5	8	2
TAXES	0	0	0	242	0	0
TC	0	0	0	0	16	0
VLR	0	0	0	0	0	257

Table 10. Accuracy metrics

	Deed	PDR	PSA	TAXES	TC	VLR
Sensitivity	0.38136	0.9383	1.0000	0.9528	0.66667	0.9625
Specificity	0.97471	1.0000	0.8940	1.0000	1.0000	1.0000
Balanced accuracy	0.67803	0.9691	0.9470	0.9764	0.83333	0.9813
Overall accuracy	88.06					

5   Conclusion and Future Scope

We have presented here an automatic multi-page multi-class scanned image document classification technique using SVM which has obtained an overall accuracy of 88.06%. This model can be further tuned and validated using other kernel functions. Also, other supervised machine learning algorithms like Gradient boosting, Random forest, Multinomial Logistic regression, K-nearest neighbor and Deep learning techniques (multi-layer perception, Convolution neural network) can be applied to validate and compare with presently obtained accuracy. Also, there are various other techniques of representing the features into n dimensional space, such as Doc2vec, Fast Text etc., can be explored individually and in combinations with each other. These techniques of embedding retain the semantic information to a large extent and can be useful in classifying documents with very similar representation in $tf-idf$ feature space.

The proposed research is expected to have an impact in the field of property data management by enabling and improving the possibility of RPA (robotic process automation). Though the research has been performed using the data from the real estate property domain but can be adopted by any field having a requirement of automated image-based document classification and management.

References

1. Jiang, Y., Shen, Q., Fan, J., Zhang, X.: The classification for e-government document based on SVM. In: IEEE - 2010 International Conference on Web Information Systems and Mining (2010)
2. Mtimet, J., Amiri, H.: Document class recognition using a support vector machine approach. In: IEEE - 2016 2nd International Conference on Advanced Technologies for Signal and Image Processing (ATSIP) (2016)
3. Awal, A.-M., Ghanmi, N., Sicre, R., Furon, T.: Complex document classification and localization application on identity document images. In: IEEE - 2017 14th IAPR International Conference on Document Analysis and Recognition (ICDAR) (2017)
4. Chagheri, S., Roussey, C., Calabretto, S., Dumoulin, C.: Technical documents classification. In: IEEE - Proceedings of the 2011 15th International Conference on Computer Supported Cooperative Work in Design (CSCWD) (2011)
5. Gordo, A., Rusiñol, M., Karatzas, D., Bagdanov, A.D.: Document classification and page stream segmentation for digital mailroom applications. In: IEEE - 2013 12th International Conference on Document Analysis and Recognition (2013)
6. Shinde, S., Joeg, P., Vanjale, S.: Web document classification using support vector machine. In: IEEE - 2017 International Conference on Current Trends in Computer, Electrical, Electronics and Communication (CTCEEC) (2017)
7. Plansangket, S., Gan, J.Q.: A new term weighting scheme based on class specific document frequency for document representation and classification. In: IEEE - 2015 7th Computer Science and Electronic Engineering Conference (CEEC) (2015)
8. Dsouza, F.H., Ananthanarayana, V.S.: Document classification with a weighted frequency pattern tree algorithm. In: IEEE - 2016 International Conference on Data Mining and Advanced Computing (SAPIENCE) (2016)
9. Joshi, S., Nigam, B.: Categorizing the document using multi class classification in data mining. In: IEEE - 2011 International Conference on Computational Intelligence and Communication Networks (2011)
10. Baygin, M.: Classification of text documents based on Naive Bayes using N-Gram features. In: IEEE - 2018 International Conference on Artificial Intelligence and Data Processing (IDAP) (2018)
11. Hsu, C.-W., Chang, C.-C., Lin, C.-J.: A practical guide to support vector classification. Department of Computer Science, National Taiwan University, Taipei 106, Taiwan (2016)
12. Hastie, T., Tibshirani, R., Friedman, J.: The Elements of Statistical Learning: Data Mining, Inference, and Prediction. (PDF), 2nd edn. Springer, New York. p. 134
13. Samanta, D., Podder, S.K.: Level of green computing based management practices for digital revolution and New India. Int. J. Eng. Adv. Technol. (IJEAT) 8(3S) (2019). ISSN 2249-8958

14. Sivakumar, P., Nagaraju, R., Samanta, D., Sivaram, M., Hindia, N., Amiri, I.S.: A novel free space communication system using nonlinear InGaAsP microsystem resonators for enabling power-control toward smart cities. Wirel. Netw. J. Mob. Commun. Comput. Inf. ISSN 1022-0038

15. Mahua, B., Podder, S.K., Shalini, R., Samanta, D.: Factors that influence sustainable education with respect to innovation and statistical science. Int. J. Recent Technol. Eng. 8(3). ISSN 2277-3878

16. Praveen, B., Umarani, N., Anand, T., Samanta, D.: Cardinal digital image data fortification expending steganography. Int. J. Recent Technol. Eng. 8(3) (2019). ISSN 2277-3878

17. Dhanush, V., Mahendra, A.R., Kumudavalli, M.V., Samanta, D.: Application of deep learning technique for automatic data exchange with Air-Gapped Systems and its Security Concerns. In: Proceedings of IEEE International Conference on Computing Methodologies and Communication, Erode, 18–19 July 2017

18. Kumar, R., Rishabh, K., Samanta, D., Paul, M., Vijaya Kumar, C.M.: A combining approach using DFT and FIR filter to enhance Impulse response. In: Proceedings of IEEE International Conference on Computing Methodologies and Communication, Erode, 18–19 July 2017

19. Ghosh, G., Samanta, D., Paul, M., Kumar Janghel, N.: Hiding based message communication techniques depends on divide and conquer approach. In: Proceedings of IEEE International Conference on Computing Methodologies and Communication, Erode, 18–19 July 2017

20. Singh, R.K., Begum, T., Borah, L., Samanta, D.: Text encryption: character jumbling. In: Proceedings of IEEE International Conference on Inventive Systems and Control @IEEE, Coimbatore, 19–20 January 2017

21. Ben-Hur, A., Horn, D., Siegelmann, H., Vapnik, V., Jerome, N.: Support vector clustering. J. Mach. Learn. Res. (2001)

22. Optical Character Recognition (OCR) – How it works. Nicomsoft.com. Accessed 16 June 2013

23. Chakrabarti, S., Samanta, D.: Image steganography using priority-based neural network and pyramid. In: Emerging Research in Computing, Information, Communication and Applications, pp. 163–172 (2016). https://doi.org/10.1007/978-981-10-0287-8_15

24. Ghosh, G., Samanta, D., Paul, M.: Approach of message communication based on twisty "Zig-Zag". In: 2016 International Conference on Emerging Technological Trends (ICETT) (2016). https://doi.org/10.1109/icett.2016.7873676

25. Hossain, M.A., Samanta, D., Sanyal, G.: Extraction of panic expression depending on lip detection. In: 2012 International Conference on Computing Sciences (2012). https://doi.org/10.1109/iccs.2012.35

26. Hossain, M.A., Samanta, D., Sanyal, G.: Statistical approach for extraction of panic expression. In: 2012 Fourth International Conference on Computational Intelligence and Communication Networks (2012). https://doi.org/10.1109/cicn.2012.189

27. Khadri, S.K.A., Samanta, D., Paul, M.: Approach of message communication using fibonacci series: in cryptology. Lect. Notes Inf. Theory (2014). https://doi.org/10.12720/lnit.2.2.168-171

28. Schantz, H.F.: The history of OCR, optical character recognition. [Manchester Center, Vt.]: Recognition Technologies Users Association. ISBN 9780943072012

29. Vamvakas, G., Gatos, B., Stamatopoulos, N., Perantonis, S.J.: A complete optical character recognition methodology for historical documents

30. van der Maaten, L.J.P., Hinton, G.E.: Visualizing data using t-SNE. J. Mach. Learn. Res. 9, 2579–2605 (2008)

A Revised Study of Stability Issues in Mobile Ad-Hoc Networks

Nismon Rio Robert[1(⊠)] and Viswambari Madhanagopal[2]

[1] Department of Computer Science, Christ (Deemed to be University),
Bengaluru 560 029, Karnataka, India
nismon.rio@christuniversity.in
[2] Department of Computer Science, Patrician College of Arts and Science,
Chennai 600 020, Tamil Nadu, India
viswambari1391@gmail.com

Abstract. Adhoc or "short live" network has developed tremendously in the recent time, which can work without any access point or mobile towers. That means it is an infrastructure less network. Mobile Ad-hoc Networks can be referred as "MANETs". The locations can be changed, and it can discover the path dynamically. In other words, the nodes move dynamically leading to the update of the topology, frequent change in topology, optimization of routing and fading the interference of multiuser are few issues connected with MANETs that affects the efficiency of the data transfer. The purpose of this survey is to reveal the various types of mechanisms which can be used to resolve the problem of routing performance related issues in MANETs. This paper also presents the classification of link stability, route repair and stable path algorithms in tabular format.

1 Introduction

Channel access to nodes must be provide by protocols, in such a way that collisions are minimum. The nodes should also be treated fairly by the protocol, with respect to bandwidth allocation. Due to broadcasting transmission, collisions may occur. The mobile nodes are moving frequently. Therefore, nodes must be distributed and scheduled for accessing the channel. MANETs never uses any centralized device for coordinating the mobile nodes. The mobility factor into consideration, by the protocols as the nodes is movable. The network provider provides a guarantee known as QOS in order to meet a performance standard for bandwidth, jitter, delay, etc. Since nodes are having mobility nature in most of the time, it is difficult to provide QoS support to data in wireless network. In MNAETs energy is a very big problem and it limits the capability of a routing protocol as the wireless devices are functioning based on battery power. MANETs routing protocols must have some default capabilities such as Denial of Service attacks, Brute-force attacks and other attacks. The MANETS routing protocols must be alert about any other threat and attack. The transmission might interfere with each other based on the characteristics of transmission, which is considered as one of the major problems in MANETs as links present exit and sometimes does not. The change in topology must be reflected in the Adhoc network through routing tables and

© Springer Nature Switzerland AG 2020
L. C. Jain et al. (Eds.): ICICCT 2019, LAIS 9, pp. 380–388, 2020.
https://doi.org/10.1007/978-3-030-38501-9_38

adapt routing algorithms, since the topology is not constant. For instance, the routing table in a fixed network is updated after an interval of 30 s, which is very slow in case of MANETs. The main objective of this research work is to understand the basic concept in MANETs, which is an emerging technological field in the recent times, and can be very useful for achieving future generation network. Development of MANETs protocols and applications is a very complicated task since there is a limitation in resource for bandwidth and energy consumption. The aforesaid challenges, factors, properties and related works of MANETs will pave a new way for researchers to make development in this area (Fig. 1).

Fig. 1. Mobile ad hoc networks (MANETs)

2 Advantages of MANETs

The advantages of a MANETs include the following:

- Information and services about geographical position is provided by MANETs
- Self-organized network.
- It is a scalable network, which means that it adds any number of nodes in transmission
- Robust and improved flexibility
- The network can setup in any time and at any place.

3 Classification of Routing Protocols in MANETs

The MANETs begin the transmission process by broadcasting requests to the neighboring (available) nodes during the discovery process of the route, as the routing protocol of the MANETs responsibly coordinates and controls mobile nodes for delivery of the data packets from source to destination. The process of route discovery is performed in different ways based on the functionalities of routing protocols, for which many routing protocols are developed for MANETs. Mobility of nodes, constraint of resource and channels prone to error are the basic limitations of designing routing protocols for MANETs. Authors have proposed several approaches to analyze the quality of link. In MANETs, the link quality varies frequently, hence the evaluation of link stability and its availability in wireless network is one of the most important features to be taken under consideration. Unreliable parameters are used, which are subjected to errors and variations, and this marks one of the basic limited of this work. Many researchers have focused on link availability and stability issues. The mobile nodes are moving frequently. Therefore, nodes must be distributed and scheduled for accessing the channel. MANETs never uses any centralized device for coordinating the mobile nodes. MANETs routing protocols must have some default capabilities such as Denial of Service attacks, Brute-force attacks and other attacks. The MANETS routing protocols must be alert about any other threat and attack. The rest of the section is structured as follows: Sect. 2 presents the characteristics of MANETs. Here highlights the distinct mechanisms have been used in the existing algorithms. Section 5 is the conclusion suggesting future work. Finally, references are listed (Fig. 2).

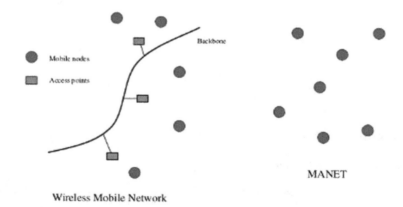

Fig. 2. Wireless mobile network vs mobile ad hoc networks (MANETs)

A table is maintained by Pro-Active protocols at each node for updating information periodically, even if no message is forwarded by the nodes, it shows the below characteristics:

- These protocols are updated version of wired network routing protocols
- Every node maintains one or more tables

- Every table stores the information of the mobile nodes
- Tables should be updated frequently.

4 Related Works

Various authors have been researched regarding routing failure process to improve the performance of MANETs as follows:

4.1 Related Works on Link Stability Algorithms

In these papers [2, 5–11, 19, 21, 23] researchers have evaluated link stability mechanisms on different kinds of parameters and they were brought some results.

4.2 Related Works on Route Repair Algorithms

Authors are in [3, 6, 12–17, 20] highlighted on route repair algorithms to recover the link from failure.

4.3 Related Works on Stable Path Algorithms

Authors have worked for finding out stable path algorithms in [4, 8, 11, 12, 18, 22, 24–27]. These algorithms are presented in tabular format and shown in below. Table 1 shows the classification of link stability, route repair and stable path algorithms.

Table 1. Classification of link stability, route repair and stable path algorithms

No.	Author(s)	Objective(s) of the work	Protocol(s) Technique(s) used	Metric(s) used	Result(s)
1	Sunita Nandgave-Usturge et al.	To increase the performance of MANET	Cross layer approach	RSS	(i) Avoids link failure (ii) Improves in congestion control
2	Devi et al.	To detect congestion in transmission	Route recovery technique	Queue length, data rate and MAC contention	(i) Increases in packet delivery ratio (ii) Minimizes in packet drop and delay
3	Jasmine Jeni et al.	To recovery route from link failure	Link failure recovery algorithm	Route reply buffer table and received signal strength (RSS)	(i) Increases in packet delivery ratio and throughput (ii) Minimizes end-to-end and routing overhead

(*continued*)

Table 1. (*continued*)

No.	Author(s)	Objective(s) of the work	Protocol(s) Technique(s) used	Metric(s) used	Result(s)
4	Adwan Yasin et al.	To enhance networks lifetime connectivity	Power efficient routing	Distance delay, RSS and mobility	Finds optimum path
5	Veerana Gatate et al.	To enhance route quality	Linear optimization	Peak Signal to Noise Ratio (PSNR), distance and delay	Increases in packet delivery ratio
6	Prasath et al.	To evaluate link quality	Dynamic source routing protocol	RSS, hop count, Round Trip Time (RTT) and delay	(i) Minimum delay (ii) High throughput
7	Prabha et al.	To improve multipath routing	Ad hoc on-demand multipath distance vector	Node's queuing delay, mobility and bit error rate	(i) Predicts the link availability (ii) Finds optimal route
8	Sujata et al.	To improve the link quality	Stable backbone based multipath routing	Residual bandwidth, residual energy and node mobility	(i) Minimize the packet drop and delay (ii) Maximize the throughput
9	Gaurav Singal et al.	To estimate link stability	Multicast routing protocol	Distance, bandwidth and SINR	(i) Improvements in packet delivery ratio (ii) End-to-end delay
10	Hui Xia et al.	To evaluate link stability	Relative stability metric and local stability metric	RSS	(i) Increases in packet delivery ratio (ii) Minimization in end-to-end delay and (iii) Routing overhead
11	Ali Moussaoui et al.	To establish stable path	Optimized link state routing protocol	RSS	(i) Minimizes end-to-end delay (ii) Minimum packet loss
12	Sharvani et al.	To enhance the reliability	Ant colony optimization	GPS, number of hops and bandwidth	(i) Increases in Throughput (ii) Reduces in end-to-end delay (iii) Reduces routing overhead

(*continued*)

Table 1. (*continued*)

No.	Author(s)	Objective(s) of the work	Protocol(s) Technique(s) used	Metric(s) used	Result(s)
13	Jyoti Jain et al.	To improve bandwidth utilization	AODV	RSS and mobility	(i) Reduces in throughput and delay (i) Reduces the chances of route break
14	Anjaneyu lu et al.	To increase the network life time	Maximum hop count routing protocol	Energy and hop count	(i) Reduces in delay (ii) Efficient bandwidth utilization
15	Senthil Kumar et al.	To reduce the link break	Cross layer scheme	RSS, bandwidth and delay	(i) Increases in throughput (ii) Increases in packet delivery ratio
16	Arjot Kaur et al.	To reduce link failure	AODV	RSS, hop count and sequence number	(i) Yields minimum packet loss (ii) Increases in throughput
17	Surabhi Purwar et al.	To predict the link stability	AODV	RSS and distance	(i) Increases in packet delivery ratio (ii) Minimum end-to-end delay and routing overhead
18	Abdulmale k Al-hemyari et al.	To increase the route lifetime	ODMRP	Distance and delay	(i) Reduces in delay (ii) Minimum overhead
19	Preetha et al.	To reduce the route failure	AODV	Node energy and delay	(i) Enhances in throughput (ii) Reduces in dropped packets
20	Srinivasan et al.	To compute the reliability factor	AODV	Residual energy, node expiration time and connection expiration time	(i) Increases in packet delivery ratio (ii) Reduces route reconstruction
21	Rekha Patil et al.	To select more stable path	AODV	Cost matrix and node energy	(i) Reduces in control overhead (ii) Minimizes in end-to-end delay

(*continued*)

Table 1. (*continued*)

No.	Author(s)	Objective(s) of the work	Protocol(s) Technique(s) used	Metric(s) used	Result(s)
22	Rajesh et al.	To find out a stable path	AODV	Periodic nodes movement and hello packets	(i) Increases in packet delivery ratio (ii) Enhances in throughput
23	Deva Priya et al.	To prolong the life time of a network	PPCLSS	Nodes energy, distance and neighbors stability	(i) Increases in high throughout (ii) Reduction in control overhead
24	Geetha Nair et al.	To reduce the packet loss	PLSS	Total mobile nodes and total network lifetime	(i) Improvements in packet delivery ratio (ii) Reduces in end-to end delay
25	Qingyang Song et al.	To estimate the link stability	RCPLE	Transmission rate, window size selection and GPS	Enhances in packet delivery ratio
26	Ravindra et al.	To meet the QoS requirements of real-time applications	QoS Protocol	Bandwidth, hello message	(i) Packet delivery ratio increases greatly (ii) Packet delay and energy dissipation decrease significantly
27	Surjeet et al.	To find a QoS constrained route from source to destination	MQAODV	Bandwidth and delay	Provides more accurate bandwidth estimation and end-to-end delay
28	Somesh Maheshwari et al.	To discover feasible and best path in a network	ACO	Bandwidth and cost	Finds best route from source node to destination node
29	Gagandeep Singh Hundal et al.	To select the secure and shortest path	AODV	Beacon frame range	Reduces packet loss

5 Conclusion

The main objective of this research work is to understand the basic concept in MANETs, which is an emerging technological field in the recent times, and can be very useful for achieving future generation network. Development of MANETs protocols and applications is a very complicated task since there is a limitation in resource for bandwidth and energy consumption. The aforesaid challenges, factors, properties and related works of MANETs will pave a new way for researchers to make development in this area.

References

1. Sharvani, G.S., Ananth, A.G., Rangaswamy, T.M.: Efficient stagnation avoidance for MANET with local repair strategy using ant colony optimization. Int. J. Distrib. Parallel Syst. **3**(5), 123–137 (2012)
2. Jain, J., Gupta, R., Bandhopadhyaya, T.K.: Performance analysis of proposed local link repair schemes for ad hoc on demand distance vector. Inst. Eng. Technol. **3**(2), 129–136 (2014)
3. Anjaneyulu, A., SitaKumari, C.H.: Preventing link failure and increasing network life time using MEMHRP and S-MAC protocol in MANET. Int. Res. J. Innovative Eng. **1**(3), 94–103 (2015)
4. Senthil Kumar, R., Kamalakkannan, P.: A review and design study of cross layer scheme based algorithm to reduce the link break in MANET. In: International Conference on Pattern Recognition, Informatics and Mobile Engineering, pp. 139–143, February 2013
5. Kaur, A., Manpreet Kaur, Er.: Improvement in AODV routing protocol to reduce link failure problem in ad hoc network. Int. J. Res. Appl. Sci. Eng. Technol. **2**(11), 300–305 (2014)
6. Patle, S., Kumar, P.: Decision making approach to prefer route repair technique in AODV routing protocol of MANET. Int. J. Res. Eng. Technol. **4**(2), 9–16 (2015)
7. Purwar, S., Prakash, S.: Reliable pair protocol for link stability in MANET. Int. J. Comput. Netw. Wirel. Commun. **2**(3), 322–327 (2012)
8. Praveen, B., Umarani, N., Anand, T., Samanta, D.: Cardinal digital image data fortification expending steganography. Int. J. Recent Technol. Eng. **8**(3) (2019). ISSN 2277-3878
9. Dhanush, V., Mahendra, A.R., Kumudavalli, M.V., Samanta, D.: Application of deep learning technique for automatic data exchange with Air-Gapped Systems and its Security Concerns. In: Proceedings of IEEE International Conference on Computing Methodologies and Communication, Erode, 18–19 July 2017
10. Kumar, R., Rishabh, K., Samanta, D., Paul, M., Vijaya Kumar, C.M.: A combining approach using DFT and FIR filter to enhance Impulse response. In: Proceedings of IEEE International Conference on Computing Methodologies and Communication, Erode, 18–19 July 2017
11. Ghosh, G., Samanta, D., Paul, M., Kumar Janghel, N.: Hiding based message communication techniques depends on divide and conquer approach. In: Proceedings of IEEE International Conference on Computing Methodologies and Communication, Erode, 18–19 July 2017
12. Singh, R.K., Begum, T., Borah, L., Samanta, D.: Text encryption: character jumbling. In: Proceedings of IEEE International Conference on Inventive Systems and Control @IEEE, Coimbatore, 19–20 January 2017

13. Al-Hemyari, A., Ismail, M., Hassan, R., Saeed, S.: Improving link stability of multicasting routing protocol in MANET. J. Theor. Appl. Inf. Technol. **55**(1), 109–116 (2013)
14. Preetha, K.G., Unnikrishnan, A., Poulose Jacob, K.: An effective path protection method to attain the route stability in MANET. Int. J. Adv. Res. Comput. Commun. Eng. **2**(6), 2343–2348 (2013)
15. Srinivasan, P., Kamalakkannan, P.: Route stability and energy aware routing for mobile ad hoc networks. Int. J. Comput. Commun. **8**(6), 891–900 (2013)
16. Patil, R., Megha Rani, R., Bainoor, R.: Link stability based on QoS aware on-demand routing in mobile ad hoc networks. IOSR J. Comput. Eng. **5**(3), 52–57 (2012)
17. Rajesh, T., Madhava Reddy, M., Kishore Babu, T.: Relational permanence routing protocol under video transmission for MANET. Int. J. Comput. Sci. Commun. Netw. **1**(3), 212–217 (2011)
18. Deva Priya, M., Priyanka, P.: Probabilistic prediction coefficient link stability scheme based routing in MANET. Int. J. Comput. Sci. Eng. Technol. **6**(4), 246–256 (2015)
19. Nair, G., Muniraj, N.J.R.: Prediction based link stability scheme for mobile ad hoc networks. Int. J. Comput. Sci. Issues **9**(6), 401–408 (2012). No. 3
20. Samanta, D., Podder, S.K.: Level of green computing based management practices for digital revolution and New India. Int. J. Eng. Adv. Technol. (IJEAT) **8**(3S) (2019). ISSN 2249-8958
21. Sivakumar, P., Nagaraju, R., Samanta, D., Sivaram, M., Hindialraj, N., Amiri, S.: A novel free space communication system using nonlinear InGaAsP microsystem resonators for enabling power-control toward smart cities. Wirel. Netw. J. Mob. Commun. Comput. Inf. ISSN 1022-0038
22. Mahua, B., Podder, S.K., Shalini, R., Samanta, D.: Factors that influence sustainable education with respect to innovation and statistical science. Int. J. Recent Technol. Eng. **8**(3). ISSN 2277-3878
23. Song, Q., Ning, Z., Wang, S., Jamalipour, A.: Link stability estimation based on link connectivity changes in mobile ad-hoc networks. J. Netw. Comput. Appl. **35**(6), 1–26 (2012)
24. Ravindra, E., Agraharkar, P.: QoS-aware routing based on bandwidth estimation for mobile ad hoc networks. Int. J. Adv. Electr. Electron. Eng. **3**(2), 11–16 (2014)
25. Surjeet, A.P., Tripathi, R.: QoS bandwidth estimation scheme for delay sensitive applications in MANET. Commun. Netw. **5**, 1–8 (2013)
26. Maheshwari, S., Bhardwaj, M.: Secure route selection in MANET using ant colony optimization. Am. J. Netw. Commun. **4**(3–1), 54–56 (2015)
27. Hundal, G.S., Bedi, R., Gupta, S.K.: Enhancement in AODV protocol to select the secure and shortest path in Mobile Ad hoc Network. Int. J. Comput. Sci. Inf. Technol. **5**(5), 6543–6546 (2014)

Compressed Sensing Based Mixed Noise Cancellation in Passive Bistatic Radar

D. Venu[1]([⊠]) and N. V. Koteswara Rao[2]

[1] Department of ECE, University College of Engineering,
Osmania University, Hyderabad, Telangana, India
`dunde.venu@gmail.com`
[2] Department of ECE, Chaitanya Bharathi Institute of Technology and Science,
Hyderabad, Telangana, India

Abstract. This paper proposes a unique compressed sensing based pathway to improve mixed noise cancellation in Passive Bistatic Radar (PBR). Mixed noise is considered as Additive White Gaussian Noise (AWGN) including Impulse Noise (IN). The proposed technique applies a best sparsifying basis that adapts to the structure of the problem and reduces the size of the measurement matrix drastically. According to simulation results, it has been confirmed that the proposed system gives higher state estimation capabilities as compared to the conventional LMS filtering techniques. Without loss of generality, the testing of the performance metric has been done over the FM signals. The paper explains the simulation methodology and the details.

Keywords: Passive Bistatic Radar (PBR) · Mixed noise cancellation · Compressed sensing

1 Introduction

PBR is a well-established system as far as surveillance is concerned. The system applies existing conventional and practical technicalities. These results in low price surveillance, covert operation and a drastic reduction in pollution attributed to electromagnetism. However, there still remains a myriad of challenges regarding both technology development and processing techniques. Due to their excellent coverage, passive radars which employ passive broadcasting signals have become very interesting [1–3]. With Passive Bistatic Radar (PBR), there is no need for dedicated transmitters, in contrast to the typical active or monostatic radars. The communication signals used by PBR, for example, FM, DAB, DVB-T, GSM, and WIFI, are completely different and exploits a transmitter [4, 5]. The high level transmits powers specifically accessible from FM broadcast transmitters empower detection ranges around 250 km [6]. For the directly transmitted signal in a standard monostatic system, the channel which receives the signal is a powerful, smart, and independent is the need of great importance. At that point, the transmitted signal is put up as a source of the perspective signal so that it can be compared with surveillance signal for ease in echo detection and analysis. Also, to acknowledge acceptable signal-to-noise ratio, transmitted signals needs long integration time and do possess levels, so that targets are accomplished [4].

© Springer Nature Switzerland AG 2020
L. C. Jain et al. (Eds.): ICICCT 2019, LAIS 9, pp. 389–404, 2020.
https://doi.org/10.1007/978-3-030-38501-9_39

However, because of the built-in transmission pattern that is not designed by the designer, there are possibilities of having variable and unforeseeable features of the resulted waveform. On the contrary, conventional radars do have well-established setup to produce ambiguity function in both range and Doppler side lobes. Keeping the above factors in view [7–9] transmitted waveform may lead to, Increase in Doppler effect, Masking of target echo due to the direct signal being received, Echo problems due to many lower level echoes. In addition, moving target indicator techniques were not able to get rid of the masking effect, therefore, novel adaptive cancellation filters were designed for this purpose [6, 10–15]. Since the purpose of this paper revolves round FM radio signals, but it can be applied to cell phone transmissions, television waveforms, digital radios, navigation satellite etc. For target detection, the selection of FM transmissions possibly leads to waveforms with worst ambiguity. Also, reference signals depict information pertaining to (i) Doppler and range determination, and (ii) The side lobe level based on PSLR. Parallel and in the meantime, the estimation of the particular noise floor is based slightly over Doppler – range surface. The calculation specifically includes the dominant 2D-CCF peak and noise floor, peak–to–noise floor ratio (PNFR).

2   Contribution

Unlike alternative systems, PBR primarily based systems have totally different techniques to enhance estimation accuracy and denoising ability. But most of those systems don't guarantee mixed noise obliteration. The mixed noise primarily consists of two parts, particularly the white Gaussian noise and impulse noise. Presence of mixed noise in radar systems will result in sub-optimal solutions to state estimation [16–18]. In compressive sensing Technique, which includes three parameters of measurement which is (Φ), dictionary (Ψ), and a vector (α). While producing a higher rate of resolution, this technique is not being affected by other factors i.e. rate or pulse length. In many papers, the technique of CS to PBR is described to gain better results for various PBR systems. Especially, object detection, and tracking was explained in [20] by utilizing a minimum range of guiding representation for DVB-T/DAB signals. In [21], compressive sensing methods are implemented for the object detection using WIFI based a PBR system which uses a familiar range of training progression to develop the basis, whereas the matrix for measurement is a part of discrete Fourier basis. However, there are very few numbers of papers which concentrate on using appropriate strategies [22, 23], for reconstruction, whereas others addressed data integration in multistate passive radar (MPR) systems [24, 25]. In multi-channel FM PBR systems to enhance the range-Doppler resolution, we have a tendency to contemplate the application of compressive sensing (CS) [26], whereas we have a tendency to create an attempt to beat a number of the above-mentioned limitations. Therefore, to attain higher resolution in FM PBR system, this paper exploits the working for utilizing compressive sensing technique for the FM based mostly multi-channel PBR systems. Thus, the main focus of our study is to avoid the problems found in the conventional PBR signal system.

Thus we have a tendency to resolve the object-scene by using a convex optimization technique; the solution given in this paper doesn't affect from instantly waveform disturbance or inter-channel position problems. Additionally, it is not mandatory to enclose all Doppler shifts because of completely distinct channels within the identical Doppler shift. By using the block (sparsity) search method, the problem described above can be solved wherever there's a direct relation between the structure of FM/GSM channels and therefore the related Doppler disparity. The solution presented here additionally gives higher resolution for the similar range of channels whereas trying to reduce the number of the unvarying pre-processing strategies (e.g., channel phase correction) as discussed in [27, 28].

3 Definitions

In PBR systems, the two main components that are required to create a radar system are the reference signal (direct signal) and another one is the surveillance signal (Fig. 1).

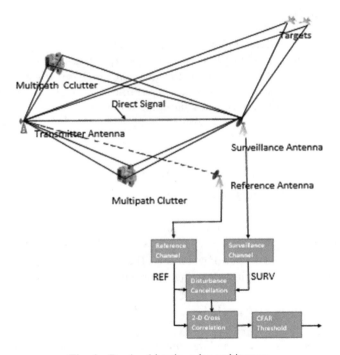

Fig. 1. Passive bistatic radar architecture

The components of the transmission system are, FM transmitter with the antenna system, Moving objects and environment model. The Receiver system, surveillance antennas, reference antennas, and amplifiers. The signal from the reference channel and the surveillance channel consists of noise which needs cancellation. A filter system is present for noise cancellation process. The schematic of the entire model has been shown below in the following Fig. 2.

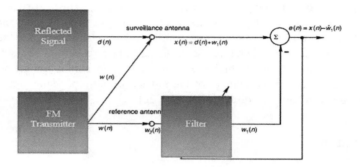

Fig. 2. Schematic of the system

4   Signal Model

Firstly, we assume an object drifting far away at a range rate that is unchanging say v, where v is less than c. Assuming that the broadcast transmitter transmits a narrowband signal: $s(t)e^{j2\pi ft}$ with s(t) being described for the complex baseband transmitted pulse-shape. The Carrier frequency is denoted by f. From a particular position target, the demodulated baseband received signal takes the following expression:

$$r(t) = s(t - \tau)e^{-j2\pi ft} + n(t) \tag{1}$$

Where:

$$\tau = 2(d_0 + vt), t_0 = 2d_0/c, f = 2v/\lambda$$

$n(t) =$ the white noise and $\lambda =$ wavelength.

In the process of sampling, we ensure that the length of the transmitted waveform $s = [s_0, s_1, \ldots, s_{L-1}]^T$ and the size of window N. The size of N is greater than the length of s, where norm of s is unity. When sampling is done, the resultant signal can be written as:

$$r = \sum \partial diag\left(\left[1, e^{j2\pi f}, \ldots, e^{j2\pi f(N-1)T}\right]\right)J_n S + n \tag{2}$$

Where n is the noise vector and the time shift matrix is denoted by J_n and s denote extended transmitted waveform vector, defined as Eq. (3).

$$J_n = \begin{cases} I_N & n = 0 \\ Z^n & 0 < n \leq P \\ (Z^{|n|})^T & -P \leq n < 0 \end{cases} \tag{3}$$

Z takes the following form:

$$Z = \begin{bmatrix} 0_{(N-1),1} & I_{N-1} \\ 0 & 0_{1,(N-1)} \end{bmatrix} \tag{4}$$

5 PBR System Parameters

The efficiency of a PBR system totally rely on its waveform

$$A(t_d, f_c) = \int_R s(t - t_d)s^*(t)e^{j2\pi f_c t}dt \tag{5}$$

R represents the range of integration. The characteristics of the waveform primarily the trade-offs between Doppler resolution and range and limit the execution of several PBR systems. The configuration of the radar system has been shown in Fig. 3 with its common notations where r_0 represents the distance between receiver and transmitter, r_1 represents the distance between target and transmitter and r_2 represents the distance between target and receiver respectively.

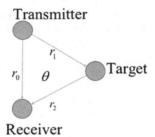

Fig. 3. Radar system

The estimates for velocity and position are calculated from the following Eqs. (6) and (7) respectively.

$$r_2 = \frac{ct_d(ct_d + 2r_o)}{2(ct_d + r_o(1 - \cos\theta))} \tag{6}$$

$$v = v_{r_1}^2 + v_{r_2}^2 + 2v_{r_1}v_{r_2}\cos\theta^i \tag{7}$$

The symbols c, f_c and t_d represents the velocity of the transmitted signal, frequency and the respective signal delay in *m/sec*, Hz and *sec* respectively. The parameters θ_v, θ and r_1 are related to the system parameters from Eqs. (8)–(12) respectively.

$$r_1 = r_o - r_2 + ct_d \tag{8}$$

$$\theta_v = \arcsin\left(\frac{v_{r_1}}{v}\sin\theta\right) \tag{9}$$

$$v_{r_1} = \frac{r_2}{r_1 + r_2}\left(\frac{cw_d}{2\pi f_c} + r_o r_2 \frac{d\theta}{dt} - r_o \cos\theta\right) \tag{10}$$

$$\theta^i = \arcsin\left(\frac{r_o}{r_1}\sin\theta\right) \tag{11}$$

The vector of state estimates r_2 and v at any instant k are denoted by $\mathbf{x}(k)$

6   Compressed Sensing

Compressed sensing (CS) is a powerful technique to solve problems which have inherent sparsity. Since the systems involved with radar applications are generally sparse in nature, CS is a substantial possibility for target estimation and white noise eradication. The CS-based strategies mostly work on two steps, namely: Encoding and Decoding.

Data Encoding
When there are n sensors in control area, the state measurements $\mathbf{x}(k)$ at any time step taken from n sensors is compressed into linear measurements totaling m through a basis $\boldsymbol{\Phi}$
Where m and n are such that m is less than n, and $\mathbf{y} \in \mathbf{R}^m$ as in (12).

$$\begin{aligned}\mathbf{y}(k) &= \boldsymbol{\Phi}\mathbf{x}(k) + \mathbf{n}(k) \\ &= \phi\varphi\theta(k) + \boldsymbol{n}(k) \\ &= \partial\boldsymbol{\theta}(k) + \mathbf{n}(k)\end{aligned} \tag{12}$$

Vector $x(k)$ has m non-sparse elements and $(k) = \varphi\theta(k)$ represents $\mathbf{x}(k)$ into its independent basis matrix and the transform coefficients vector $\theta(k) \in \mathbf{R}^n$. The additive white noise vector is denoted as $\mathbf{n}(k)$ with unit variance and zero in the signal. The measurement matrix is Φ and $\partial = \Phi\varphi \in \mathbf{R}^{m,n}$ is the corresponding sensing matrix. The above encoding introduces sparsity into the solution. For this application, we tend to construct a normalized optimal basis based on the positional index of the points where the local maximum of the ambiguity function occur (13). Since points of local maxima on the ambiguity function curve affects compression, this technique ensures that those corresponding indices are sparsified.

$$\varphi = \max_{i,j}|A(iT, jf)|$$

$$s.t \quad \|\varphi\|_2 = 1 \tag{13}$$

Data Decoding

After transmission, the following step is to recover the signal from the sparse solution. The following Decoding problem (14), is an NP-hard problem with the l_0-norm minimization being involved. The solution involves solving its convex approximation, the l_1-norm minimization problem, given by

$$\hat{\boldsymbol{\theta}}(k) = arg\,min_{\boldsymbol{\theta}(k)}\|\boldsymbol{\theta}(k)\|_1 \tag{14}$$

$$s.t. \quad \mathbf{y}(k) = \partial\boldsymbol{\theta}(k) + \mathbf{n}(k).$$

Once the transformation coefficients vector (k) has been recovered, then the following equation recovers the state data vector:

$$\hat{\mathbf{x}}(k) = \boldsymbol{\Psi}\hat{\boldsymbol{\theta}}(k) \tag{15}$$

Where the recovered state measurements vector is denoted as $\hat{x}(k)$. The whole formulation has been shown in Fig. 4. The performance of Compressive Sensing can be calculated by the Peak Signal to Noise Ratio (PSNR) which is given by Eq. (16).

$$PSNR = 20\log\left(\frac{MAX_I}{\sqrt{MSE}}\right) \tag{16}$$

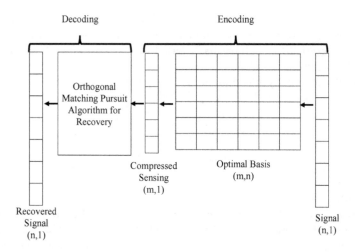

Fig. 4. Compressed sensing framework

7 Filter Formulation

Adaptive mixed noise filtering is a technique to remove the effect of mixed noise from a signal [29]. Matrix Y columns have a vector representation defined as $y = vec(Y)$. We consider two kinds of noise-namely Gaussian Noise and Sparse Noise. A data set for the signal of dimension m × n can be represented as $X = [x_1, x_2, ..., x_d]$ where each $x_i \in R_{mn} \times 1$ (mn = m × n) has been obtained by concatenation of all the columns of the set. The noisy model Y of the signal in presence of Gaussian (G) and sparse noise (S) can be expressed as in Eq. (17).

$$Y = X + S + G \tag{17}$$

In this paper, we consider a two-dimensional signal where Z is its sparse representation. Now $Z = D_1 X D_2$. Here, D_1 is a 2-D sparsifying transform applied along the spatial dimension, and D_2 is a 1-D sparsifying transform applied on the spectral dimension. Hence the denoising operation of the filter can be described as an optimization problem to minimize the total Gaussian noise present in the signal under the presence of an upper limit of the sparse noise S. The problem also considers a hard constraint on the bounds of the sparsifying differential operators in both directions. The whole problem has been described below.

$$\min_{Z,S} \|Y - X - S\|_F^2 + \lambda \|S\|_1$$

$$P = D_h x D \tag{18}$$

$$Q = D_v x D \tag{19}$$

The whole problem can be converted into an unconstrained optimization problem as (19).

$$\min_{Z,S} \|Y - X - S\|_F^2 + \lambda\|S\|_1 + \mu\|P\|_1 + \mu\|Q\|_1 \tag{20}$$

Where μ is the regularization parameter. The same parameter is used for both horizontal and vertical total variation regularization terms for considering equal contribution in both directions. We solve the following problem by using split-Bregman approach [30] which breaks down the optimization problem into smaller sub-problems. By using the Bregman variables B_1 and B_2, we can rewrite the problem as

$$p_1 : \min_P \mu\|p\|_1 + v\|p - D_h XD - B_1\|_F^2 \tag{21}$$

$$p_2 : \min_P \mu\|Q\|_1 + v\|p - D_v XD - B_2\|_F^2 \tag{22}$$

$$p_3 : \min_S \lambda\|S\|_1 + v\|Y - X - S\|_F^2 \tag{23}$$

$$p_4 : \min_X v\|Y - X - S\|_F^2 + v\|p - D_h XD - B_1\|_F^2 + v\|Q - D_v XD - B_2\|_F^2 \tag{24}$$

The sub problems p_1, p_2, and p_3 are of the form

$$\arg\min_X \|Y - X\|_F^2 + \beta\|X\|_1 \tag{25}$$

The sub-problems can be solved using the Orthogonal Matching Pursuit algorithm [31].

8   Computer Program: Validation and Verification

By using the CS based adaptive filter the problem of noise eradication from the transmitted signal can be solved. The role of the CS is to first encode the signal with the help of a best basis as derived in Eq. (7). The mixed noise is then removed by passing the signal through the filter by solving the optimization problem stated in Eqs. (13)–(16). It should be noted that the decoding part of the CS is embedded inside the filter where the signal is recovered using an orthogonal matching pursuit (OMP) based route. The complete process flow has been shown in the flowchart in Fig. 6 and the schematic has been shown in Fig. 5 respectively. The results of the presented scheme in this paper has been found by the root mean square value of deviations of the estimated states from the target states as given by Eq. (25).

$$x_{rms} = \sqrt{\frac{1}{N}\sum_{i=1}^{N}\left(|x_{\text{estimated}}| - |x_{\text{target}}|\right)^2} \tag{25}$$

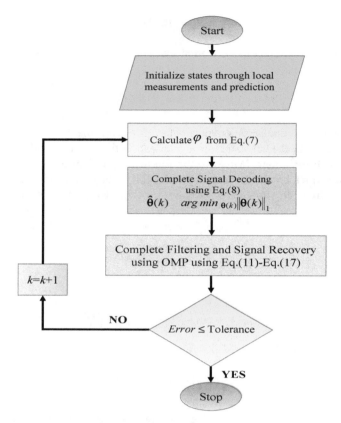

Fig. 5. Flowchart

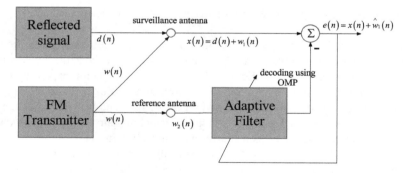

Fig. 6. Schematic of the proposed system

9 Results and Discussion

The ambiguity plots show the denoising characteristics of the algorithm when compared to that of the LMS filter. A characteristic significance of FM signals in the presence of sidelobes in the ambiguity function. It is shown from Fig. 7b that the ambiguity function has a lot of sidelobes especially due to the hyperspectral noise component of the noise in the signal. The proposed CS based noise filtering scheme decreases the sidelobes of the system as is evident from Fig. 7a. This shows the fact that the methodology is successful in denoising the FM signal. The quality of the signal is judged by PSNR in Eq. 16 and therefore the results for the estimated PSNR has been plotted for both the systems in Fig. 8. The PSNR for the signal is much higher in case of the CS based filtering than when compared to the LMS approach. The denoising technique eliminates the mixed components along with the Gaussian noise when compared to the strict LMS filtering technique which additionally establishes the fact that proposed strategy is better than the common techniques of filtering (Fig. 11).

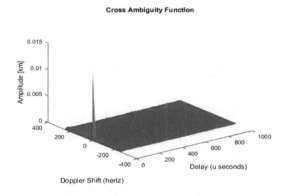

Fig. 7. Ambiguity function in case of CS based filter.

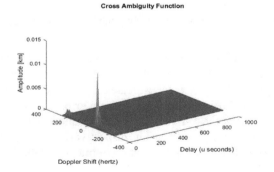

Fig. 8. Ambiguity function in case of normal LMS filter

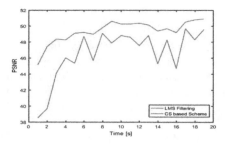

Fig. 9. PSNR comparison of the two techniques

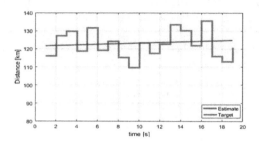

Fig. 10. Estimated target distance.

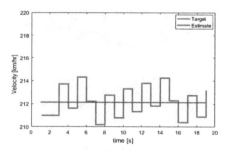

Fig. 11. Estimated target velocity with the proposed algorithm

The estimation results for the target distance and the target velocity for the given algorithm, under the FM class of signals, has been given shown in Fig. 9a and b respectively. The graphs show that the estimated signals are equally distributed along the target signals. The CS based scheme reduces the error in estimation of the states of the system. Fig. 10 shows the variation of error calculated with Eq. 25 with sample time with and without compressed sensing respectively. The Fig. 12 shows that the recovery system keeps the error of the signal within acceptable limits removing the noise from the signal.

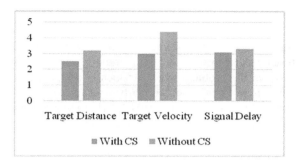

Fig. 12. Percentage estimation error for FM signal

The ambiguity function waveform for the case of the FM signal as a function of the estimated velocity and also the signal delay has been shown in Fig. 8. The CS based filter helps in removing the sidelobes of the ambiguity function at completely different velocities and signal delay of the target. This suggests the ability of the filter when it comes to efficient noise removal under the different state of conditions of the target (Fig. 13).

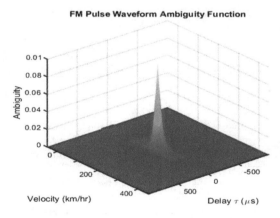

Fig. 13. Ambiguity function in case of estimated velocity [km/hr] and signal delay without compressed sensing.

Moreover, the scheme expanded the PSNR ratio by 14.25% for FM signals when compared to the generic techniques of filtering. The novel CS based scheme is computationally cheap, efficient and scalable for applications in giant systems which can guarantee efficient target detection under operational constraints.

10 Conclusion

This paper examines the application of a centralized novel CS based filtering technique on the passive bistatic radar system. An ideal route towards selecting the scarifying basis was considered in the problem which drastically decreased the impact of noise. The system was useful and performed similarly well for over the FM signal. When compared to generic filtering techniques like LMS Filtering both signal types the estimation errors for velocity and target distance detection were reduced to 2.5 and 3.67%. Thus the results prove that the proposed solution has the capability for mixed noise reduction when compared to generic filtering systems.

References

1. Special issue on passive radar systems. IEE Proc. Radar. Sonar Navig. **152**(3), 106–223 (2005)
2. Griffiths, H.D., et al: Television based bistatic radar. IEE Proc. F (Commun. Radar Signal Process.) **133**(7), 649–657 (1986)
3. Howland, P.E.: Target tracking using television-based bistatic radar. IEE Proc. Radar Sonar Navig. **146**(3), 166–174 (1999)
4. Griffiths, H.D., Baker, C.J.: Passive coherent location radar systems. Part 1: performance prediction. IEE Proc. Radar Sonar Navig. **152**(3), 153–159 (2005)
5. Zemmari, R., Nickel, U., Wirth, W.-D.: GSM passive radar for medium range surveillance. In: Proceedings of the European Radar Conference (EuRAD 2009), pp. 49–52, October 2009
6. Howland, P.E., et al.: FM radio based by static radar. IEE Proc. Radar Sonar Navig. **152**(3), 107–115 (2005)
7. Baker, C.J., et al.: Passive coherent location radar systems. Part 2: waveform properties. IEE Proc. Radar Sonar Navig. **152**(3), 160–168 (2005)
8. Griffiths, H.D., et al.: Measurement and analysis of ambiguity functions of off-air signals for passive coherent localization. Electron. Lett. **39**(13), 1005–1007 (2003)
9. Lauri, A., et al.: Analysis and emulation of FM radio signals for passive radar. Presented at the 2007 IEEE Aerospace Conference, Big Sky, MT, pp. 3–10, March 2007
10. Kulpa, K.S., et al.: Ground clutter suppression in noise radar. Presented at the International Conference on Radar Systems (Radar 2004), October 2004
11. Axelsson, S.R.J.: Improved clutter suppression in random noise radar. Presented at the URSI 2005 Commission F Symposium on Microwave Remote Sensing of the Earth, Oceans, Ice, and Atmosphere, April 2005
12. Gunner, A., Temple, M.A., Claypoole Jr., R.J.: Direct-path filtering of DAB waveform from PCL receiver target channel. Electron. Lett. **39**(1), 1005–1007 (2003)
13. Kulpa, K.S., et al.: Masking effect and its removal in PCL radar. IEE Proc. Radar Sonar Navig. **152**(3), 174–178 (2005)
14. Cardinali, R., et al.: Comparison of clutter and multipath cancellation techniques for passive radar. Presented at the IEEE 2007 Radar Conference, Boston, MA, March 2007
15. Colone, F., et al.: Cancellation of clutter and multipath in passive radarusing a sequential approach. In: IEEE 2006 Radar Conference, Verona, NY, 24–27 April 2006, pp. 393–399 (2006)
16. Gini, F.: Sub-optimum coherent radar detection in a mixture of K-distributed and Gaussian clutter. IEE Proc. Radar Sonar Navig. **144**(1), 39–48 (1997)

17. Narayanan, R.M., et al.: Doppler estimation using a coherent ultrawide-band random noise radar. IEEE Trans. Antennas Propag. **48**(6), 868–878 (2000)
18. Sangston, K.J., et al.: Coherent radar target detection in heavy-tailed compound-Gaussian clutter. IEEE Trans. Aerosp. Electron. Syst. **48**(1), 64–77 (2012)
19. Herman, M.A., et al.: High-resolution radar via compressed sensing. IEEE Trans. Signal Process. **57**(6), 2275–2284 (2009)
20. Berger, C.R., et al.: Signal extraction using compressed sensing for passive radar with OFDM signals. In: 2008 11th International Conference on Information Fusion, pp. 1–6. IEEE (2008)
21. Maechler, P., et al.: Compressive sensing for WiFi-based passive bistatic radar. In: Proceedings of the 20th European Signal Processing Conference (EUSIPCO), pp. 1444–1448. IEEE (2012)
22. Sevimli, R.A., et al.: Range-doppler radar target detection using denoising within the compressive sensing framework. In: 22nd IEEE European Signal Processing Conference (EUSIPCO), pp. 1950–1954. IEEE (2014)
23. Misiurewicz, J., et al.: Compressed sensing algorithms performance with superresolution in passive radar. In: 14th International Radar Symposium (IRS) (2013)
24. Ender, J.H.: A compressive sensing approach to the fusion of PCL sensors. In: Proceedings of the 21st European Signal Processing Conference (EUSIPCO), pp. 1–5. IEEE (2013)
25. Subedi, S., et al.: Motion parameter estimation of multiple targets in multistatic passive radar through sparse signal recovery. In: IEEE International Conference on Acoustics, Speech and Signal Processing (ICASSP), pp. 1454–1457. IEEE (2014)
26. Candes, E.J., et al.: An introduction to compressive sampling. IEEE Signal Process. Mag. **25** (2), 21–30 (2008)
27. Olsen, K.E., et al.: Performance of a multiband passive bistatic radar processing scheme—Part I. IEEE Trans. Aerosp. Electron. Syst. **27**(10), 16–25 (2012)
28. Samanta, D., Podder, S.K.: Level of green computing based management practices for digital revolution and New India. Int. J. Eng. Adv. Technol. (IJEAT) **8**(3S) (2019). ISSN 2249–8958
29. Sivakumar, P., Nagaraju, R., Samanta, D., Sivaram, M., HindiaIraj, N., Amiri, S.: A novel free space communication system using nonlinear InGaAsP microsystem resonators for enabling power-control toward smart cities. Wirel. Netw. J. Mob. Commun. Comput. Inf. ISSN 1022-0038
30. Mahua, B., Podder, S.K., Shalini, R., Samanta, D.: Factors that influence sustainable education with respect to innovation and statistical science. Int. J. Recent. Technol. Eng. **8** (3). ISSN 2277-3878
31. Praveen, B., Umarani, N., Anand, T., Samanta, D.: Cardinal digital image data fortification expending steganography. Int. J. Recent. Technol. Eng. **8**(3) (2019). ISSN 2277-3878
32. Dhanush, V., Mahendra, A.R., Kumudavalli, M.V., Samanta, D.: Application of deep learning technique for automatic data exchange with air-gapped systems and its security concerns. In: Proceedings of IEEE International Conference on Computing Methodologies and Communication, Erode, 18–19 July 2017
33. Kumar, R., Rishabh, K., Samanta, D., Paul, M., Vijaya Kumar, C.M.: A combining approach using DFT and FIR filter to enhance Impulse response. In: Proceedings of IEEE International Conference on Computing Methodologies and Communication, Erode, 18–19 July 2017
34. Ghosh, G., Samanta, D., Paul, M., Kumar Janghel, N.: Hiding based message communication techniques depends on divide and conquer approach. In: Proceedings of IEEE International Conference on Computing Methodologies and Communication, Erode, 18–19 July 2017

35. Singh, R.K., Begum, T., Borah, L., Samanta, D.: Text encryption: character jumbling. In: Proceedings of IEEE International Conference on Inventive Systems and Control, Coimbatore, 19–20 January 2017. IEEE (2017)
36. Olsen, K.E., Woodbridge, K.: Performance of a multiband passive bistatic radar processing scheme—Part II. IEEE Trans. Aerosp. Electron. Syst. Mag. **27**(11), 4–14 (2012)
37. Aggarwal, H.K., et al.: Hyperspectral image denoising using spatio-spectral total variation. IEEE Geosci. Remote Sens. Lett. **13**(3), 442–446 (2016)
38. Goldstein, et al.: The split Bregman method for L1-regularized problems. SIAM J. Imaging Sci. **2**(2), 323–343 (2009)
39. Pati, Y.C., et al.: Orthogonal matching pursuit: recursive function approximation with applications to wavelet decomposition. In: Conference Record of the Twenty-Seventh Asilomar Conference on Signals, Systems and Computers. IEEE (1993)

E-Class Education Model in Modern Educational Technology-Based Approach

Trung Tran[1(✉)], Thanh Xuan Pham[2,3], and Thao Thi-Thanh Vu[2]

[1] Vietnam Academy for Ethnic Minorities, Hanoi, Vietnam
trungtl978@gmail.com
[2] Hochiminh City University of Technology and Education,
Ho Chi Minh City, Vietnam
[3] Dong Nai University, Bien Hoa, Vietnam

Abstract. eClass education model inherits the view of approaching modern instructional technology as the orientation basis for digital teaching organizations with personal development orientation in the current education ecosystem. This paper presents the actual implementation of eClass education model in teaching students of Elementary Education major (school year: 2018–2019) of Dong Nai University, Vietnam. T-test survey results with WIHIC classroom assessment tool show positive improvements when applied in practice.

Keywords: eClass model · Education ecosystem · Teaching technology · WIHIC assessment tool

1 Introduction

A model of teaching is a description of a learning environment that helps students to acquire information, ideas, skills, value, way of thinking and means of expressing themselves as well as efficient learning capacity; it includes teachers' behavior when that model is used [9].

So far, there have been many forms that change the models of teaching from traditional classroom to computer-assisted classroom, then to online learning (eLearning and mLearning) or personal learning environments (PLE) on mobile applications in the context of today's digital knowledge society.

However, the organization of digital learning environment in Vietnam is separating components of the educational environment: technology, pedagogy/education - psychology, sociology [4, 6, 7, 11] … Particularly with the trend of Education 4.0, the lack of coherent foundation from approaching modern teaching technology will make the teaching model at schools becomes outdated, not in line with the current trend of education development - towards personal development [2].

This paper deals with eClass education model in the new educational context - meeting the flexibility, open - personalized approach on the basis of modern teaching technology platform; and thereby practical applications in organizing the teaching Pedagogy2 subject at Dong Nai University for students of Elementary Education major (schoolyear 2018–2019).

© Springer Nature Switzerland AG 2020
L. C. Jain et al. (Eds.): ICICCT 2019, LAIS 9, pp. 405–416, 2020.
https://doi.org/10.1007/978-3-030-38501-9_40

2   eClass Education Model: Modern Educational Technology-Based Approach

2.1   Educational Technology and Features of Modern Educational Technology:

Educational technology scientifically organizes the educational process through identifying the exact educational objectives, learners' subjects, educational content, etc. in order to achieve educational objectives with optimal costs and time [3] (Fig. 1).

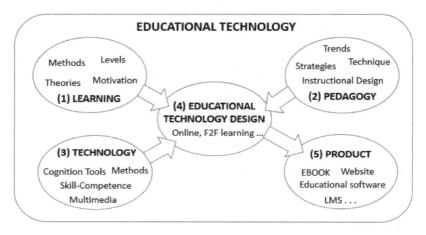

Fig. 1. Educational technology components (*source:* [1])

Approach to Educational Technology is to demonstrate the following characteristics of Educational Technology:

- Integrity, mutuality - expressed through almost all aspects of education according to Educational Technology process, etc.
- Cohesion of technology with education: Inheriting scientific achievements in fields related to education, etc.
- Emphasis on software element: clear and creative process of organizing educational activities - based on educational design towards personalization in practical contexts and suitable with educational objectives.
- Foundations based on cognitive sciences -specifically based on learning theory, learning styles, etc., in the context of education in the digital knowledge era.
- Openness with the ability to respond quickly and flexibly to changes in social practice.
- Standardization of conditions that can be implemented according to Educational Technology process includes: human resources (experts, teachers, managers, students), infrastructure/equipment, information systems and appropriate educational technology policies.

2.2 Technology 4.0 Context and Education Ecosystem

Technological transformation is happening so fast that education models need to be linked to research, technology development and direct participation in production and life [14].

Teachers' role in the 21st century become more complex which requires them to have orientation in technology and be responsible not only for teaching but also for students' learning. The role of teachers has changed and continues to change from being an instructor to becoming a constructor, facilitator, coach, and creator of learning environments. Today teachers are required to be facilitators helping learners to make judgments about the quality and validity of new sources and knowledge, be open-minded and critical independent professionals, be active cooperators, collaborators, and mediators between learners and what they need to know, and providers to scaffold understanding [13].

Education Ecosystem: The educational ecosystem shows the close links between the learning components with each other, and with the external learning environment (larger learning ecosystem) through the movement of connected knowledge and technology environment; it shows personalization through establishing relationships in order to create an educational connection environment for personal development in accordance with the trends/motivations of that educational ecosystem.

The structure and model of development of an education/learning ecosystem includes the following components: (1) Individuals/groups of learners with the role of exploiting, using, recreating and creating knowledge; (2) Teachers, along with other educational support resources; (3) Personal environment with the ability to connect

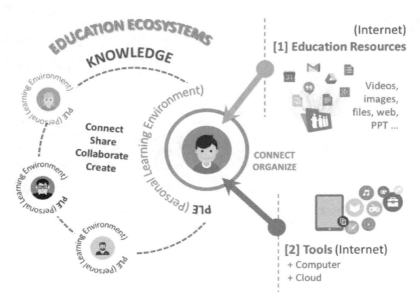

Fig. 2. PLE system with the ability to connect and integrate tools and Education resources PLEs connect, share, collaborate, create together (in Education ecosystems)

education on the basis of information and communication technology (ICT); (4) Open educational resources, Massive Online Open Courses (MOOCs); (5) The ability to connect knowledge between components within the ecosystem and outbound connections to the larger ecosystem [5, 10, 12] (Fig. 2).

2.3 eClass Education Model from the Perspective of Modern Educational Technology

From the above analysis, it is possible to identify the following basic elements of eClass education model in modern educational technology-based approach on the basis of the education ecosystem (Fig. 3).

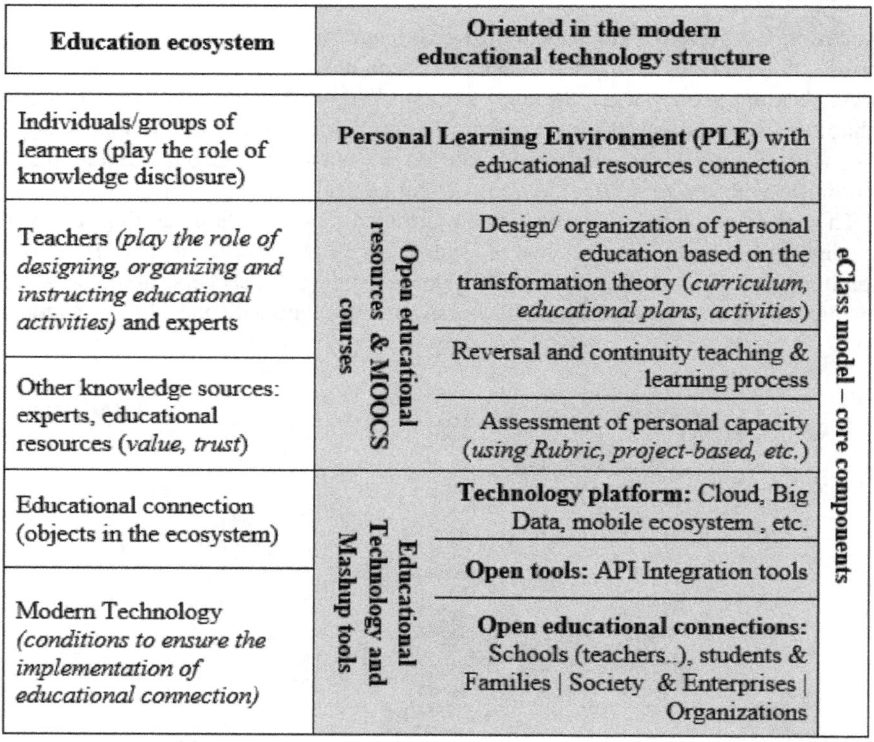

Education ecosystem	Oriented in the modern educational technology structure		
Individuals/groups of learners (play the role of knowledge disclosure)	Personal Learning Environment (PLE) with educational resources connection		
Teachers *(play the role of designing, organizing and instructing educational activities)* and experts	Open educational resources & MOOCS	Design/ organization of personal education based on the transformation theory *(curriculum, educational plans, activities)*	eClass model – core components
		Reversal and continuity teaching & learning process	
Other knowledge sources: experts, educational resources *(value, trust)*		Assessment of personal capacity *(using Rubric, project-based, etc.)*	
Educational connection (objects in the ecosystem)	Educational Technology and Mashup tools	Technology platform: Cloud, Big Data, mobile ecosystem , etc.	
		Open tools: API Integration tools	
Modern Technology *(conditions to ensure the implementation of educational connection)*		Open educational connections: Schools (teachers..), students & Families | Society & Enterprises | Organizations	

Fig. 3. eClass model with core components (*Source:* [2])

In a nutshell, it can be said that eClass education model as a specific product of educational technology is suitable to the current educational context and composed of the following elements:

- Personalization: creating a personal learning environment (PLE)
- Open technology platform and API integration tools
- Open educational connections: schools (teachers, managers) - students - students' parents - social organizations, enterprises, etc.
- Open source of digital knowledge, also known as open educational resources - including Massive Online Open Courses (MOOCs)
- The process of education design/organization based on connected learning theory is ongoing/continuous in combined learning method (in classroom and outside classroom) with the assurance of corresponding implementation conditions of the curriculum, teacher quality, facilities/equipment, etc.
- Combined assessment for personal progress and development.

3 Deploying the eClass Model for Teaching Pedagogy2

The eClass model shows the basic factors: (1) technology-based environment/tools for PLE; (2) transformative instructional design … This section below shows more details about this solution:

3.1 The Choice of Technology Solution

Tuan and Thanh [2], listed the popular applications for individual learners can participate and well organize their learning process on the mobile platform are identified through the Table 1 below:

The learning & teaching process is in 3 stages (Fig. 4):

- **PreClass:** providing the lectures (videos, materials) for students through Google Classroom system
- **inClass:** organizing practical experience activities in class (at a high level) to help students build their competence
- **AfterClass:** providing guidance, strengthening exercises for students.

Table 1. Analysis table for identifying common applications for mobile learning organizations:

Grouping as required	Mobile Apps	Explanation for the choice of applications
Connection, contact notificaton, personal information …	Email, Zalo, Viber, SMS, Call Facebook và trên Google Classroom	**Gmail:** popular, 100% students use **Zalo, Facebook, Viber** (for chat, discuss, share information in group, if needed)
Educational Resources: videos, materials …	Google Drive, OneDrive Youtube…	**Google Drive, Youtube:** popular, easy to integrate into Google Classroom
Organize and manage the learning process in Class	Google Classroom OneNote, Edmodo…	**Google Classroom:** free, simple, fast and stable system. Easy to manage class; helping teachers to mark papers quickly; Many teachers join the class, integrate other open services of Google such as: Drive, Youtube, etc.
Student note and organize information from the lectures.	OneNote, SketNote … Google Docs, Google Drive	**Camscan + Google Drive:** easy to manage learning notes, safe and collaboration well.
(In class) students connect mobile phones to projectors	ezCast, Camscan MS Powerpoint Google Classroom	**Camscan:** scan document (pdf) from groupwork to Google Classroom **eZCast:** connect the projector and present the report on the spot (others can see result)
Survey and evaluation	Google Form Google Classroom	The combination is applied for surveying, testing and evaluating results (Rubric, score)
Online interaction	Google Calander Meet, Youtube Google Classroom MS Office, Google Drive	Set up appointments (livestream) with **Meet, Youtube** with students **Google Drive:** collaborate online (doing homework/projects in group)
Personal learning organization on mobile - *With normal applications built on mobile (or installed more)*	**File manager:** store personal resources on mobile and Internet Drive Camera, Record videos and Contacts **Chrome:** its compatibility with many add-ons **Google Translate:** use if reading english materials **Google Apps** (Drive, Docs, Slide, etc.) collaboration and online working (in group)	

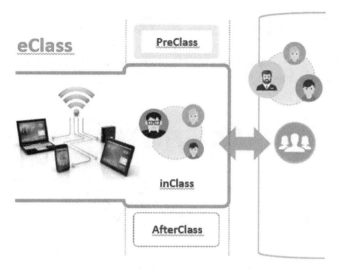

Fig. 4. The learning & teaching process

3.2 Transformative Instructional Design for the Personal Competence Development

Thanh [15] proposed a personal competence development model (Fig. 6) from inheriting the main point of view, including: reference frame transformation (as assumptions for understanding personal experience; it includes awareness, will, attitude, feelings and beliefs, etc.) all due to the learner's previous assimilative and transformative process of experiences when placed in real-life situations [8]; and personal transformative process is described by experiential learning phases (experience, reflection, etc. positive experiment) and levels of competence (Knows, Knows How, Show How, Does) [16].

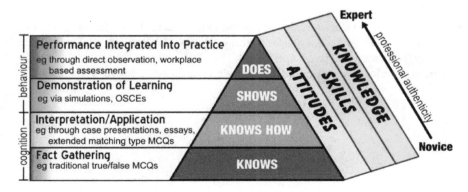

Fig. 5. Model of competence and performance assessment aspects (*Miller 1994*)

The framework of transformative instructional design is determined according to the process in the Table 2: (1) Transforming into the individual competence development course/program in the business/work context; (2) Determining the transformative level and (3) Designing transformative instructional design for teaching-learning activities (Fig. 5).

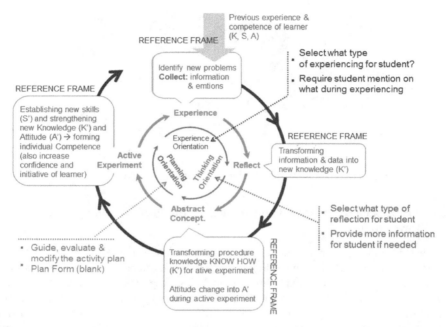

Fig. 6. Model of development individual's competence based on transformative learning theory [15]

Explaination:

- **Inner Circle:** teacher activities to help/support student during learning phases
- **Middle Circle:** the experience learning phases (Experience, Reflect, Abstract Conceptualization, Active Experiment)
- **Outter Circle:** Transforming the learner's competence throught changing (K', S' and A')

Table 2. Summary of the transformative instructional design framework

TRANSFORMATIVE INSTRUCTIONAL DESIGN						
(1) COURSE		(2) COMPETENCE LEVEL		(3) TEACHING - LEARNING ACTIVITIES DESIGN		
TEACHING CONTENT (Design of individual competence development approach)	SUBJECT 2	**LEVEL OF COMPETENCE TRANSFORMATION** (Determining the levels and results of individual competence transformation)	Continuous and reverse class	**LEARNING ACTIVITIES DESIGN** (Designing teaching and learning activities under transformational approach)	**INTERACTIVE SUPPORT** (getting guidance and organization from teachers)	**ICT TOOLS** (Mobile Apps) Determining suitable applications on Mobile (related to learning phases)
SUBJECT 1:		**KNOWS?** (Please list here)		**Concrete Experience**	Experience Orientation	Tools/Apps for concrete experience activities
SUBJECT 2:		**KNOWS HOW?** Table of the action process **Table of Process /Instructions:** **SHOW HOW** - Demo Actual action (according to teacher's instructions with forms)		**Reflection Observation & Abstract Conceptualiation**	**Thinking Orientation**	Tools/Apps for concrete experience Reflect & Show How activies
SUBJECT X: .... SUBJECT Y:		**DOES (action)** - Perform and practice regularly List of actual practice and exercises		**Ative Experimentation**	**Planning Orientation**	Tools/Apps supporting experiment activities

Table 3. Example for instructional design for pedagogy2

TABLE 2 - EXAMPLE: INSTRUCTIONAL DESIGN FOR COURSE - PEDAGOGY2						
COURSE DESIGN		COMPETENCE LEVEL	FLIPPED	TEACHING - LEARNING ACTIVITIES DESIGN		
TEACHING CONTENT (Competence Development)	**SUBJECT 2**	**COMPETENCE TRANSFORMATION** (Determining the levels of competence transformation)	Flipped, Continous applied in class	**LEARNING ACTIVITIES DESIGN** (Designing Learning activities under transformational approach)	**INTERACTIVE SUPPORT** (getting guidance and organization from teachers)	**ICT TOOLS** (Mobile Apps
SUBJECT 1: PREPARATION **SUBJECT 2: IDENTIFY THE NATURE OF THE EDUCATIONAL-TEACHING PROCESS IN THE CURRENT CONTEXT**		**KNOWS?** Know what ??? (Please list here)		GUIDE: **Students need to apply personal experience to try to solve the specific situation required:** How to teach effectively? (Student experiencing) **The process of organizing teaching in class and with online interaction** Students discuss in teamworks to deep understanding about **The Teaching Process Nature.**	Prepare education resources for the lesson (video, infographics, ebooks ...) Organize activities & guide: *(experience, teamwork & discussions in class)* Oriented questions about understanding the nature of the teaching process	**Google Classroom** **MS Office,** Google Drive *(video, PDF, hình ảnh ...)* Camscan Google Form ezCast, **Meet** OneNote
SUBJECT 3: (BEGIN) ORGANIZING TEACHING FOR COMPETENCE DEVELOPMENT SUBJECT 4: USING TECHNOLOGY IN CLASS		**KNOWS HOW?** Procedure guidance table (student build it themselves with teacher helps) [Steps | manipulation | Condition | Criteria] **SHOW HOW** - Actual action (with instructions from teacher)		**Determine the information in the table of process:** [Aspects | Education | Teaching: Objective, Content, Methods, Organization, Evaluation, Motivation, Principles] Discuss for clarifying the nature of the education and teaching process	Modeling the nature of education & teaching and gudie student identify information for textbook for the left table. Prepare questions in details to help student understand more Adjust the information table for students.	**Google Classroom, MS Office,** Google Drive, OneNote, Camscan, Google Form ezCast **Zalo, Viber, Facebook**
SUBJECT 5: SUMMARIZE AND REPORT ON STUDY - EXAMINATION		**DOES (action)** - Thực hiện, rèn luyện thường xuyên Danh sách bài tập rèn luyện, thực hành thực tế		**Practice more:** Please complete the above information table for the types of lessons: **theory, practice, integration and life skills education.**	Provide practical exercises: **theory lessons, exercises, integrated lessons and life skills education**	**Google Classroom** **MS Office,** Google Drive

3.3   Experiment and Evaluate Effective with WIHIC

- Subject: Pedagogy 2 (Lesson design: Table 3 for example design)
- Participant: Student of Elementary Major (schoolyear 2018–2019)
- Class 1: 137 students (traditional class)
- Class 2: 63 students (experiment class)
- Blue illustrates for traditional Class (Class 1)
- Red illustrates for the experiment class in which teacher use multiple apps on (Class 2)
- Evaluation tool WIHIC: It includes 7 scale and 56 items, 8 items per scale comprising of Student Cohesiveness, Teacher support, Involvement, Investigation, Task Orientation, Cooperation, Cooperation and Equity (Fig. 7).

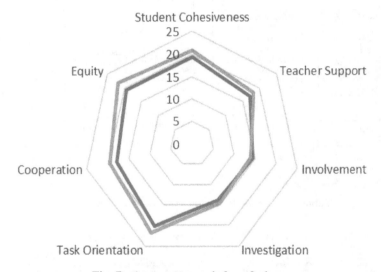

Fig. 7.   Assessment result from 2 classes

Using T-test to check the difference between 2 classes ($\alpha = 0{,}05$, $t\alpha = 1{,}660$) (Table 4).

Table 4.   The result of T-test

	Student Cohesive ness	Teacher Support	Involve ment	Investig ation	Task Orientat ion	Coopera tion	Equity
Class 1	19,31	17,15	14,45	13,89	20,15	17,95	19,35
N1 = 137	5,78	6,45	6,12	5,91	5,76	5,62	7,38
Class 2	20,82	18,27	13,95	14,67	21,78	19,45	21,8
N2 = 63	5,51	5,74	5,54	5,93	5,22	5,87	6,5

The result indicates the difference between two classes is right, Student Cohesiveness (t = −1.77), Task Orientation (t = −1.983), Cooperation (t = −1.76); Equity (t = −2.312), so it proves the effectiveness of teaching and learning in classroom using eClass model (Table 5).

Table 5. T-test (Checking the difference)

CR1 - CR2	0,85	0,93	0,89	0,89	0,84	0,85	1,06
T-test	-1,77	-1,2	0,56	-0,87	-1,95	-1,76	-2,3

4 Conclusion

The article presented an overview of educational technology and educational ecosystems in the current context; and thereby, affirming the role of educational technology as the basis for orienting the development of future education ecosystems - concretized through the eClass model. Besides, it also proved the effectiveness of teaching through the environment assessment tool of WIHIC class with T-test (with $\alpha = 0,05$ & checking the value table $t\alpha = 1,660$).

References

1. Hanh, N.V., Hop, N.H.: Textbook of Education Technology, p. 27. VNU Press, Vietnam (2016)
2. Tuan, N.A., Thanh, P.X.: Designing connected learning environment based on common mobile applications. In: ICOE 2019, pp. 108–119 (2019)
3. Tuan, N.A.: Textbook of Education Technology, Vietnam, p. 30 (2013)
4. Schneckenberg, D.: Educating Tomorrow's Knowledge Workers, p. 126 (2008)
5. Altınpulluk, H., Kesim, M.: The Future of LMS and Personal Learning Environments (2013)
6. Land, S.M., Hannafin, M.J.: Student-Centered Learning Environments: Foundations, Assumptions and Implications (1996)
7. Hannafin, M.J., Land, S.M.: The foundations and assumptions of technology-enhanced student-centered learning environments (1997)
8. Mezirow, J.: Transformative Learning, Theory to Practice. Jossey-Bass Publishers (1997)
9. Joyce, B., Weil, M., Calhoun, E.: Models of Teaching, 9th edn. Pearson, Boston (2014)
10. Chatti, M.A.: Toward a personal learning environment framework (2010)
11. Hiemstra, R.: Creating Environments for Effective Adult Learning, p. 6 (1991)
12. Van Harmelen, M.: Personal learning environments. In: Proceedings of the Sixth International Conference on Advanced Learning Technologies (ICALT 2006), pp. 1–2. IEEE Computer Society (2006)

13. Weinberger, A., Fischer, F., Mandl, H.: Fostering individual transfer and knowledge convergence in text-based computer-mediated communication. In: Stahl, G. (ed.) Computer Support for Collaborative Learning: Foundations for a CSCL Community. Proceedings of CS (2002)
14. Weller, M., Anderson, T.: Digital resilience in higher education. Eur. J. Open Distance E-Learn. **16**(1), 53 (2013)
15. Thanh, P.X.: Developing individual competence in integrated teaching based on transformative learning theory. In: Conference: Enhancing the lecturers' competence in Pedagogy Universities, University of DaNang - University of Science and Education, pp. 714–724 (2015)
16. Miller G.E.: The assessment of clinical skills/competence/performance. Acad. Med. **65**(9 Suppl), S63–S67 (1990)

An Approach to Introduce Mobile Application Development for Teaching and Learning by Adapting Allan's Dual Coding Theory

R. Gobi$^{(\boxtimes)}$

Department of Computer Science, Christ (Deemed to be University),
Bangalore, India
gobi.r@christuniversity.in

Abstract. This paper reveals around methodology that could be powerful to teach mobile applications in a class by including a decent utilization of innovative technology alongside a system. Appraisal instruments within the cloud were utilized to encourage this sort of methodology toward teaching application development. The new methodology is executed by teaching in the lab with desktops or in the classroom with student's laptops. It adapted Allan's dual code Theory. The adequacy of this methodology is obvious through an examination of outcomes.

Keywords: Android studio · Application development · Java · XML · Google classroom · Assessment methodology · Dual code theory

1 Introduction

Learning about mobile App development moves at a lightning pace. According to a report by Touch Point, today as many as 2.1 billion people in the world have a Smartphone. Mobile App is part and parcel of our life style today. People deceive themselves with the thought that creating a Mobile Applications is a tough job. Even though the knowledge of it is very easy to acquire, people are terrified to attempt. They think it is just another programming language where they'll have to study the syntax and theory. What they don't realise is that it is easy as most of the code is already written by the options themselves. We live in another millennial and need to be adaptive to the new tech advancing around us and leave our old orthodox ways behind. The class is dull and uninteresting when mobile applications are taught as a programming language without trying to practically implement them. Hence, to make learning interesting and retainable is the goal. It is imperative to bring your laptops to every class to apply the subject. The objective to learning mobile applications isn't all in the theory and syntax of Java or XML but understanding how they can use it to make a difference in this world, giving the students confidence of making use of this skill. If the courses are not well organized, learners may prefer to learn on their own and absence of concern will lead only in copying code and passing the exams. This proposed research focuses on offering measures to support learners develop smartphone apps. The tool that was being considered was Android Studio, where several

© Springer Nature Switzerland AG 2020
L. C. Jain et al. (Eds.): ICICCT 2019, LAIS 9, pp. 417–431, 2020.
https://doi.org/10.1007/978-3-030-38501-9_41

progressive topics like SQLite, Shared Preferences and firebase were made known to the students. At present, many universities have included Mobile Applications in their curriculum for the department of computer science. This research paper is organized as follows: Sect. 2 presents related works of mobile application development. Section 3 gives a description of the present teaching methodology. Section 4 explains the methodology used in creating a new framework to teach using Android Studio. Research Population of the proposed method is carried out Mobile Applications with Android Studio and it is represented in Sect. 5. Section 6 gives a list of issues in the gradle and their versions. Section 7 gives an overall summary of the proposed model.

2 Literature Review

A Clark and Paivio state that the science and training for knowledge are inclined towards cognitive issues [1]. Computer languages like Java and XML are quite like human languages as they both aim for communication. According to Ahmed [2] language should be taught through computer-assisted language learning aids. Using visuals and audio can greatly enhance attention of the listeners. People have selective attention as stated by Saul [3] and focus on what is important to them, catching up to the millennium is paramount. With the rise in new technology and its rapid growth it is only fair if it is put to good use. Harris, Gliksberg and Sagi [4] say that Perceptual learning is accomplished by repeated performance with simple visual stimuli and is easier with computers. Allan mentioned in the dual code theory that the student stores the ideas as pictures and verbal signs which expands the opportunity that when the student later attempts to review the data it will either show up as the picture or word, exclusively or together [5]. The Net Gen are more proficient than the past ages as many communicate utilizing pictures [6]. Afra and Rizwan stated that today's generation of students have extensive experience in social media [7, 8]. Allan recommends that execution in memory and other psychological assignments is interceded by phonetic procedures as well as by a nonverbal symbolism model of thought also [9]. Inspite of instruments, technological advancements, and student standards keep on transforming, one thing that is stagnant is the requirement for student commitment and learning assets that support this [10].

3 Present Teaching Methodology

The established way to teaching mobile applications consists of a one-hour lectures held three to four times a week in lecture halls. Usually, the classroom comes with a projector and a screen, the students are faced towards the teacher with his laptop connected to it. The course begins numerous PowerPoints filled with the concepts, syntax and lots of snippets of code. A few fragments of code from the created application are displayed and described in detail. The Core controls are supposedly covered in the classroom itself without any alteration or experiment. The typical lab comprises of many desktop computers where the students are required to complete the same old experiments utilizing android studio. The lab assistants oversee the two hours with or without prior knowledge in the subject, which makes it easier for the students to give up and lose their drive the first time thy fail or encounter an error. The lab experiments are designed long ago by one of the faculty and is passed down many generations, this isn't a feasible solution as we all know that technology keeps updating itself rapidly and it is our job to keep up with it. The inability of students to recollect controls is prevalent as the given experiment has no relation with what was taught as the syllabus is updated but the lab experiments lag. The only aid are the PowerPoints and the lab assistants, they prove to be of not much help in the long run. There are a lot of issues relating to the gradle like the changes in configuration, version control, etc. This causes chaos as the assistants don't know what to do and the teacher is usually engaged. The chaos further leads to the inactivity of the students as they start to question their abilities and slowly lose all motivation. The experiments are required to be complete at the end of the two hours, which makes the students compromise on the user interface design. We know that the assessment instruments for this subject are dependent on each other. When it comes to the creation of mobile applications, it is important to build on the language being widely used, in Android Studio Java is prominent. Java is a highly rated language at present which is given Fig. 1 [11].

Fig. 1. Top languages

4 Proposed Teaching Methodology

XML is a metalanguage which lets its users define their own customized mark-up languages, in most cases to display documents on the Internet but in our case to design the User interface of an application. XML is often used in a manifest document which identifies elements such as package names, operations, recipients, providers and authorizations needed by our implementation. For example, we must first classify online authorization in the document in order to use another web in the app.

Other usages of xml are colours, styles, drawable files, strings and dimensions. Java is the platform for app development with a library. Gradle is a build system based on java virtual machine with the best features. Google created Firebase which is a mobile app development platform that provides various online services such as storage, processing, real-time database, authorisation of user and the like. With an everyday analysis of the usage of these services along with the details of their users. User interface design like the colours and the minimalistic appearances are very important as they tend to attract humans making them keen on using the application. People are visual creatures and what you choose to represent your product [12]. Presently, the application's user interface (UI) and user experience (UX) are significant considerations in assessing whether or not this application will succeed in learning aspect. [13]. The easy way around is scaling down the UX features of a web page, but it is in no way the same as designing an app that is built specifically for a mobile platform [14]. To have a complete knowledge of Android studio it is essential to start teaching from the basics which is layouts that include views then the various input controls, explicit and implicit intents with values passed and how to create toast. Later menus, fragments, map drawer notifications, music players, games and location-aware applications are created to locate landmarks but also resident facilities [15]. Google Play is a global online software shop for systems built exclusively on Android created by Google [16]. Google play allows the searching, uploading, downloading, reviewing and installation of applications. To launch your app, you need to create its Android Application Package (apk) file that can be added to the play store so that others discover or distribute the file themselves. Students have trouble retaining the controls taught in lectures as they are based on mere observation the teacher going on and on with the

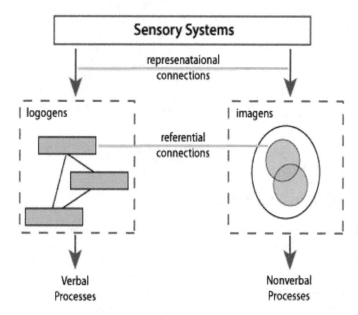

Fig. 2. Allan Pavio's dual code theory

chalk and the black board or a PowerPoint presentation. It doesn't captivate their attention like getting their hand dirty, working. Describing Allan's Dual Code Theory in the Fig. 2 below. Theory of dual coding consists of three kinds of processing: (1) representational, immediate application of verbal or nonverbal portrayals, (2) referential, or otherwise application of verbal or nonverbal portrayals, and (3) associative processing, execution of portrayals in the same oral or nonverbal scheme (Fig. 3).

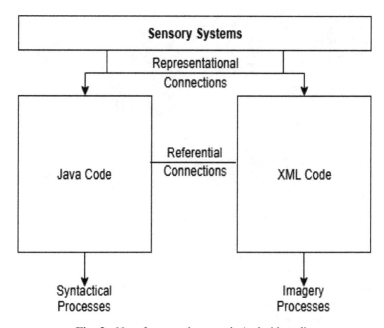

Fig. 3. New framework to teach Android studio

Allan sees learning as an embedded system whereby each stage supports and feeds to the next stage [17].

These different phases are given in Fig. 4.

1. Class (Phase A) – Representing code
2. Class work (Phase B) – Referring code
3. Submissions (Phase C) – Associating code

In order to make this new approach a success, certain additions to the regular assessments were made like daily submissions, quizzes, mid-semester test, project and end-semester test. According to this approach, the lectures are to take place inside the lab, with the teacher projecting his own new application for every control he plans on teaching his students who are seated in front of desktops with RAM over 8 GB. The new methodology comprises of three phases, they are as follows.

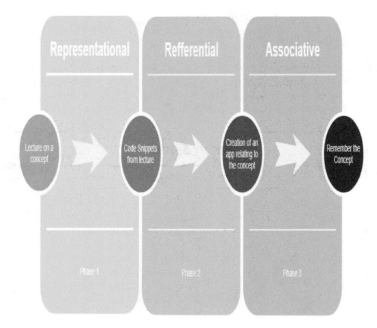

Fig. 4. Phases to reaching the goal

Phase A: The instructor is viewed as a facilitator who structures the principle target to cover from a concept and a couple of vital pieces of code from a working application. The teacher utilizes his not only his PowerPoint but also his application for a better understanding of the controls by the students. As the students are observing and simultaneously altering and experimenting his code on their own desktops, they will have examined the ideas planned to be taught through the session.

Phase B: The students work one next to the other in the lab under the guidance of the instructor and are relied upon to finish the class work lab assignment and execute the program on their phones for the lab assistant to see and evaluate in those two hours. They now attempt the concept individually, referring to the snippets uploaded by the teacher on the google classroom and make a report with their own code, API and screenshots that have to be uploaded to the link with the question on google classroom too.

Phase C: Evaluation tools intended for this methodology are altogether different from that utilized in the conventional methodology. The application is emulated on the laptop itself or shown on the phone. The new methodology approach opens students to hands-on application building following a concept. Each concept taught has an assignment to be completed in the lab. This causes the teacher to survey whether the students have comprehended the concept instructed during the lecture. This enables him to plan more difficult errands for the lab session in the next week.

Sample java code to build a music player:

```java
public class Music extends AppCompatActivity {
    boolean playing = false;
    ImageButton play, pause, rewind, forward;
    SeekBar seekbar;
    private MediaPlayer mp;
    private Handler myHandler = new Handler();
    int forwardTime = 5000, backwardTime = 5000;
    double startTime = 0, finalTime = 0;
    static int oneTimeOnly = 0;

    @Override
    protected void onCreate(Bundle savedInstanceState)
{
    super.onCreate(savedInstanceState);
    setContentView(R.layout.activity_music);
    forward = (ImageButton)
findViewById(R.id.imageButton);
    rewind = (ImageButton)
findViewById(R.id.imageButton2);
    play = (ImageButton)
findViewById(R.id.imageButton4);
    pause = (ImageButton)
findViewById(R.id.imageButton3);
    final MediaPlayer mp = MediaPlayer.create(this,
R.raw.song);
    seekbar = (SeekBar) findViewById(R.id.seekBar);
    pause.setEnabled(false);
    pause.setVisibility(View.GONE);

    play.setOnClickListener(new View.OnClickListener()
{
    @Override
    public void onClick(View v) {
    if (!playing) {
    mp.start();
    finalTime = mp.getDuration();
    startTime = mp.getCurrentPosition();
    if (oneTimeOnly == 0)
  {
    seekbar.setMax((int) finalTime);
    oneTimeOnly = 1;
    }
    seekbar.setProgress((int) startTime);

myHandler.postDelayed(UpdateSongTime, 100);
    pause.setEnabled(true);
    pause.setVisibility(View.VISIBLE);
    play.setEnabled(false);
    play.setVisibility(View.GONE);
                }
            }
        });
```

```
        pause.setOnClickListener(new View.OnClickListener()
{
            @Override
            public void onClick(View v) {
                if (playing)
        {
                    mp.pause();
                    pause.setEnabled(false);
                    pause.setVisibility(View.GONE);
                    play.setEnabled(true);
                    play.setVisibility(View.VISIBLE);
                }
            }
        });
    }

    private Runnable UpdateSongTime = new Runnable() {
        public void run() {
            startTime = mp.getCurrentPosition();
            seekbar.setProgress((int)startTime);
            myHandler.postDelayed(this, 100);
        }
    };
}
```

Sample XML code:

```xml
<?xml version="1.0" encoding="utf-8"?>
<android.support.constraint.ConstraintLayout
xmlns:android="http://schemas.android.com/apk/res/android
"
    xmlns:tools="http://schemas.android.com/tools"
    xmlns:app="http://schemas.android.com/apk/res-auto"
    android:layout_width="match_parent"
    android:layout_height="match_parent"
    tools:context=".Music">

<ImageView
        android:id="@+id/imageView"
        android:layout_width="wrap_content"
        android:layout_height="wrap_content"
        android:layout_marginBottom="8dp"
        android:layout_marginEnd="8dp"
        android:layout_marginLeft="8dp"
        android:layout_marginRight="8dp"
        android:layout_marginStart="8dp"
        android:layout_marginTop="8dp"
        app:layout_constraintBottom_toBottomOf="parent"
        app:layout_constraintEnd_toEndOf="parent"
```

```
            app:layout_constraintStart_toStartOf="parent"
            app:layout_constraintTop_toTopOf="parent"
            app:layout_constraintVertical_bias="0.09"
            app:srcCompat="@drawable/gly" />

<TextView
            android:id="@+id/textView"
            android:layout_width="248dp"
            android:layout_height="26dp"
            android:layout_marginBottom="8dp"
            android:layout_marginEnd="8dp"
            android:layout_marginLeft="8dp"
            android:layout_marginRight="8dp"
            android:layout_marginStart="8dp"
            android:layout_marginTop="8dp"
            android:text="Girls Like You (feat. Cardi B)"

        android:textColor="@android:color/background_dark"
            android:textSize="18sp"
            android:textStyle="bold"
            app:layout_constraintBottom_toBottomOf="parent"
            app:layout_constraintEnd_toEndOf="parent"
            app:layout_constraintStart_toStartOf="parent"
            app:layout_constraintTop_toTopOf="parent"
            app:layout_constraintVertical_bias="0.699" />

<SeekBar
            android:id="@+id/seekBar"
            android:layout_width="306dp"
            android:layout_height="25dp"
            android:layout_marginBottom="8dp"
            android:layout_marginEnd="8dp"
            android:layout_marginLeft="8dp"
            android:layout_marginRight="8dp"
            android:layout_marginStart="8dp"
            android:layout_marginTop="8dp"

            app:layout_constraintBottom_toBottomOf="parent"
            app:layout_constraintEnd_toEndOf="parent"
            app:layout_constraintStart_toStartOf="parent"
            app:layout_constraintTop_toTopOf="parent"
            app:layout_constraintVertical_bias="0.574" />
</android.support.constraint.ConstraintLayout>
```

In Fig. 5, a sample music player is presented. The Input Controls which are used to create the above Music Player are given in Table 1.

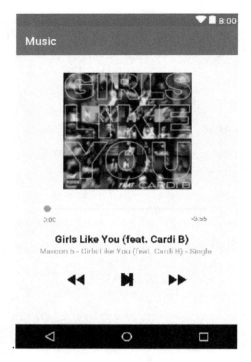

Fig. 5. Music player with input controls

Quizzes conducted indicated that students were well-versed with the tools android studio had to offer. This indicates that the students have been able to remember what was taught in class.

Table 1. Input controls

Controls	Description
Image buttons	Image that acts like a button and can be clicked, pressed, etc. to perform the given action
Image view	Displays an Image, the size can be enlarged or compressed
Text views	Display of text, manipulation of font, size, colour, etc.
Seek bar	Unlike the progress bar, it can be dragged to the left or the right

This proposed system is designed incorporate real-world objects in the implementation to undertake the intensity as well as tasks on the entities. Functions are divided into tiny goals and steps. Object – Action model is selected to operate the application [18]. Tasks in designing the application is divided into 3 levels as High level, Mid level and Low level.

- High level task is design an application with real world entities,
- Mid level task is to map the objects and
- Low level task is to check for error at the user level.

Tasks that are completed or studied by user should be calculated to find the amount of time a user spends for learning in a mobile application. The aggregate of task to be completed, particularly by a mobile users or mobile application in a period of time can be calculated to achieve better result [19]. For this purpose, the following formula can be used.

$$Task \times Time \times Frequency = Total\ Workload$$

Assume, a user is performing 1 Task in 1 h for a 365 days in a year means,

$$1 \times 1 \times 365 = 365\ h$$

So, based on this formula, a developer or a designer can calculate the amount of time a user is spending for learning with this application.

The following are the values considered in the design of the mobile application's user interface [20, 21]. These rules helps designers to create a unique design for their users. Additionally, this will also makes users to spend more time for learning by using the application. They are:

A. Classification of user actions in related circumstances.
B. Using shortcuts to help regular users like using acronyms, special keys and also by using hidden commands
C. Collecting informative feedback from the user after completing their learning by modest or significant model.
D. Dialog boxes are used to yield closure and to elicit a response from the user
E. Unnecessary prompt messages are avoided to make strive for minimalism and to make visual consistency
F. Automatic error prevention and error handling techniques are used in application
G. When an user deletes a notes or information in the screen accidently means, easy reversal of actions are permitted in the application
H. An art of correlating internal focus of control and Decreases the temporary memory load for the user.

5   Research Population to Execute This Idea

While designing a mobile application for the educators, a complex real world issue in teaching-learning should be identified and decomposed into small tasks. In order to execute this idea, a concept of Divide-and-Conquer approach is used for the problem solving. A tree structured approach should be followed for both objects and their respective actions to complete the task. End users can try to use the application in the task domain. Users can learn the interface and the tasks by just seeing the demonstrations and design itself. Some traditional model fails due to missing of metamorphic representation in their design. To identify how effective this new idea is, student's gradles and performance will be continuously evaluated for the core 'Mobile Applications using Android Studios'. Information will likewise be gathered from performances of students in numerous multiple-choice queries led on different platforms like kahoot and the university moodle. The quizzes for both the groups will be compared. The examining of the code is to be finished on their own laptops at the time of the lecture, followed by submitting a word document on google classroom. The teacher must be inventive and able to create intuitive exercises aimed towards teaching a concept. These exercises are centred around finding the correct solution and simultaneously in helping students explore new experiences, making it fun. The results of these changes were unsurprisingly high.

6   Challenges Faced for Both Learning and Commercial Aspects

Many issues with the gradle and their versions came up. The students may not be able to complete the given task within the two-hour time limit and hence, focused only on the functionality and have compromised on the User Interface and its design. The preceding are the guidelines proposed for the development of the mobile app's user interface [22].

Diversity of the users can be managed by handling the usage and task properly. For example, a single user can use one application for learning or a business user can use one application for both learning and commercial aspects.

The following issues should be concentrated at the usage aspects [23].

A. Nature of the user
B. Objective of the user
C. Educational experience of the user
D. Training resources available for the user
E. Physical abilities of the user
F. Expert reviews for an application design

The following issues should be taken into consideration at the task aspects [24].

A. Goals and subsequent tasks
B. Supporting tasks should be acknowledged
C. Functionality
D. Occurrence of action

Every single frequent user actions are designed by invoking the special keys. For the proposed framework to be successful, it requires a huge amount of time invested by the teacher as he would have to make his own fully working application based on the concept of the day to have an effective session. The students needed one on one time with the teacher to answer their queries, but there being thirty students, it was harder for the one teacher to impact them. The comparison of different challenges faced are given in Fig. 6.

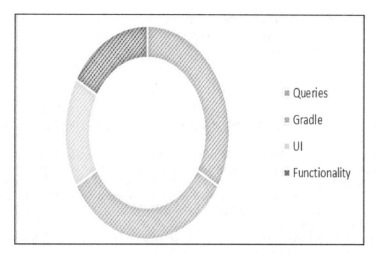

Fig. 6. Challenges faced

7 Conclusion

Different techniques have been executed in the past by educators teaching mobile application development, huge numbers of which pursued the notable instructing style. Organization of data to display to the user is carefully analysed based on consistency of data like abbreviations and colours used in the application. Information is displayed with proper alignment, numbering and proper spacing during display. Workload on the learners are reduced by using proper labels and buttons in the design. Correct information is displayed to user with neat formatting option and it can also be editable to ensure the compatibility of the information shown to the user. User control of the data organization is don based on the flexibility concept like ordering of data in terms of columns and rows. When a user is trying to entry data or for searching an information in the application, format of data is matched with the data displayed to the user. By receiving fewer input from the user and also by avoiding redundant information, the learner can use the application effectively. Time delays, unanticipated interfaces, feedback from multiple unwanted resources, incomplete use feedback and time delays are avoided by placing the icons and images at appropriate position. Moreover, the design is analysed and tested in visual programing environments to make ease in learning. It can be concluded that, the traditional lectures comprising solely of PowerPoints and no opportunity for the students to execute that code left a hole between learning the control and its use. The new methodology demonstrates a way of empowering students to effortlessly grasp the concepts in the lectures by allocating inventive assignments to students after each class to profoundly inspire them. Enabling them to recollect the controls taught all through the semester by making use of it in their final project. Instructors are recommended to embrace the proposed structure to teaching Android Studio. This methodology will give students new knowledge, as the instructors design class assignments which include the controls not only taught within the walls of the classroom.

References

1. Clark, J.M., Paivio, A.: Dual coding theory and education. Educ. Psychol. Rev. **3**(3), 149–170 (1991)
2. Alduais, A.M.S.: Integration of language learning theories and aids used for language teaching and learning: a psycholinguistic perspective. Integration **2**(4) (2012)
3. https://www.simplypsychology.org/attention-models.html
4. Harris, H., Gliksberg, M., Sagi, D.: Generalized perceptual learning in the absence of sensory adaptation. Curr. Biol. **22**(19), 1813–1817 (2012)
5. https://www.instructionaldesign.org/theories/dual-coding/
6. Oblinger, D.G., Oblinger, J.L.: Educating the net generation, is it age or IT: first steps toward understanding the net generation, pp. 12–13 (2005)
7. Alabbadi, A.A., Qureshi, R.J.: The proposed methods to improve teaching of software engineering. Int. J. Mod. Educ. Comput. Sci. (IJMECS) **8**(7), 13–21 (2016). https://doi.org/10.5815/ijmecs.2016.07.02
8. Paivio, A.: Imagery and Verbal Processes. Holt, Rinehart & Winston, New York (1971)
9. Paivio, A.: Mental Representations. Oxford University Press, New York (1986)
10. Walker, L.: My teacher is an Android: engaging learners through an Android application. In: Williams, G., Statham, P., Brown, N., Cleland, B. (eds.) Changing Demands, Changing Directions. Proceedings Ascilite Hobart 2011, pp. 1270–1274 (2011)
11. www.tiobe.com
12. https://www.forbes.com/sites/karstenstrauss/2014/09/23/how-to-get-your-app-noticed-wisdomfrom-app-annie/#316e98cd3675
13. Cohen, R., Wang, T.: GUI Design for Android Apps. Apress.open
14. UI and UX design. http://colure.co/the-importance-of-ux-and-ui-in-mobile-app-design/
15. Darwin, I.: Android Cookbook-Problems and Solutions for Android Developers. O'Reilly Media, Sebastopol (2017)
16. https://en.wikipedia.org/wiki/Mobile_app
17. www.simplypsychology.org
18. Korhan, O., Ersoy, M.: Usability and functionality factors of the social network site application users from the perspective of uses and gratification theory. Qual. Quant. **50**(4), 1799–1816 (2016)
19. Wu, D., Rosen, D.W., Wang, L., Schaefer, D.: Cloud-based design and manufacturing: a new paradigm in digital manufacturing and design innovation. Comput. Aided Des. **59**, 1–14 (2015)
20. Dinh, T.Q., Tang, J., La, Q.D., Quek, T.Q.: Offloading in mobile edge computing: task allocation and computational frequency scaling. IEEE Trans. Commun. **65**(8), 3571–3584 (2017)
21. Kim, H.K.: Designing of domain modeling for mobile applications development. In: 3rd IEEE/ACIS International Conference on Big Data, Cloud Computing, and Data Science Engineering, pp. 71–79. Springer, Cham (2018)

22. Mehrotra, A., Musolesi, M., Hendley, R., Pejovic, V.: Designing content-driven intelligent notification mechanisms for mobile applications. In: Proceedings of the 2015 ACM International Joint Conference on Pervasive and Ubiquitous Computing, pp. 813–824. ACM, September 2015

23. Ali, A., Alrasheedi, M., Ouda, A., Capretz, L.F.: A study of the interface usability issues of mobile learning applications for smart phones from the users perspective (2015). arXiv preprint arXiv:1501.01875

24. Tran, T.X., Hajisami, A., Pandey, P., Pompili, D.: Collaborative mobile edge computing in 5G networks: new paradigms, scenarios, and challenges (2016). arXiv preprint arXiv:1612.03184

CPAODV: Classifying and Assigning 3 Level Preference to the Nodes in VANET Using AODV Based CBAODV Algorithm

N. Arulkumar[1]($\boxtimes$), Mohammed Gouse Galety[2], and A. Manimaran[3]

[1] Department of Computer Science, Christ (Deemed to be University),
Bangalore, India
arul.kumar@christuniversity.in
[2] Department of Computer Network, Lebanese French University,
Erbil, KR, Iraq
[3] Department of Computer Applications, Madanapalle Institute of Technology
and Science, Madanapalle, Andhra Pradesh, India

Abstract. Vehicles communicate with nearby vehicles to share high routing and traffic information in Vehicular Ad hoc Networks (VANETs) environment. Congestion and Delay in the transmission may occur due to the density of the nodes in the network. Traffic condition depends on the vehicles in Rural and Urban environment. Increase or Decrease in vehicle's speed makes significant network changes when compared to the MANET environment. Road Side Terminals (RSTs) plays a major role in bridging the connection between the sender and the receiver nodes. The traditional AODV algorithm performs better when there are shortest path and link lifetime between the nodes in VANET. Giving 3 Level Preference to the nodes as High Preference (HP), Average Preference (AP) and Less Preference (LP) gives chances to nodes that have High Preference when compared to Less Preference. CPAODV model is proposed by implementing Classifying and giving preference to the RREQ to mitigate latency to the nodes. RST sends RREQ wisely based on the early model of Route Discovery stage itself. NS2 Simulator is used to analyze the strength of the proposed algorithm using QoS metrics like Throughput, Packet Delivery Ratio and End to End delay. This proposed CPAODV method performs better when compared to traditional AODV and CBAODV algorithm.

Keywords: AODV · CBAODV · VANET · RREQ · Categorization · Prioritization · RST · Network latency

1 Introduction

Sharing routing information with other vehicles to avoid road accidents is the primary job of any Vehicular Ad hoc Network (VANET). This network provides on-demand services to other nodes based on concepts of Ad hoc environment. Like routing information sharing, distribution of alert messages like emergencies and road accidents are the other essential services of VANETs. Alert messages are broadcasted to neighbor nodes to avoid congestion in the network [1]. The information of any anonymous node

© Springer Nature Switzerland AG 2020
L. C. Jain et al. (Eds.): ICICCT 2019, LAIS 9, pp. 432–445, 2020.
https://doi.org/10.1007/978-3-030-38501-9_42

is shared entirely to the entire network. This method of connecting nodes is of three types. They are Vehicle to Vehicle (V2V) communication, Vehicle to Infrastructure (V2I) communication and Infrastructure to Infrastructure (I2I) communication [2]. Road Side Terminals (RSTs). The RST has been equipped with internet services in order to back up routes on the remote server. To save resources, several resource management techniques are implemented with the help of RSTs. Resources are wasted when there was improper management of nodes. During congestion, the number of nodes or a blacklisted node is the primary reason for the data consumption in the network. Node's movement and congestions created are the natural incidents of the VANETs [3]. The primary work for a network is the right network design and resource management. The best approach for congestion using the reactive mechanism that responds on-demand queries is to categorize and prioritize the nodes in one network. After completing the literature review, Classifying and giving Preference to the nodes in network design are identified as the best solution to support more bandwidth reservation and link breakages. AODV routing is selected as the best model for routing and CBAODV for link breakages [4]. Finally, this technique is named as Classifying and Giving Preference to the nodes in VANET using CBAODV algorithm (CPAODV). This paper is organized as follows: The related works of the Wormhole attack are presented in Sect. 2. Section 3 describes the CPAODV algorithm proposed. Section 4 describes the way the CPAODV algorithm is generated. CPAODV's performance analysis is done using AODV algorithm and is shown in Sect. 5 as graphs. In Sect. 5.1, the proposed CPAODV algorithm is summarized in full and is constrained.

2   Related Works

Al Mallah et al. proposed a Distributed Classification of Urban Congestion Model using VANET to detect congestion [5]. This model is developed based on both recurrent and non-recurrent congestion model (NRC). Events like work zones and adverse weather are performed based on the NRC method. A real-time framework is proposed to distribute the congestion among the heterogeneous environment. The proposed model produces better accuracy in terms of Bayesian classifier, deterministic classification, and random forest and also in boosting technique. The proposed model helps to reduce congestion by finding the root cause of congestion. Jawhar et al. developed a classification design for Unmanned Aerial Vehicle (UAV) based systems [6]. Different network architectures, frameworks, and data traffic requirements are analyzed to communicate UAV node among communication links and network layers. Middle layer services are discussed to provide unbroken data transmission. Additionally, UAV model is used for data collection in Wireless Sensor Networks (WSNs). Energy consumption is reduced by implementing the UAVs for this function in the network.

The dynamic spatial partition model is proposed by Rayeni et al. for density-based emergency message dissemination in VANETs [7]. It is developed as a reliable time-efficient and multi-hop broadcasting model for both dense and light traffic situations. This proposed model is named as Dynamic Partitioning Scheme (DPS) for VANET. Moreover, this model outperforms efficient broadcasting protocols in VANETs when compared to delay and reliability during emergency message broadcasting. Santamaria et al. created a partition based multicast tree protocol for data dissemination in VANET environment [8]. This model is named as PAMTree, and it is suitable for distributing the services in the network and also to increase the reliability of users to access those services. Here, characteristics of the VANET architecture are used effectively to share services among other nodes.

Selvi et al. developed and proposed an efficient message prioritization and scheduled partition model for emergency message broadcasting in VANET [9]. This prioritization model is used to give importance to emergency messages, general messages and also for entertainment messages. SMTP based metrics are used to locate the message in the network. Adaptive scheduling partitioning is used for message broadcasting at an emergency. The hybrid partitioning approach is developed by Arora et al. for emergency message dissemination in VANET [10]. In order to reduce delays in the network, relay vehicles are addressed in this proposal. Several problems generated by hidden nodes and packet collisions in the networks are addressed and simulated using NS2. This proposed model provides feasibility and effectiveness when compared to other traditional models.

3   Description of CPAODV Algorithm

The main concern of every VANET architecture lies in the forwarding of Data management, connection management, node classification, and resource allocation of nodes. The assignment of resources generates a traffic-free network as well as offers the destination node with much more network traffic. RSTs regulate the whole system and assign the connection speeds based on the primary consideration of a vehicle in the freeway environment in order to give the available bandwidth throughout the system [11]. Warning notifications would be sent to vehicles of speedy vehicle directional movement. The bandwidth-consuming node should also be described as well as indicated in the successive interaction as a blacklisted node. The AODV routing algorithm is handpicked for short route selection, CBAODV routing technique for connectors and RSTs are given preference for Classifying and assigning preference to the nodes in order to save bandwidth management in order to effectively allocated bandwidth.

The vehicle users have to be controlled by RSTs by allocating the route. This provides its subscribers on the allocated manner as well as route an anticipated system. For all the distribution of packets of data between vehicles is done based on the network topology and the routing protocol is used. Although the nodes are rapidly moving as well as RSTs could not control the vehicles delicately. There will be examined and publicized to all the other RSTs, the node which brings the collision in the network. The node will be temporarily banned in the future to prevent congestion issues. VANET invites in specific methods such as forwarding routing control, leakage early detection as well as clustering route allocation.

Particularly in comparison with reactive routing guidelines on roads and highways, the reactive protocols perform better results. Ad-hoc on-demand Distance Vector (AODV), amongst these systems, is selected to become the most robust routing protocol for load balancing in VANET. Even though there are substantial node ratios, the existing AODV routing protocol works better. Connection configuration takes just a little less time.

AODV retains only the current routing information as a routing table at each participating node. Neighbor node information is available to all the participating nodes. RREQ is used by a Source Node (SN) to make a Route Request based on flooding technique. RREQ is forwarded until it finds the Destination Node (DN). DN sends a Route Reply (RREP) to SN based on the reverse path technique. Intermediate Node (IN) helps to forward the RREQ and RREP information, and finally, each intermediate node makes an entry in their routing information table. In connection breakage, this AODV algorithm suffers a lot, and it resends the RREQ to maintain the connection. CBAODV algorithm is chosen to avoid link breakage issue in the VANET.

3.1   Classifying and Partitioning the Users

Connection breakage is a tedious task in VANET environment. AODV performance lacks when there is a connection breakage during data transmission [12]. Congestion creates a collision in the network because of numerous generations of RREQ and RREP data packets for a single communication. In such case, RSTs are configured based on the concepts of CBAODV algorithm. Whenever a connection breakage occurs, RSTs act as an intermediate node to continue further communication without regeneration of RREQ and RREP. It leads to reducing the random occurrence of multiple RREQ and RREP. Finally, this CBAODV could be strengthened by partitioning the users in the network. The proposed vehicle movement in the Freeway model for CPAODV algorithm is designed, and it is also illustrated in Fig. 1.

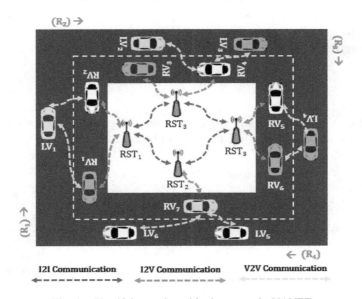

Fig. 1. Classifying and partitioning users in VANET

Vehicular user travels in their desired speed, direction, and distance from source to destination. The role of RST is to help all the users until they reach destination [13]. In general VANET communication, a single road is allocated to all the users. Some users during their travel, they move in an average speed or rapid speed, or they stop at the unwanted location. This situation creates a collision and leads to data loss. The traditional VANET communication is designed in Freeway Movement Model. RSTs are allocated in their transmission range to cover the single road with different lanes. All the moving vehicles travel at their desired speed until they reach destination. The number of generated RREQ and RREP packets are more because of without any control mechanism. RSTs, in this scenario, forwards the data packets either by RST to RST communication (I2I) or RST to Vehicle communication (I2V) or Vehicle to Vehicle communication (V2V). Vehicle movements in the roads are controlled by assigning the preference for the vehicles using Vehicle identification (V_{id}).

In the proposed model, a single city is divided into four roads as R1, R2, R3, and R4. Each road is further divided into different lanes as Left and Right lanes. RST_s are assigned to partition (P_n) the users based on Preference like P_1, P_2, and P_3. Also, the category of the vehicles as C1, C2,...Cn.

Here, RST_1 allocates and controls the lanes for each vehicle in (R_1) until the communication ends. RST_2, RST_3, and RST_4 are assigned in R_2, R_3 and R_4 roads within their transmission range to support for the V_n vehicles. The Vehicles LV_1, LV_2, LV_3, LV_4, LV_5, and LV_6 are assigned to travel as Left side path in the road. The Vehicle RV_1, RV_2, RV_3, RV_4, RV_5, RV_6, and RV_7 are assigned to travel in Right side path. Vehicles in Yellow color (LV_1, LV_2, and RV_7) are assigned as Emergency vehicles with High Preference (HP), Vehicles in blue color (RV_1, RV_3, LV_4, RV_6) are assigned as Average Preference (AP), Vehicles in white color (RV_2, RV_4, RV_5, LV_5, LV_6) are assigned as Less Preference (LP). The Total Network Capacity T_{NC} is shared with all the active nodes in the network, and it is termed as the Balanced Network Capacity (B_{NC}).

The formula to calculate BNC is

$$B_{NC} = \frac{T_{NC}}{A_{NN}}$$

Based on the above formula, route allocation using preference for vehicles with detailed information in Table 1. This table is generated based on Fig. 1.

Table 1. Assigning preference for vehicles in the routes

Node preference (NP_n)	Total network capacity (T_{NC}) = 40 Mb	Active nodes in network (A_{NN}) = 13	Balanced network capacity (B_{NC})
HP	20 Mb	3	6.66 Mb/node
AP	15 Mb	4	3.75 Mb/node
LP	5 Mb	5	1 Mb/node

In Table 1, Node Preference are considered as HP, AP, and LP. Total Network Capacity (Bandwidth) T_{NC} available to the nodes in the entire network is 40 Mb. Total Active Nodes in the Network are 13 Nodes. Based on the T_{NC}, entire 40 Mb is divided and assigned to all the nodes. At HP level, 20 Mb is divided for 13 Nodes. So, each node in HP will receive 6.66 Mb bandwidth and nodes at high speed in the network. At AP level, each node is assigned with 3.75 Mb/node. At last, 5 Mb in LP is divided, and 1 Mb is assigned to each node. The implemented preference technique is categorized as:

- Node Preference (NP_n) are assigned as HP, AP, and LP
- HP routes are assigned for high preference nodes like emergency vehicles
- AP routes are assigned for medium preference nodes like general vehicles
- LP routes are assigned for fewer preference nodes like heavy loaded vehicles
- HP nodes are in Yellow Color
- AP nodes are in Blue Color
- LP nodes are in White Color

3.2 Distance Calculation to Avoid Collision

Collision avoidance is a major concern in any VANET environment. So, to avoid collision in the network, distance between the vehicles is taken into consideration. The formula that is used to calculate distance between the vehicles is

$$Distance\ (VN_{n1} : VN_{n2}) = \sum_{i=1}^{n} D_Z$$

Here, VN_{n1} is the sender node, and VN_{n2} is the receiver node. n is total neighbor nodes between RST_{n1} and RST_{n2}. D_Z is the difference between a pair of VNs in VN_{n1} and VN_{n2}.

The Minimum Distance for each vehicle in HP is assigned as Min_{Dist} = 10 m and these vehicles V_1 and V_2 travels at a distance = 10 ms to avoid the collision. This same method is used for AP vehicles as Min_{Dist} = 10 m. However, Min_{Dist} = 20 m is assigned for vehicles at LP level. Nodes in the LP level are heavily loaded and big vehicles. Vehicles at HP level are assigned at Min_{Speed} = 70 km/h as Minimum Speed, and Maximum Speed is Max_{Speed} = 80 due to high preference nodes. In this same way, Min_{Speed} = 60 km/h and Max_{Speed} = 70 km/h are given for preference AP nodes. For preference LP nodes, Min_{Speed} = 50 km/h and Max_{Speed} = 60 km/h are used for node movement inside the network. The distance calculation is considered to avoid collision between vehicles, and it is given in Table 2.

Table 2. Distance calculation in CPAODV

Node preference (NP)	Minimum distance (Min_{Dist})	Minimum speed (Min_{Speed})	Maximum speed (Max_{Speed})
HP	10 m	70	80
AP	10 m	60	70
LP	20 m	50	60

The proposed CPAODV algorithm

RST_n is the number of RSTs in the network
V_n is the total number of vehicles
RP_n is the preference of the nodes
AN_n is the active nodes in the routes
L_n is the lanes created based on RP_n
Min_{Speed} is the Minimum Speed
Max_{Speed} is the Maximum Speed
BV_n is the total Blacklisted node
BV_i is a Blacklisted node

- RST_1 checks the road (route) availability
- RST_1 makes partition of the route as $L_n = L_1$, L_2, L_3,,,L_m
- RST_1 receives the V_n vehicles
- RST_1 check the (BV_i) details to assign preference NP
- RST_1 assigns
    - RP_1 to L_1, RP_2 to L_2 and RP_3 to L_3
- RST_1 checks (T_{NC}) to assign (A_{NC})
$$Distance\ (VN_{n1}:VN_{n2})\ = \sum_{i=1}^{n} D_z$$
- RST_1 forwards route preference information to all the availableRST_n
- Nodes Preference (NP) is calculated and sent to all the available RST_n
- Available Bandwidth (AB) is calculated
- Bandwidth is Allocated (BA) based on (AB)
- Bandwidth Reserved for future vehicles
    $(BR) = (AB) - (BA)$
- V_n starts transmission based on the direction of RST_1
    V_n moves in the order as V_i, V_{i+1},,,, V_{m+1}
- Distance between vehicles is calculated to avoid the collision
    $$Distance\ (RST_{n1}:RST_{n2})\ = \sum_{i=1}^{n} D_z$$
- BNC is calculated and assigned based on

$$B_{NC} = \frac{T_{NC}}{A_{NN}}$$

- RST checks the distance as
    - Minimum Distance (Min_{Dist})
    - Minimum Speed (Min_{Speed})
    - Maximum Speed (Max_{Speed})
- if$((V_i<Min_{Speed})$ && $(V_i>Max_{Speed}))$
    - RST_i sends the alert message to V_i
    - RST_i makes V_i as Blacklisted Node (BV_i)
- Hello messages are broadcasted periodically.
- RSTi checks vehicle status (V_{st})
    - If$((V_{st}=0)$ or $(V_{st}=1))$

    o   RST$_i$ sends V$_{st}$ message to the entire network
- If V$_i$ sends a message to another V$_m$
  - RST$_i$ forwards data packets based on AODV
- If V$_i$ faces connection breakage between V$_i$ and V$_m$
  - NearbyRST$_i$ helps to avoid breakage based on CBAODV algorithm

3.3   Advantages of CPAODV Model

CPAODV could be implemented with any routing algorithm with the help of RSTs. Collision in the network is reduced due to the proper allocation of bandwidth in the network. Vehicular nodes are strictly controlled based on the preference assigned in routes. Whenever a node seeks to route support from multiple nodes or RSTs, the request is solved based on providing proper bandwidth and speed. Assigning preference for routes and nodes are simple and effective technique for each node to avoid collision. The main advantages of the proposed CPAODV are:

- Freeway model is chosen to limit the vehicles to its lane
- RSTs control Vehicle's Speed (VS)
- Traffic status is shared with the entire network
- Tracking a vehicle location is an easy task
- The scenario is designed as an obstacle-free road
- Nodes Preference and Routes Preference are calculated effectively
- Bandwidths are reserved for future usage
- Vehicles are blacklisted as (BV$_n$) when they violate rules
- Queue model is used to allocate vehicles to the Lanes
- Resource utilization is done effectively with the help of RSTs and Vehicles

4   Performance Analysis of AODV, CBAODV, and CPAODV

The performance of the proposed CPAODV algorithm is implemented by using NS 2.35 simulator. The strength of the proposed algorithm is calculated based on the QoS parameters like Packet Delivery Ratio (%), End to End Delay (Sec.) and Throughput (Bits/Sec.). The ratio of data delivered from the source vehicle to the destination vehicle without loss is termed as Packet Delivery Ratio (%). The delay occurred in sending a data packet from source node to destination node is called End to End Delay (Sec.). The number of successful delivery of data packets in a definite time is referred to as Throughput (Bits/Sec.). The proposed CPAODV algorithm is compared with the AODV algorithm concerning number of nodes like 7, 12, 17, 22 and 27 in a VANET environment. The short information about the simulation parameters used in experimental setup as Table 3.

The simulation of the traditional freeway movement model using the CPAODV algorithm is given in Fig. 3. Nodes are tested with different sizes in the network as 7, 12, 17, 22 and 27 nodes respectively. RSTs are used in the same environment without performing any network changes in the network design to analyze the results. RST_1, RST_2, and RST_3 are considered as the essential nodes to decide and to control the data transmission. Vehicles are identified as V_1, V_2, V_3, V_4, V_5, V_6 and V_7 in the travel without any order until its destination. It is found that the vehicles travel at random speed and results in congestion in the network.

Table 3. Simulation parameters

Parameter	Value
Tool	NS 2.35
Data packet	512 bytes
MAC type	IEEE 802.11P
Channel type	Channel/Wireless
Mobility model	Freeway mobility model
Propagation model	Two ray ground
Network interface type	Phy/Wireless Phy
Interface queue type	Queue/Drop Tail/PriQueue
Routing protocol	AODV, CBAODV, CPAODV
Channel and capacity	Wireless 10 Mb/s
Buffer size	1000 packets
Highway length	8 km
Number of nodes	7, 12, 17, 22, 27
Vehicle speed limit	60 km/h, 70 km/h, 80 km/h
Simulation time	900 s
Traffic rate	CBR
Simulation area	1000 m × 1000 m
Transmission range	100 to 700 m

In Fig. 2, to control and to streamline the participating nodes in the network, RST_1 is assigned as the primary node to assign preference and route allocation for V_n nodes. RST_2 and RST_3 are assigned to help the data transmission. These two RSTs are used as a secondary node for the data transmission, which helps to avoid connection breakage also. During the performance analysis, the X-axis is measured as the number of nodes and Y-axis is the parameter used for the routing protocols.

In Fig. 3, the Throughput (Bits/Sec.) of AODV vs. CBAODV vs. CPAODV algorithm is calculated based on the changing the number of nodes as 7, 12, 17, 22 and 27 correspondingly. Throughput is calculated by diving the overall message received at the destination node by the total number of messages sent from the source node to destination node for all the nodes.

Packet Delivery Ratio (%) of AODV vs. CBAODV vs. CPAODV is measured by the total number of data packets sent successfully carried to the total number of data

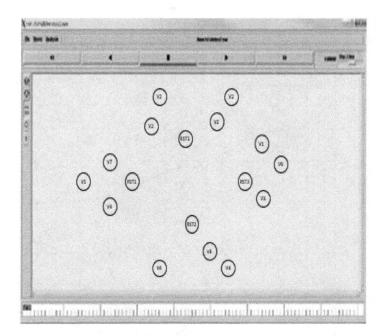

Fig. 2. Vehicle movement in CPAODV model

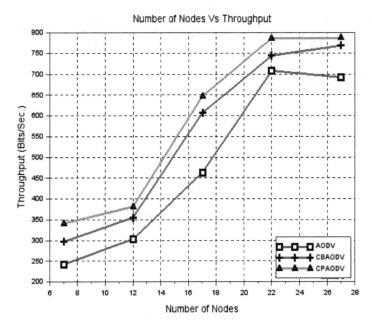

Fig. 3. Throughput (Bits/Sec.) Analysis of AODV vs. CBAODV vs. CPAODV

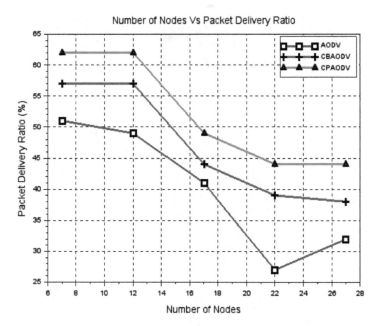

Fig. 4. Packet Delivery Ratio (%) Analysis of AODV vs. CBAODV vs. CPAODV

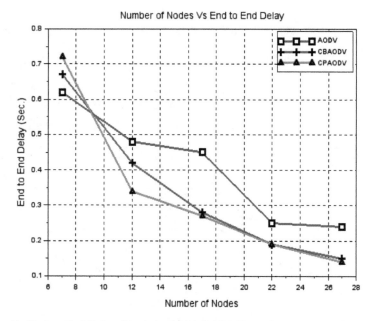

Fig. 5. End to End Delay (Sec.) Analysis of AODV vs. CBAODV vs. CPAODV

packets sent out by a sender node based on the nodes 7, 12, 17, 22 and 27. The ratios are analyzed and given as a graph in Fig. 4.

End to End Delay (Sec.) is considered as the subtraction of the arrived data packet to the originated data packet time through the sending node in the network. It is the average time taken for a data packet to transverse the network. It is also called as latency in the data transmission. This calculation is beneficial when measuring the efficiency of the proposed CPAODV algorithm. The delay calculated is given as a graph in Fig. 5.

4.1 Findings and Interpretation

Figure 3 gives the value of average Throughput (Bits/Sec.) generated for AODV, CBAOD, and CPAODV for nodes 7, 12, 17, 22 and 17. The difference in the Throughput values for the proposed CPAODV increases by delivering data packets using RSTs without any delay in the network. For instance, when the number of nodes is 7, Throughput (Bits/Sec.) of AODV is 242, CBAODV is 297, but CPAODV is 341.

The Packet Delivery Ratio (%) is given in Fig. 4 is done by comparing the AODV, CBAODV, and CPAODV for nodes 7, 12, 17, 22 and 17. The number of data packet delivered increases when the number of nodes increases in the network. Here, CPAODV delivers number of data from Source to Destination nodes. For an illustration, the Packet Delivery Ratio (%) of CPAODV for seven nodes has a more significant value of 62(%), CBAODV has a value of 57(%), but the AODV has the result of 54 (%) only. This ratio is high because in CPAODV the RSTs assign preference for both route and nodes and also it controls the flow of nodes until destination.

End to End Delay (Sec.) is the average delay that occurred in the network due to the delay occurred in the transmission. These details are given in Fig. 5. RSTs are responsible for reducing this delay occurred between the nodes network. These values are calculated without changing changes in the simulation environment. The result shows that the delay in CPAODV is less when compared to the delay occurred in AODV and CBAODV algorithm. After performing preference calculation, Routing and Connection breakage are concentrated to reduce the delay with the help of RSTs. For example, when the number of nodes is 12, the delay for CPAODV is 0.34 (Sec.) only, but the delay for AODV is 0.48 (Sec.), and CBAODV is 0.42 (sec.).

5 Summary

The proposed CPAODV algorithm is used to assign preference to nodes and routes using RSTs. RSTs are placed in the network for assigning and reserving bandwidths in the network. The node which consumes more bandwidth is identified and blacklisted in the future. Alert messages are sent periodically to backup the routing information in the network by vehicles and RSTs. For implementation purposes freeway model is used to analyze the results for nodes in highway environment. Min_{Dist} and Max_{Dist} are calculated to avoid the distance between two vehicles. All the participating nodes travel in the Queue model to avoid collision in the network. By designing the network carefully using some concepts from AODV and CBAODV algorithm, End to End Delay (Sec.)

values are decreased in CPAODV. Moreover, the Packet Delivery Ratio (%) and Throughput (Bits/Sec.) values are increased with the help of RSTs in CPAODV algorithm.

5.1 Limitation of CPAODV

Though, the proposed CPAODV algorithm performs better when compared to AODV because of categorizing and by assigning 3 level preference for nodes and routes in VANET, other parameters apart from Classifying and giving preference to the vehicles which travel in both directions for highway environment. Route allocation by partitioning is done with the help of RSTs in highway environment only, and this approach fails when RST_1 is idle due to unavoidable circumstances [14, 15].

References

1. Hasrouny, H., Samhat, A.E., Bassil, C., Laouiti, A.: VANet security challenges and solutions: a survey. Veh. Commun. **7**, 7–20 (2017)
2. Gong, H., He, K., Yingchun, Q., Wang, P.: Analysis and improvement of vehicle information sharing networks. Phys. A: Stat. Mech. Appl. **452**, 106–112 (2016)
3. Zhang, E., Zhang, X.: Road traffic congestion detecting by VANETs. In: 2nd International Conference on Electrical and Electronic Engineering (EEE 2019). Atlantis Press (2019)
4. Arulkumar, N., George Dharma Prakash Raj, E.: CBAODV: an enhanced reactive routing algorithm to reduce connection breakage in VANET. In: Suresh, L., Dash, S., Panigrahi, B. (eds.) Artificial Intelligence and Evolutionary Algorithms in Engineering Systems. Advances in Intelligent Systems and Computing, vol. 325. Springer, New Delhi (2015)
5. Al Mallah, R., Quintero, A., Farooq, B.: Distributed classification of urban congestion using VANET. IEEE Trans. Intell. Transp. Syst. **18**(9), 2435–2442 (2017)
6. Jawhar, I., Mohamed, N., Al-Jaroodi, J., Agrawal, D.P., Zhang, S.: Communication and networking of UAV-based systems: Classification and associated architectures. J. Netw. Comput. Appl. **84**, 93–108 (2017)
7. Rayeni, M.S., Hafid, A., Sahu, P.K.: Dynamic spatial partition density-based emergency message dissemination in VANETs. Veh. Commun. **2**(4), 208–222 (2015)
8. Santamaria, A.F., Sottile, C., Fazio, P.: PAMTree: partitioned multicast tree protocol for efficient data dissemination in a VANET environment. Int. J. Distrib. Sens. Netw. **11**(5) (2015). https://doi.org/10.1155/2015/431492
9. Selvi, M., Ramakrishnan, B.: An efficient message prioritization and scheduled partitioning technique for emergency message broadcasting in VANET. In: 2018 3rd International Conference on Communication and Electronics Systems (ICCES), pp. 776–781. IEEE (2018)
10. Arora, P.O.: Reliable emergency message dissemination in VANET using hybrid partitioning approach. In: 2018 International Conference on Communication and Signal Processing (ICCSP), pp. 0672–0677. IEEE (2018)
11. Ali, G.G.Md.N., Chong, P.H.J., Samantha, S.K., Chan, E.: Efficient data dissemination in cooperative multi-RSU vehicular ad hoc networks (VANETs). J. Syst. Softw. **117**, 508–527 (2016)

12. Samanta, D., Podder, S.K.: Level of green computing based management practices for digital revolution and new India. Int. J. Eng. Adv. Technol. (IJEAT), **8**(3S) (2019). ISSN 2249–8958

13. Sivakumar, P., Nagaraju, R., Samanta, D., Sivaram, M., Hindialraj, N., Amiri, S.: A novel free space communication system using nonlinear InGaAsP microsystem resonators for enabling power-control toward smart cities. Wirel. Netw. J. Mob. Commun. Comput. Inf. ISSN 1022-0038

14. Mahua, B., Podder, S.K., Shalini, R., Samanta, D.: Factors that influence sustainable education with respect to innovation and statistical science. Int. J. Recent. Technol. Eng. **8** (3). ISSN 2277-3878

15. Praveen, B., Umarani, N., Anand, T., Samanta, D.: Cardinal digital image data fortification expending steganography. Int. J. Recent. Technol. Eng. **8**(3) (2019). ISSN 2277-3878

16. Dhanush, V., Mahendra, A.R., Kumudavalli, M.V., Samanta, D.: Application of deep learning technique for automatic data exchange with air-gapped systems and its security concerns. In: Proceedings of IEEE International Conference on Computing Methodologies and Communication, Erode, 18–19 July 2017

17. Kumar, R., Kumar, R.: Samanta, D., Paul, M., Vijaya Kumar, C.M.: A combining approach using DFT and FIR filter to enhance impulse response. In: Proceedings of IEEE International Conference on Computing Methodologies and Communication, Erode, 18–19 July 2017

18. Ghosh, G., Samanta, D., Paul, M., Kumar Janghel, N.: Hiding based message communication techniques depends on divide and conquer approach. In: Proceedings of IEEE International Conference on Computing Methodologies and Communication, Erode, 18–19 July 2017

19. Singh, R.K., Begum, T., Borah, L., Samanta, D.: Text encryption: character jumbling. In: Proceedings of IEEE International Conference on Inventive Systems and Control, Coimbatore, 19–20 January 2017. IEEE

20. Kumar, N.A., George Dharma Prakash Raj, E.: VCS-RSCBAODV: vehicular cloud storage concepts for RSCBAODV protocol to reduce connection breakage in VANET. Adv. Comput. Sci. Int. J. **4**(3), 147–153 (2015)

21. Choksi, Y., Tandel, P., Khatri, T.: AODV modification to address link breakage issue: a review. In: 2017 International Conference on Trends in Electronics and Informatics (ICEI), pp. 576–578. IEEE (2017)

22. Wang, S., Yao, N.: A RSU-aided distributed trust framework for pseudonym-enabled privacy preservation in VANETs. Wirel. Netw. **25**(3), 1099–1115 (2019)

23. Nabil, M., Hajami, A., Haqiq, A.: A stable route prediction and the decision taking at sending a data packet in a highway environment. In: Proceedings of the 2nd International Conference on Big Data, Cloud and Applications, p. 66. ACM (2017)

TSA – QoS: Throughput Aware Scheduling Algorithm to Enhance the Quality of Service in Mobile Networks

Ramkumar Krishnamoorthy[1(✉)] and P. Calduwel Newton[2]

[1] Department of Information and Communications Technology, Villa College,
Malé 20373, Maldives
kramdharma@gmail.com
[2] Department of Computer Science, Tiruchirappalli 620 022, India
calduwel@yahoo.com

Abstract. Mobile Networks are one type of wireless networks and provides the service to the subscribers regularly. Sometimes, it struggles to offer the expected Quality of Service (QoS) to the users at all times because it contains different kinds of users and variable application needs. It needs to focus many parameters with several technologies. Call Admission Control (CAC) is the one among them and having the capability to control that who can get admission and who should be rejected. Similarly, several reasons exist for rejecting and accepting the calls. Bandwidth is one of the main parameters for measuring the network capacity in Mobile Networks. It should be used in effective manner the networking system might be well upgraded when bandwidth is properly assigned and goes wasted when it is assigned poorly. It leads to drop many ongoing calls and refuse newly requesting calls by CAC. Then, at last it reduces the call acceptance probability. So, there is high demand to design proficient CAC algorithm to upgrade or improve the system capacity and use the bandwidth efficiently. The main aim of this paper is to propose a new CAC algorithm TSA – QoS that uses the bandwidth effectively and also increases the throughput. Finally, it improves the QoS in Mobile Networks.

Keywords: Scheduling · TACA · Call Admission Control · Bandwidth · Throughput · AMR codec · Quality of Service

1 Introduction

Mobile users count has been increasing astonishingly day-by-day without any limit because Long Term Evolution (LTE) [1] technology got more attention in usages. It makes the people to bring their interests in out of the box that is beyond the limit usage. Every time the users are always connected with the network in order to share information with other people. Additionally, 4G handsets are available in cheaper cost than previous generations and also the service provider giving more high speed data to access by paying in a lesser amount. LTE technology fulfills most of the users by adopting some advanced components based architecture. It provides service to different types of network users and it is why called Heterogeneous networks. Generally, in

© Springer Nature Switzerland AG 2020
L. C. Jain et al. (Eds.): ICICCT 2019, LAIS 9, pp. 446–454, 2020.
https://doi.org/10.1007/978-3-030-38501-9_43

networking the system can be determined by reckoning the number of users presently accessing the service. To keep on maintaining this, the particular network provider must provide truthful service continuously even if any obstacles exist in the network.

4G has Multi-Input-Multi-Output (MIMO), Orthogonal Frequency Division Multiple Access (OFDMA), interactive Multimedia Service, portable service, Easy and Fast access, highly scalable, easy adjustment with its prior networking protocols including Hyper LAN, Wireless LAN, 2.5G, 2.75G and 3G networks etc. It refers to Heterogeneous Networks [2] and briefly termed as 'HetNet'. It also offers the break-free connection in the connected users and provides many applications for targeting more participants into the networks. Due to its diversified application requirements, it is somewhat difficult to provide an appropriate/better QoS [3] to all the users at a time. To avoid this situation, the 'HetNet' uses an efficient Radio Resource Management (RRM) framework which periodically offers the resources to the users by looking at the expected channel feedback values. For this, CAC [4] plays a crucial role in RRM functionalities. Both CAC and RRM determine in improving the network capacity as well as increasing the system throughput.

CAC is specially framed only for voice call operations and not for data access which maintains a list for giving permission to the new user and accepts the existing calls who are trying to connect to the new network. It admits the new user only who resolves certain constraints and similarly it accepts a fewer number of handoff calls. It can be done only when the CAC system has sufficient bandwidth to provide continuous service to the requested call. In general, the Base Station usually reserves some amount of bandwidth for giving services to the handoff calls. Entering newly arrived call must not violate the speech quality of ongoing voice call. It should be done effectively by the CAC.

This paper is structured as follows. Section 2 deals with previous works and Sect. 3 shows the proposed TSA – QoS Algorithm in detail. Section 4 displays the Scenarios and Discussions for TSA – QoS. Section 5 covers the Simulation Results and Sect. 6 presents the Conclusion. Finally, References got placed in Sect. 7.

2   Previous Works

Many CAC algorithms were designed for enhancing the QoS in wireless networks during admission. Calduwel et al. [5] developed a priority based resource block assignment algorithm to the real-time services in wireless networks. The further work is then extended in the name of DSA – QoS [6] to analyze that priority and it is verified through a mathematical model. AlQahtani et al. [7] proposed a new CAC algorithm for enhancing the QoS in LTE – A based network. Delay is considered as a main parameter to assign the resources efficiently. Users are classified by taking application type without degrading the QoS performance.

Ivesic et al. [8] released an improved scheduling algorithm for providing services to multimedia applications in LTE networks. They also enhanced the Quality of Experience (QoE) of users by performing cross-layer approach. Results show that, it reaches higher throughput even by having limited number of resources in the scheduler. Carvalho et al. [9] implemented an algorithm for solving the network selection issue in next generation wireless networks. They also formed an optimization problem in CAC by using Semi-Markovian for optimally assigning the resources. It works jointly based CAC for addressing the inter-RAT issues.

Ovengalt et al. [10] incorporated Fuzzy logic theory to analyze the features of CAC in LTE. It predicts both certainty and uncertainty to reduce the Call blocking ratio and increase Call acceptance ratio using Fuzzy. Additionally, it concentrates on QoS parameters for improving QoS performance while admitting the users. Sonia et al. [11] developed a novel downlink based CAC algorithm for Random Way Point mobility scenarios. It produces good results in less number of calls dropping ratio and more number in acceptance ratio. Finally, it increases the system throughput and allows more admission to real-time users.

Calduwel et al. [12] introduced a modified CAC algorithm by only looking at the real-time users while giving admission into the network. AHP is utilized for allowing the users (Codecs) in proper order. In order to make that, three alternative criteria have been set in that algorithm for finding the importance value among them. And then, WSM is adapted for calculating the weight value of each criterion. Ghosh et al. [13] proposed Heterogeneity based admission control scheme for enhancing the QoS in HetNets. Received Signal Strength (RSS) and Bandwidth are mainly considered and incorporated into one novel hybrid CAC to take scheduling decision in an effective manner. It is verified by implementing nested cell in a particular environment.

Maharazu et al. [14] created reservation oriented CAC algorithm to focus on Best Effort (BE) service users. This is can be done by having an adjustable threshold factor which balances the throughput and number of users in admission. They verified by using analytical model which shows an improved performance in BE traffics. Ramkumar et al. proposed an algorithm for pushing out the fake users from the networking system which ultimately admits only the genuine users in the cell [15–20]. It increases the throughput then it further enhances the QoS also in Mobile Networks [21, 22].

3 TSA – QoS: A Proposed Algorithm

TSA – QoS is mainly designed for real-time users and is an updated version of TACA [12] algorithm. It is developed to utilize the available bandwidth effectively in mobile networks. It also offers services to more number of users than TACA. Additionally, it sets 20 ms as a threshold for restricting the bandwidth damage. It assigns portion of the channel to other users who initially gets rejected from the algorithm.

Input: Number of Real-Time Users, Number of RBs, Threshold and Share.

Initialization: $N_{RB}=\Phi$; $N_{RT}=\Phi$; $\Theta=20$ms; $\psi = C_{NB}/40$;

1. Identify the Available bandwidth in network
2. Observe the number of users and number of resource blocks available at the scheduler
3. Invoke TACA Procedure
4. Set the threshold time 'Θ' during admission
5. Admit the previously unaccepted users until it meets the new threshold value
6. Assign the resources, $\psi = $ (No. of bytes / 40) to each users
7. Repeat steps 1 to 6 for all the users until all the users are scheduled

> *Procedure TACA()*
> *Begin*
> - Calculate the Total bandwidth of each requesting calls
> - Use Analytic Hierarchy Process and Weighted Sum Method for finding the rank of the codecs and admit them one by one
> - Restore the leftover bandwidth
>
> *End*

Where,

TACA – Throughput Aware CAC

ψ – Share of each user

Output: Allow (Admission) the users one by one According to TACA rank.

It is done repetitively until all the incoming calls are admitted without violating the bandwidth constraints. According to that Medium Access approach [16], each resource block can carry 40 bytes of payload during transmission. By applying this, TSA – QoS identifies the share of each and every user who seeks resource blocks.

4 Scenarios and Discussions

At First, the Network is configured with 512 Kbps bandwidth and 15 calls for 1 s simulation time. 4 users in G.711 codec, 2 users in G.729 codec, 3 users in G.723 codec, 2 users in AMR codec and iLBC codec 1 user are assumed for network scenario. Packetization delay 20 ms is treated for all types of codec users. TSA – QoS sets 20 ms as a threshold time for admitting the users in the network. TACA assigns the whole bandwidth to each user permanently. In general, packets have been produced by the codec for every 20 ms exactly. This time is utilized by the proposed TSA – QoS for saving the bandwidth wastage and admits the users effectively. Table 1 shows the performance comparison of both TACA and TSA – QoS algorithms. Initially, for the first case 512 Kbps is taken and it produces 1042 Packets Per Second (PPS) by TACA. But, TSA – QoS transfers 1166 PPS in the same bandwidth. It continues until the users are satisfied by using the resource assignments.

Table 1. Utilized bandwidth and PPS

TACA		TSA – QoS	
Bandwidth	PPS	Bandwidth	PPS
512 Kbps	1042	512 Kbps	1166
1024 Kbps	2332	1024 Kbps	2532
2048 Kbps	5726	2048 Kbps	6826
5120 Kbps	18844	5120 Kbps	21344

Figure 1 shows the throughput comparison of both TACA and TSA – QoS algorithm. It is tested for 1 s (60 ms) with 512 Kbps bandwidth with 15 users.

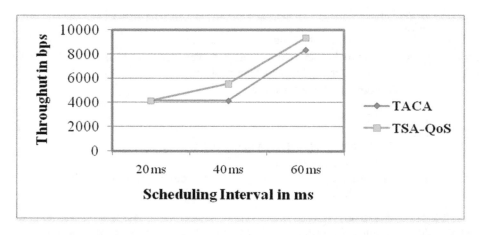

Fig. 1. Throughput results for one sec

Figure 2 displays the increased throughput for various bandwidths to different number of users in the network. The users range is taken like 50, 100, 200 and 500 by setting the similar type of mobility and various bandwidths like 2048 Kbps and 5120 Kbps respectively.

First 20 ms, the users got admission like TACA. The remaining rejected calls have to wait for its next turn (20 ms). It causes delay and wastes bandwidth. In order to mitigate these issues, TSA – QoS exactly admits the users consecutively. Wireshark application [17] is used to collect the data sets of each codecs in TSA – QoS.

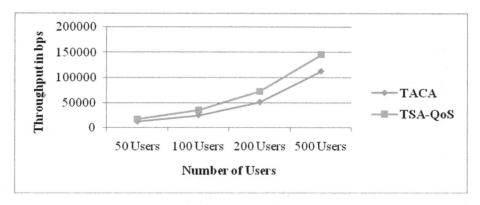

Fig. 2. Throughput comparisons for various users

5 Simulation Results

The proposed TSA – QoS is analyzed by taking the data sets that are collected using Wireshark. It is identified individually before making TSA – QoS algorithm values are tested. After that, the code is developed using MATLAB and results are observed. Based on that, the TSA – QoS saves the bandwidth and offer more number of admissions to other users. It tends to increase the throughput and enhances the QoS. But, TACA wastes the bandwidth frequently, because it assigns the bandwidth in static manner. Suppose, if the assigned user did not speak or be idle means, assigned bandwidth goes wasted. It is observed and utilized properly to other users for admission. TSA – QoS is configured by using the following parameters and shown in Table 2.

Table 2. Simulation parameters

Parameters	Values
Codec	G.711, G.729, G. 723, G.726, iLBC and EVRC
Bandwidth	512 Kbps, 1024 Kbps, 2048 Kbps and 5120 Kbps
Simulation time	1 s
Packetization delay	20 ms
Packet size	160, 20 and 30 Bytes
Header size	40 Bytes

New CAC is built as mentioned in TSA – QoS for validating the performance of the throughput and bandwidth per call. Based on this, the results are shown in Fig. 3. It depicts that, TSA – QoS throughput got improved while on admission. It also allows many other users to make and receive the calls.

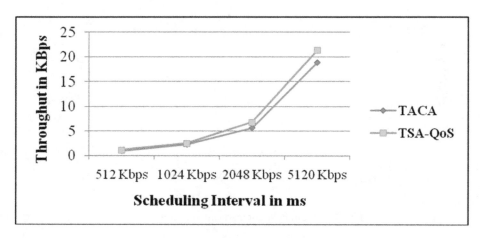

Fig. 3. Throughput results of four cases

TSA – QoS not only increases the Throughput but also provides enhanced results in the acceptance ratio while on user admission. It gives more number of opportunities to several users than TACA. It accepts only limited users and wastes the bandwidth in an altering the assignments, these wasting of bandwidth can be effectively assigned to the set of users in network. Figure 4 illustrates the Acceptance Ratio in % for various users in the network dispersed coverage.

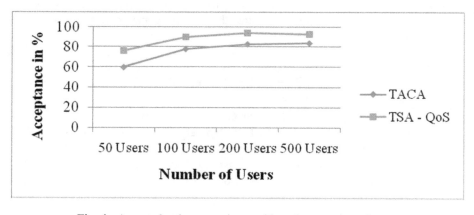

Fig. 4. Accepted ratio comparisons with various number of users

According to the above figure, TSA – QoS yields better results by accepting more number of users in various mobility ranges. Obviously, the proposed TSA – QoS increases the throughput, offers more admission, less dropping ratio for with different setup configurations even more number of users present in the network.

6   Conclusion

Developing QoS enhanced Scheduling algorithm is not at all easy in today's networking environment because of limited amount of available spectrum and frequently changing user mobility. Especially, the real-time applications need to reserve sufficient bandwidth for preparing the voice or video calls to connect remote people. Some decision making algorithm causes to withdraw certain ongoing connections or reject newly requesting connections due to poor management of bandwidth assignments. Therefore, an effective approach is much needed for accommodating many users, not degrading the voice quality, improving the call acceptance ratio and reducing the dropping ratio. TSA – QoS performs better than existing TACA algorithm for saving the used bandwidth and increasing the throughput. Finally, it enhances the QoS in Mobile Networks.

References

1. Capozzi, F., Piro, G., Grieco, L.A., Boggia, G., Camarda, P.: Downlink packet scheduling in LTE cellular networks: key design issues and a survey. IEEE Commun. Surv. Tutorials **15**(2), 678–700 (2013)
2. Ramkumar, K., Calduwel Newton, P.: Interference aware downlink scheduling algorithm in mobile networks. In: 11th International Conference on Intelligent Systems and Control (ISCO), pp. 282–286. IEEE Xplore (2017)
3. Calduwel Newton, P., Arockiam, L.: A novel prediction technique to improve quality of service (QoS) for heterogeneous data traffic. J. Intell. Manuf. **22**(6), 867–872 (2011)
4. Erik, C., Marcel, J., Tomas, B.: Admission control methods in IP networks. Adv. Multimed. **2013**, 1–7 (2013)
5. Calduwel Newton, P., Ramkumar, K.: An innovative technique (ANINTECH) to reduce delay of VoIP downlink packet scheduling in 4G technology. Int. J. Appl. Eng. Res. (IJAER) **10**(72), 363–368 (2015)
6. Ramkumar, K., Calduwel Newton, P.: DSA - QoS: delay based scheduling algorithm to enhance the QoS in mobile networks. Int. J. Pure Appl. Math. **118**(9), 93–101 (2018)
7. AlQahtani, S.A.: Users' classification-based call admission control with adaptive resource reservation for LTE-A networks. J. King Saud Univ. – Comput. Inf. Sci. **29**(1), 103–115 (2017)
8. Ivesic, K., Skorin-Kapov, L., Matijasevic, M.: Cross-layer QoE-driven admission control and resource allocation for adaptive multimedia services in LTE. J. Netw. Comput. Appl. **46**, 336–351 (2014)
9. Carvalho, G.H.S., Woungang, I., Anpalagan, A., Coutinho, R.W.L., Costa, J.C.W.A.: A semi-Markov decision process-based joint call admission control for inter-RAT cell reselection in next generation wireless networks. Comput. Netw. **57**(17), 3545–3562 (2013)
10. Ovengalt, C.B.T., Djouani, K., Kurien, A.: A fuzzy approach for call admission control in LTE networks. In: 5th International Conference on Ambient Systems, Networks and Technologies (ANT-2014) (2014). Procedia Comput. Sci. **32,** 237–244
11. Sonia Ben, R., Nidal, N., Sami, T.: A novel resource allocation scheme for LTE network in the presence of mobility. J. Netw. Comput. Appl. **46**, 352–361 (2014)

12. Calduwel Newton, P., Ramkumar, K.: TACA-throughput aware call admission control algorithm for VoIP users in mobile networks. In: Advances in Computer and Computational Science, vol. 1, pp. 259–270. Springer (2017). ISBN 978-981-10-3770-2

13. Ghosh, A., Misra, I.S.: A joint CAC and dynamic bandwidth allocation technique for capacity and QoS analysis in heterogeneous LTE based BWA network: few case studies. Wirel. Pers. Commun. **97**, 2833–2857 (2017)

14. Chakrabarti, S., Samanta, D.: Image steganography using priority-based neural network and pyramid. In: Emerging Research in Computing, Information, Communication and Applications, pp. 163–172 (2016). https://doi.org/10.1007/978-981-10-0287-8_15

15. Ghosh, G., Samanta, D., Paul, M.: Approach of message communication based on twisty "Zig-Zag". In: 2016 International Conference on Emerging Technological Trends (ICETT) (2016). https://doi.org/10.1109/icett.2016.7873676

16. Hossain, M.A., Samanta, D., Sanyal, G.: Extraction of panic expression depending on lip detection. In: 2012 International Conference on Computing Sciences (2012). https://doi.org/10.1109/iccs.2012.35

17. Hossain, M.A., Samanta, D., Sanyal, G.: Statistical approach for extraction of panic expression. In: 2012 Fourth International Conference on Computational Intelligence and Communication Networks (2012). https://doi.org/10.1109/cicn.2012.189

18. Khadri, S.K.A., Samanta, D., Paul, M.: Approach of message communication using fibonacci series. In: Cryptology. Lecture Notes on Information Theory (2014). https://doi.org/10.12720/lnit.2.2.168-171

19. Maharazu, M., Zurina, M.H., Azizol, A., Abdullah, M.: An adaptive call admission control with bandwidth reservation for downlink LTE networks. IEEE Access **5**, 10986–10994 (2017)

20. Ramkumar, K., Calduwel Newton, P.: ESA – QoS: energy aware scheduling algorithm for enhancing the QoS in mobile networks. Int. J. Pure Appl. Math. **119**(7), 127–132 (2018)

21. Gürsu, M., Vilgelm, M., Fazli, E., Kellerer, W.: A medium access approach to wireless technologies for reliable communication in aircraft. In: Wireless Sensor Systems for Extreme Environments, in Book, Wireless Sensor Systems for Extreme Environments: Space, Underwater, Underground, and Industrial, pp. 431–451. Wiley (2017)

22. Lamping, U., Sharpe, R., Warnicke, E.: "Wireshark 1.12", NS Computer Software and Services P/L. http://wiki.wireshark.org

Applying Virtual Reality Technology to Biology Education: The Experience of Vietnam

Phuong Le Thi[(✉)] and Linh Do Thuy

VNU University of Education, Vietnam National University, Hanoi, Vietnam
{phuongle,dothuylinh}@vnu.edu.vn

Abstract. Virtual Reality Technology (VRT) is built with hardware and digital formats in order to create a realistic immersive environment. VRT is most suitable for subjects that require labs and real-life experiments such as nature science, biology, chemistry, physics. In Vietnam, VRT has been implemented in education, albeit limited because of many reasons. In this study, the authors suggest a lesson design procedure to integrate VRT into education, specifically designed for teaching and learning Biology subject in high school in Vietnam with 5 steps and several contents that are suitable for applying VRT. This study set up a foundation base for the other deeper studies in using VRT in Biology in Vietnam.

Keywords: Virtual Reality Technology · Biology teaching · Information communication technology

1 Introduction

The use of information communication technology (ICT) has been increasing rapidly, especially in the field of education to make students "future-ready" for the digital world. ICT plays an important role in enhancing teaching and studying, also it helps teachers and students interact with a larger networked community [1, 3]. The need of using ICT has grown quickly in order to have such an effective teaching training program [4, 5].

Virtual Reality Technology (VRT) is built with hardware and digital formats to create a realistic immersive environment [6, 7]. This technique has supported lots of fields, such as engineering, medicine, tourism, education and many other aspects of life [8]. Especially, it offers a powerful tool to facilitate and provide a lot of privileges to education [9]. When interacting with VRT, learners showed a statistically improvement in knowledge achievement and also positive attitude towards VRT as trying VRT is an interesting and "joyful" experience [7, 10, 12]. VRT is most suitable for subjects that require labs and real-life experiments such as nature science, biology, chemistry, physics [7] and medical education. It gives students an opportunity to learn the topics or practice experiments in a virtual environment that is inappropriate to carry out in conventional environment [8, 11].

© Springer Nature Switzerland AG 2020
L. C. Jain et al. (Eds.): ICICCT 2019, LAIS 9, pp. 455–462, 2020.
https://doi.org/10.1007/978-3-030-38501-9_44

Visualization of different biological things, phenomena and processes is an information carrier channel for Biology teaching and learning as they would directly affect students' perception [14, 15]. Quite a few biological virtual environment has been developed, such as models and processes in molecular biology [12], model of cell biology as muscular, intestinal and nerve cell [16], drosophila's fertilization-related objects [17], plant cells-organelles and photosynthesis process [18], research on structure and function of the eye [11]. Experiences in three-dimensional space are proved effective to significantly enhance biological knowledge achievements while increasing interest and engagement [11, 12].

According to General Education Renovation Project of Vietnam Ministry of Education and Training, students must meet the capacity requirements of using ICT by taking the advantages of ICT in their learning progress as well as enhancing students' capacity of information sharing, communication and cooperation. Virtual reality, which is a breakthrough technology would be a strong integrated approach to education and particularly Biology education in particular. In this paper, actual situation of applying VRT in Vietnam is reviewed and also a procedure of integrating VRT into Biology education is designed.

2 Virtual Reality Technology in Education

2.1 General Education Renovation Project – the Education Reform in Vietnam

In 2018, programs and textbooks for general education were all renovated in the direction of developing quality and competence of learners in order to transform a heavy education on knowledge to a fundamental and comprehensive development of education. To develop learners' competence, the 2018 general education curriculum has innovated educational contents and methods concerning differentiated teaching, integrated teaching and teaching through active activities of learners.

The common feature of the educational methods applied in the 2018 general education curriculum is positive activity of learners, in which the teachers play the role in organizing and guiding students through creating friendly learning environment and problematic situations to encourage students to actively participate in learning activities, discover their abilities, aspirations, train their habits and ability to self-study.

In the process of renewing educational methods, and particularly the method for testing and evaluating indispensable role of technology, technology could participate in all stages of the teaching process, from lesson design to organizing teaching activities in classroom [21].

In the 2018 General Education Program, the method for teaching Biology subjects is conducted according to the following orientations [21]:

– Promote students' activeness and creativity; avoid memorizing mechanically; focus on fostering self-reliance and self-study so that students are able to expand their intellectual capital and advance the necessary qualities as well as competencies after they graduate from schools.

- Practice skills of applying biological knowledge to detect and solve practical problems; encourage and create opportunities for students to experience and be creative.
- Apply educational methods in a flexible and creative way that is suitable for educational objectives, contents of the lesson, characteristics of students and other specific circumstances. Teachers could combine of several teaching methods in one subject determining by lesson requirements. Traditional methods of teaching (presentations from teachers, one-way conversations, …) focus attention on promoting students' activeness and initiative. Modern teaching methods emphasize the role of students in their learning process (teaching through hands-on activities, problem solving teaching method, project-based teaching, experiential teaching, discovering teaching, differentiate teaching, … with appropriate teaching techniques).
- Forms of lesson organizing are implemented in a diversified and flexible manner; combining individual study, group study, classroom learning, reverse learning, online learning, … 2018 General Education Program promotes the application of ICT in teaching process, appreciate resources outside textbooks and school teaching equipment, fully exploit the advantages of ICT in teaching through using knowledge storage facilities, multimedia, electronic materials (such as experimental films, virtual experiments, simulated test, …).
- Apply integrated teaching approach. In order to teach these topics, teachers have to develop situations that students' mobilization knowledge and skills to find an answer to cognitive, practical and technological problems.

2.2 Biology Teaching and Learning in Vietnam and the Lack of Applying ICT

Biology is a specific science that studies the whole living world, so the forms of learning are extremely diverse. Alongside with the fast development of science and technology, many teaching aids have been brought into teaching process, helping teachers to guide students to obtain knowledge most effectively. From electronic lesson plans, in recent years, more and more modern teaching and learning technology devices have appeared, such as smart electronic boards, electronic textbooks, E-learning lesson design software … By using images, diagrams, video clips …, teachers attract attention and raise interest for students, assist them to acquire knowledge, reduce the abstraction of the lesson content, create necessary conditions for students to practice and formulate skills, contribute to innovating teaching methods and assessing students' learning results.

However, the use of ICT in teaching in general and teaching Biology in particular is very limited in schools. This is due to the rigorous requirements of using technology in biology teaching:

- It requires teachers to have the skills of designing lesson plans and adapting new technology to teach biological contents.
- Teachers need to be proficient in using presenting software, accessing the Internet to find documents, using several different support tools. Teachers also must have

creative, sensitive and aesthetic thinking to find teaching materials and insert multimedia forms into lessons properly.

– Each teacher must have creative and diverse adaptation when using visual aids in lessons. There is a fact that many teachers nowadays are too abusive to use electronic visual aids in teaching causing a rampant teaching that does not transfer enough information to students. To solve this problem, teachers must rely on pre-set contents and skills to achieve the set goals. Therefore, teachers need to carefully research the lesson and choose appropriate electronic visual aids to be integrated into the lesson,

– Teachers need to avoid giving out too much flashy color pictures or too long illustrative videos that distracting students but not promoting the initiative and positive thinking.

– While demonstrating videos, a simple, mellow audio presentation is required to help students understand the lesson better.

– When using electronic visual equipment, it is necessary to combine many teaching methods accompanied by lively presentation of teachers. It would deliver great impact on the acquisition of knowledge by the students.

In addition, since Biology in Vietnam is still considered as a minor subject, students often value Math, Literature, Foreign Language, Physics and Chemistry more. This mentality may cause a decrease of student attention to this subject, so teachers do not invest enough time to use technology in the lesson design.

2.3   The Need of Applying VRT in Biology Teaching and Learning in Vietnam

VRT is the most immersive type of reality technology that educators and learners are willing to use in education. In Vietnam, VRT has been implemented in education, albeit limited.

The Infantry shooting training simulated system (Virtual shooting range) is researched, designed and manufactured by Institute of Simulation Technology – Military Technical Academy. The system simulates the target object as well as the field in three-dimensional environment, also simulates sound, images and the recoil of a gun like when firing real bullets.

In medical field, Faculty of Medicine – Duy Tan University has applied VRT to simulate virtual body for teaching, learning and research of Anatomy subject. Major parts of a human body such as skeletal, muscular, nervous, digestive-system has been simulated. Through interactive hardware, students get acquainted with endoscopy and anatomy instead of studying directly on the real corpses, templates of pictures.

CDIT Institute of Information and Communication Technology which belongs to Posts and Telecommunications Institute of Technology is one of the leading institutions in VRT application research in Vietnam. They have built and brought this tool into support lectures teach photography and video recording technique. This model has been applied since 2016 with the simulation of real devices as virtual devices that is interactive and be able to run on a PC, thereby allowing students to practice before working on real devices.

VRT raise students' enthusiasm, interest and curiosity about lesson knowledge. VRT also solve the shortage of practice equipment for students. Besides, a number of basic knowledges is digitized into the application, so teachers do not need to repeat the knowledge, but only focus on explaining and answering students' questions raised.

Nevertheless, there are still some limitations that need to be overcome to make integration of VRT more successful in Vietnam.

Firstly, VRT is an indisputably powerful tool that has tremendous impact on education allowing student to discover environment that would be otherwise inaccessible [11, 19]. However, the application of VRT in education filed in Vietnam is constrained because of high market price of great-resolution computing engine and peripheral devices requirements, typically a head mounted display. Even designing 3D model for VRT is tremendously expensive.

Secondly, there is a fact that most of consumers do not have full awareness of what VR could offer. VR companies in Vietnam just focus mainly on video-games users or business users, but not for educational purposes.

Thirdly, for less-developed and developing countries the biggest concern is government must consume a large financial budget for the most advanced technology and equipment without considering how to make the most of their advantages [20]. Teachers and educational staffs in Vietnam schools need to be highly-professional trained to use VRT most effectively.

2.4 Lesson Design Procedure Integrating VRT into Biology Education

The authors suggest a lesson design procedure to implement VRT in education, specifically designed for teaching and learning Biology subject in high school in Vietnam (Fig. 1).

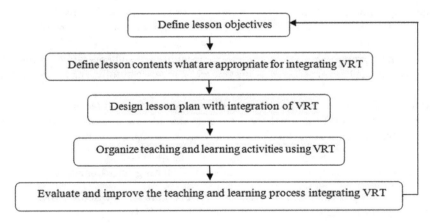

Fig. 1. Process diagram of integrating VRT into biology teaching in high school in Vietnam

Define Lesson Objectives: teachers need to rely on latest update of Vietnam General Education Program to clarify lesson objectives. The lesson targets must follow the general aim as developing learners' competency including knowledge attainment, skill and attitude improvement, especially information technology skill.

For integrating VRT into lesson, students are expected to explore a new 360^0 world and thereby promote their curiosity and wonder towards biological science.

Define Lesson Contents that Appropriate for Integrating VRT: teachers need to take a look constantly to all the contents and find out what are suitable for integrating VRT. These contents may include a lot of images, models or experiments which are able to be converted into virtual model due to empirical nature of Biology.

Design Lesson Plan with Integration of VRT: teachers may choose to use several positive methods of teaching such as group teaching, problem solving learning, blended learning to create an active classroom atmosphere. More importantly, since students would be entirely immersed in a simulated world contributed by VRT, it is essential to model the environment with high quality textures and 3D characters to make VRT the best experience for the students.

Organize Teaching and Learning Activities Using VRT: VRT offers an interesting and unlimited way to visualize objects, processes, location or historical events depends on the content of the lesson. However, there are some points that do need attention in order to use VRT most properly and effectively.

– VRT is quite new technology for Vietnamese students, so they may feel over-excited about exploring a novel world, thereby distracting their attention from the main content of the lesson.
– Since the students are totally immersed in simulated world, a serious supervision is needed to assure the safety of the students. Besides, academic staffs must involve actively while designing virtual scenarios in order to use VRT most effectively.
– There is time limit of each lesson (around 45 min/lesson in Vietnam) and thus teachers should control the class to ensure lesson time.
– There are several researches report on cybersickness of students while testing VRT, so teachers and academic staffs must watch students carefully during lesson time.

Evaluate and Improve the Teaching and Learning Process Integrating VRT: after trying to integrate VRT into the lesson, teachers should look back at the whole process to assess the pros and cons, then revise the process to make it improved.

2.5   Examples of Applying VRT to Biology Teaching and Learning

The authors suggested several contents of the upcoming High School Biology Education Program in Vietnam are suitable for applying VRT:

Brief contents in biology curriculum of Vietnamese high schools	VRT applied contents
– Cytology (The Cell)	– Cell structure (Eukaryotic and Prokaryotic cell) – Structure and function of membrane – Respiration of the cell – Cell cycle and Cell division
– Microorganism	– Structure of bacteria, viruses – Virus replication in host cells
– Plant form	– The transport of water and minerals in plants – Respiration and photosynthesis in plants – Growth and reproduction in plants
– Animal form	– Metabolism of matter and energy in animals – Induction in animals – Nervous systems
– Genetic	DNA Duplicating, genetic transcription and translation
– Evolution	– The evolutionary history of biological diversity
– Ecology	– Ecosystem – Relationships in communities

3   Discussion

VRT offers many advantages over other teaching techniques, hence it attracts the attention from educators and learners from around the world, including Vietnam. Vietnam has been gradually adapting VRT to education to meet the General Education Program's student quality requirements. The most outstanding characteristic of VRT is allowing users to interact with a cyberspace in the real-time. Students show a positive attitude towards integrating VRT into their learning process with reference to relaxation and no difficulty in perceiving biological contents. Additionally, VRT could motive biology learning of students by stimulating multi-sensory organs of students, making them interested in learning activities. Students also could learn through activities by their personal comfort and speed, thereby increase autonomy and initiative of the students.

Beside all advantages of VRT in education, it still has some disadvantages need to be addressed such as high cost for hardware and digital formats, incomplete awareness of using VRT and lack of research on VRT in the field of education. VRT should be considered as a tool to stimulate education process but not completely replace traditional education. The efficiency of lesson would reach the highest when combining both methods.

Vietnam is a developing country with a huge number of students (over 23.5 million students in 2018–2019 school year). There is a lot of burden on the government and educators in creating breakthroughs in education, especially in integrating ICT into teaching. Even though the proportion of central budget expenditure tends to increase over time, government, educators and learners in Vietnam need to put more effort to bring VRT into practice in high schools and other levels as well.

References

1. Jung, I.: ICT-pedagogy integration in teacher training: application cases worldwide. J. Educ. Technol. Soc. **8**(2), 94–101 (2005)
2. Wang, Q.: A generic model for guiding the integration of ICT into teaching and learning. Innov. Educ. Teach. Int. **45**(4), 411–419 (2008)
3. Chen, T.: Recommendations for creating and maintaining effective networked learning communities: a review of the literature. Int. J. Instr. Media **30**(1), 35 (2003)
4. Ratheeswari, K.: Information communication technology in education. J. Appl. Adv. Res. 45–47 (2018)
5. Anderson, J., van Weert, T., Duchâteau, C.: Information and communication technology in education: a curriculum for schools and programme of teacher development (2002)
6. Martín-Gutiérrez, J., et al.: Virtual technologies trends in education. EURASIA J. Math. Sci. Technol. Educ. **13**(2), 469–486 (2017)
7. Hussein, M., Nätterdal, C.: The benefits of virtual reality in education-a comparision study (2015)
8. Alhalabi, W.: Virtual reality systems enhance students' achievements in engineering education. Behav. Inf. Technol. **35**(11), 919–925 (2016)
9. Youngblut, C.: Educational uses of virtual reality technology. Institute for Defense Analyses, Alexandria, VA (1998)
10. Kaufmann, H., Schmalstieg, D., Wagner, M.: Construct3D: a virtual reality application for mathematics and geometry education. Educ. Inf. Technol. **5**(4), 263–276 (2000)
11. Shim, K.-C., et al.: Application of virtual reality technology in biology education. J. Biol. Educ. **37**(2), 71–74 (2003)
12. Tan, S., Waugh, R.: Use of virtual-reality in teaching and learning molecular biology. In: 3D Immersive and Interactive Learning, pp. 17–43. Springer (2013)
13. Ammanuel, S., et al.: Creating 3D models from radiologic images for virtual reality medical education modules. J. Med. Syst. **43**(6), 166 (2019)
14. Camp, J.J., et al.: Virtual reality in medicine and biology. Futur. Gener. Comput. Syst. **14**(1–2), 91–108 (1998)
15. Asenova, A., Reiss, M.: The role of visualization of biological knowledge in the formation of sets of educational skills (2011)
16. Gay, E., Greschler, D.: Is virtual reality a good teaching tool. Virtual R. Spec. Rep. **1**(4), 51–59 (1994)
17. Karr, T.L., Brady, R.: Virtual biology in the CAVE. Trends Genet. **16**(5), 231–232 (2000)
18. Mikropoulos, T.A., et al.: Virtual environments in biology teaching. J. Biol. Educ. **37**(4), 176–181 (2003)
19. Bell, J.T., Fogler, H.S.: The application of virtual reality to (chemical engineering) education. VR **4**, 217–218 (2004)
20. Hamidi, F., et al.: Information technology in education. Procedia Comput. Sci. **3**, 369–373 (2011)
21. Vietnam Ministry of Education and Training, Vietnam General Education Program (2018)

Improved Scheme Securing Voting System for Elections

Jabbar El-Gburi[✉]

Faculty of Informatics, University of Debrecen, Debrecen 4028, Hungary
enjabbar.enjabbar@yahoo.com

Abstract. In this paper, we provide an improved Electronic Voting System (EVS) for political and social elections based on known Cryptographic schemes. In cryptography, the ElGamal encryption system is an asymmetric key encryption algorithm which is based on Diffie-Hellman key exchange. We will use the ElGamal algorithm to generate and encrypt random keys for the voters. This scheme is designed in a way that the communication channels are anonymous and in the meantime privacy, eligibility and fairness are applied to the entire system. Encrypted receipt-free transactions are provided to the voters after they submit a vote successfully. After the voting completes the voters can check on voting online using the encrypted receipt, only to confirm their participation in the election. The voters will not be aware of the chosen candidate to prevent cheating and vote selling, common issues in elections these days. The name of the chosen candidate will be confidential. The proposed voting system efficiently gives the opportunity for people to vote via their own PCs/laptops, thus decreases the queues accumulated up at voting centers. In addition, it offers a highly dependable authentication approach by national ID or biometrically which leads to overcoming electoral fraud.

Keywords: Secure voting · Online elections · Encryption · Cryptography

1 Introduction

The old traditional way of casting votes by manually marking ballot paper is yet to revolutionize in this digital age in many countries [1]. To keep voting and the election process secures in any given country, the need to develop a different scheme for Political and Educational selection process exists. Electronic voting and counting technologies are being increasingly used around the world with much success [3, 4]. This type of scheme can be used for political, social and educational matters. So, for that purpose, we propose a secure electronic voting system. If we look deeper into classic voting systems, it is clear that time, security and cost are all at compromised levels [5]. The fields of election technologies relevant to counting are changing day by day and emerging frameworks are being considered for future use. There are other major problems in current voting systems such as losing voting papers or environmental concerns (fire, flood, etc.). It is a common occurrence as well that voter turnout is low, as people do not take the time to vote in elections with current voter turnouts averaging in the 10–20% range in some countries [2].We propose a system here where

© Springer Nature Switzerland AG 2020
L. C. Jain et al. (Eds.): ICICCT 2019, LAIS 9, pp. 463–476, 2020.
https://doi.org/10.1007/978-3-030-38501-9_45

voters will be at ease to cast a vote more securely by using their own credentials. This can be done using a one-way return channel using Cyber Space. There are some benefits of Electronic Voting:

- Our scheme uses an improved mechanism for the identification of the voter.
- The most important benefits of E-Voting, it allows the disabled persons to vote at their highest ease, they would not have to suffer any difficulty of moving.
- It will be a completely developed system that will also increase the voting speed; there will be no issue of ballot papers, inks, marking, identifications, etc.
- In this process, there will be no chances of Vote Rejection, no wrong votes will be cast.
- This system will also reduce the number of staff required for election.
- This technology will also increase the voting turnout.
- It will improve trust in the electoral process.
- No Logistical Arrangements will be required, i.e. printing, designing, etc. and their shifting more precisely.
- This system is developed especially for handling complex problems of the election.
- No possibilities of changing the ballots.

1.1 Challenges

The act of conducting an election is one of the most complex operations in a country. The Election Management bodies in a country understand that when an election is called all of the staff has to go through a great operational challenge. It becomes their foremost duty to ensure the smoothness of the election process [6]. After observing such a difficult way of conducting the election, the Election Administrators should get to know about the latest technologies to lessen their burden of conducting the election, counting of votes and result declaration as well [7]. Candidates will be unable to check who has voted for him/her as there will be an encrypted result declared. All the information will be kept secret even the author will not be allowed to disclose who this participant has voted [8, 9]. After the voting, the participant can see his/her name with his serial number (issued by the system) online. The participant will be able to view his/her voting details, just during the specific time, after which there will be no extra details regarding the candidate to whom he/she voted [13]. We will use a secure mechanism and terminologies like Mix-net code and re-encryption in order to make sure that this code is secure and it'd be difficult to guess. This scheme has been adopted in several companies because of good results in practical work [10, 11]. The people or the election staff who manages voting processes nowadays always potentially mistake-prone to counting or any other relevant activity [12]. The Electronic Voting System (EVS) will minimize the chances of mistakes in all facets. In this scheme, the voter will receive an automatic message after casting the vote to selected member and further-more a receipt will also be generated by the system for the voter [14, 15].

2   Traditional Voting Prior to Electronic Voting

The participant has to visit the voting center to pass through the voting process, for that all of the participant usually uses the traditional way of voting like using pen and paper to mark the box with the pen for the chosen candidate and after this process, some authorized staff members have to sort out all the hard copies of the papers, after sorting, there starts the counting process to finalize the results for each candidate. The staff members are directed to make a detailed list separately for each candidate with their achieved number of votes. This work is done very carefully in order to lessen the chances of counting mistakes. If we take an example that there are a hundred of participants for two candidates; Alice and Bob.

One of these two who get the highest number of votes will win the election, it will require a lot of time and hard work to get it done at the right time, in other words, it is really a time-consuming and a sort of hectic job. In order to improve this voting system, there are several scientists and researcher who did a lot of work on this concept just for its improvement and security. In this scheme we are going to use a secure channel called One Way Channel Encryption, using the mix-net server, this protocol will be simple encryption not having the complex cryptographic algorithms. The main benefit of this protocol will be that we will not require a massive fast computer to run this algorithm. we will be able to run this on simple computer systems with its high-speed functionalities without any delay. This system will be completely confidential with encryption of data. The voter will be confirmed about the casting of vote by message (sent by the system) with relevant information.

The structure of this paper is as the following Section one introduction. This part provides a concise introduction about the provided work by focal pointing on the essential themes, methods, benefits, and challenges. Section two describes Traditional Voting Prior to Electronic Voting. The third section summarizes the algorithm and the proposed method. This part provides the methods and the used algorithm for structured evaluation and analysis for the authentication process. The following subjects are discussed, proposed algorithm preliminaries, materials and methods, logical view of the scheme, the main work of the voting scheme, Verification Message, Receipt-freeness, how homomorphic utilized within this research, how receipt-freeness really works, attacker in E-voting, tallying after the election ends. The fourth section discusses the results and discussion are given. The fifth section shows a comparison with other related works. The last section debates the current objectives achieved in this research paper.

3   Proposed Algorithm

In this section, we give our proposed algorithm as shown graphically in Fig. 1 and in the Proposed Voting Algorithm.

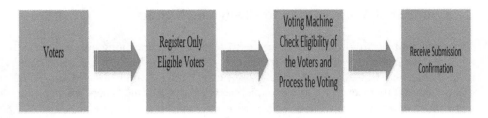

Fig. 1. Electronic voting scheme

The authority members authorized themselves to the system using the private key to authorize the voters, without learning about their chosen candidate. Only to make sure this is an authorized participant and this are his/her first vote in this election period. The encrypted code that voter gets it at the registration time can work only one time, after the participation for the first time this encrypted code is expired [16].

The voter can log-in with their data to do the voting and get the receipt to confirm that the voting has done successfully, besides this receipt is encrypted and no one can learn about it even the voter. This scheme works as follows; each participant will be having a Chip Card on which his/her all the personal data is stored. This Card is issued to every person on the time of registration and each person is issued a unique Registration Number. After getting logged in on the system, the voter will be given a list of candidates with their details from which he/she can choose one whom he/she wants to vote. The voter will get a serial number for the proof that he made the voting. So, for the security reason, every participant will be issued a unique number in order to avoid selling and cheating of the votes. Since our main priority is the security so the issued encrypted number will be unable to be decrypted by the user. This system will be developed by keeping in view all the aspects of security, privacy, and safety.

3.1   Materials and Methods

In this section, we describe the scheme. Each participant has a chip card on which his/her personal data is stored. This card is issued to every person at the time of registration as a voter and each person is issued a unique Registration Number (RN). After getting logged into the system, the voter will be given a list of candidates with their details from which he/she can choose who they wish to vote for. The voter will get an encrypted serial number as proof of voting.

For security reasons, every participant will be issued a unique number in order to avoid selling and cheating votes. The issued encrypted serial number will be unable to be decrypted by the user. The system's main objective is the aspects of security, privacy, and safety.

Proposed Voting Algorithm

Input: voter credentials.

Output: voting confirmation and receipt.

if voterid = =0 then

  Return invalid voter.

  if voterid != 0 then

    if Check voter id = = valid then

      Generate token as VTOK

      if VTOK is valid then

        Increment the count

        Repeat

      else

        Generate an error

      end if

    else

        Invalid voter

      end if

    end if

end if

3.1.1   Logical Scheme View

This scheme is based on Homomorphic Encryption. Let P is the plain text space, and the C is the ciphertext space, P works with them $\oplus$, and the C works also with the $\oplus$ Enc1(m1) is the encryption of the message using number r, the encryption scheme is $(\oplus, \oplus)$-homomorphic

$$Cl = Enc(m1) \ and \ C2 = Enc\,2(m2)C1 \oplus C2Enc(m1 \oplus m2) \tag{1}$$

Choosing the two numbers, a prime number (p) and element number (g) $Z/pZ\ V_{ID}$, Sk_v, e, f, c) $Sk_v V_{ID}$ is the ID of the voter, Sk_v is the secret key of the voter, e is the input of the voter request, f is the output of the request of the voting it will encrypt it, V_{ID}"r"y,

$$R \equiv \text{Enc Sk}_v(g) \tag{2}$$

$$Y \equiv g^x \pmod{p} \tag{3}$$

3.1.2   The Registration Phase

Voter has to register before the election time in the database of the country or any organization that organizes the voting process. The authority (A) members have to create the public (Pkv) and private (Skv) Key for the voters based on El-Gamal algorithm. Generate (Ge) the random numbers for the voters and encrypt it. In the registration time, the authority can register as well to be part of the voting process. The author public key (Pk_A), and private key (Sk_A), that these authorities can manage the voting as well, so here they can be an author and at the same time they are voter as well, they can participate in this E-Voting process. This phase is considered as an identification phase, where the authority committee is engaged with the central role, as it corroborates that the citizen is eligible to vote. As elucidated Fig. 2 the citizen has two options in order to acquire the election right, the first one comprises that he/she could go the Department of Civil Status (Authority Committee) to validate his/her data, while the second option to go with the process of biometric authentication by sending his/her scanned fingerprints. If the citizen request was accepted and verified, then he/she turns into a certified voter. While after the phase of verification, the voter data shall be saved into election server and in a later phase will be moved a mobile cellular telephone company server for further processing illustrated in Fig. 3 which shows the Authentication Phase, this phase takes place during the period time preceding the voting process.

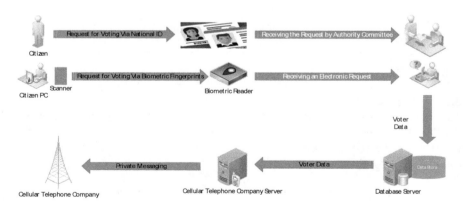

Fig. 2. Identification process

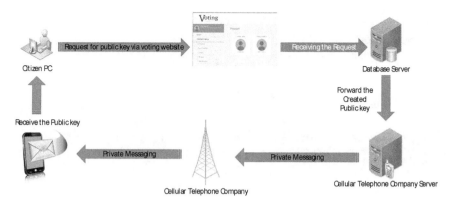

Fig. 3. Authentication process

We can notice that there is a circumstance, fact, that influence and contributes to a result of organizing the process of E-Voting which include hardware and software associated impact. At this stage, the voter has the ability to access the voting system straightforward via his own laptop (see Fig. 3). Furthermore, more clarification is represented in Fig. 3, the verified voter login to the system site via the national ID. At the moment that the voter is signed to the system, the election database server shall produce a public key through El-Gamal algorithm, this key shall be subsequently directed across the mobile company via short message service and reach the voter, bear in mind that, acquiring the public key specifies the verified voters.

3.1.3 The Voting Phase

Usually, the process of voting is considered as the most sensitive phase, as the voter will not be able to logout subsequent to obtaining the public key. In spite of that, the voter at the present time is capable to choose his/her candidates that exist in the list within the system website. In order to improve the dispatch of voting speed approach, the system website shall just show the candidates names based on associated geographical location (refer to Fig. 4). After choosing the wanted candidates and after the received public key has been inserted via the website, then the voter must press on the button of submitting in order to forward his/her election selection to the server. For that reason; the encryption algorithm of ElGamal is put into effect to encrypt the generated data of the voter which shall be afterward delivered as ciphertext to the election database server.

This scheme will provide all the functions required for the security and privacy of E-voting System. Only the registered participants will be authorized to cast the vote on E-voting System. One of the main security features is that neither the voter nor the author will be able to recover the confidential details that who has cast the vote against which candidate. We can explain using the following example:

Joh, vote number = 298675 for Alice
Richard, vote number = 467598 for Alice
Tomi, vote number = 689032 for Bob

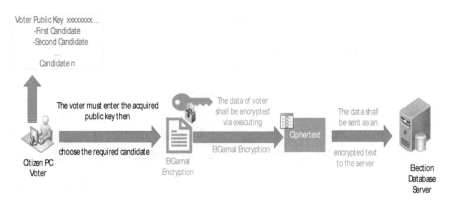

Fig. 4. The process of voting

Since there are different participants for the same contestant, but still they are issued a different serial number after submission the vote. The reason for that security system is just to avoid cheating and selling of the vote. Each person can participate in one time. During the voting time, no one can have any idea about the result until the end of voting time. So, there will be a free choice for every voter to choose the candidate that suits him/her best, that he/she is the best person for this community or the company.

3.1.4   Verification Message
Each vote verifies the result to the participant. That means after submitting the vote, the voter can get verification serial number that he/she can later check online. The voter will receive a message of "Successful Vote Submission" without any further details for the confidentiality of the data to keep the election secure and honest.

Participant gets the authority to participate

Register **V:** (V_{ID}, Sk_v, Pk_A)
Register **A:** (V_{ID}, Pk_v)

3.1.5   Receipt-Freeness
The participant will be unable to give proof of voting to anyone. Because his/her serial number will not show to which consistent he voted for. So, avoiding the people to sell their ballots. Similarly, after the participant has voted he/she will get an encrypted number, ciphertext number that no one can tell to whom he/she has voted. Since the result will be decrypted, in a way to convince the people that this is the result, here we can use homomorphic encryption, for this we can multiply ciphertext together, like adding the plain-text to get the result for Alice.

3.1.6   Homomorphic Encryption Example
Let us go through an example of homomorphic encryption. First, pick a "universe" and operation (R, *). Then we define a function f:

$$f(x) = |x| \tag{4}$$

We then check

$$f(xy) = f(x)\,(f(y)|x.y| = |x||y| \tag{5}$$

From this we get,

$$(R, *) : f(x) = x \wedge 2 \tag{6}$$

and,

$$(Z, +) : f(x) = 3x \tag{7}$$

where,

$$(Z(modp), +) : f(x) = x + p \tag{8}$$

To build an encryption function that is homomorphic, we can make from both addition and multiplication Enc(7) + Enc(2) = Enc(28), encrypt one bit at a time, we want E(key, message) to be homomorphic. So, we have:

$$(Z\ mod\ 2, +)\ \&\&\ (Z\ mod\ 2, *) \tag{9}$$

Using E, we then get:

$$E(KEY, m1 * m2) = E(KEY, m1) * E(key, m2) \tag{10}$$

And

$$E(key, m1 + m2) = E(key, m1) + E(key, m2) \tag{11}$$

Using Key(*Gen*), we Encrypt(*key, m*) Homomorphic and then Decrypt(*key, c*). From this E is homomorphic and

$$E(key, 1 + a * b) = E(key, 1 + E(key, a) * E(key, b) \tag{12}$$

Using *N AND*, as well as *!(AND)* to write a program that is homomorphic, we run this function to encrypt the data and send it to the cloud, *F(x)* Evaluate (key, function), f (x) encrypted data, example how to search data in the text, *F(x)* = evaluate (key, search, _forewords), $y = f(x)$, Result = Decrypt (key, y), The homomorphic algorithm is a bit slow encryption function that is why it's not so popular compared to other encryption algorithms like, RSA and El-Gamal algorithms. By using this algorithm, we don't need to download the whole data from the cloud to modify it and then upload it to the server, we can modify the data in the server and re-encrypt it back.

Our scheme is secure as compared to traditional voting. It is even faster and cheaper, more secure and easy to do with user-friendly functions. There will be the authority of the voting; only the authorized voter can participate in the voting. Security of the vote: in the voting time neither the author nor the voter himself can learn about the vote; this is for the security reason so the voter will not be involved to sell vote to others.

3.1.7   Receipt-Freeness Methods

There is a connection between V, C. It is the data of the participant stored already in the database of the participant, before participation in this election program, he/she has to be registered in the system and authorized by the author of the election. Then, when this participant logs in with his/her credentials and this data matches in the system he/she gets the authority to participate in the election, then he/she will choose the favorite competitor and will vote for him/her. C is the competitor ID stored in the system; with one condition the participant can choose his favorite competitor to vote for him/her. $Cf = Cj$ is the favorite competitor.

3.1.8   Attacker in E-Voting

The vote will be fully encrypted. The attacker cannot have any idea to whom this voter has voted. So here, if the attacker tries to cancel the vote from the system this is the middle attack between the starting point of the voting and the server to store the result. There is supposed to be another type of attacker, the attacker can bribe the voter to buy his/her vote in order to win the election. This is supposed to be happened by having all the personal information from the voter so he/she can do all this on behalf of a person to vote the candidate. For protection, to avoid this attacker we can use extra security in the voting center to make sure this is the owner of the voting card he/she can walk in to give the vote to a favorite candidate. This scheme prevents all types of attack by securing and encrypting the voting information. Then providing the voter an encrypted serial number, a message confirming that voting submitted successfully, this number is encrypted and no one can tell to whom the vote is conducted. An attacker cannot track the voters' vote to get any information about the vote/voter.

3.1.9   Tallying After the Election Ends

Let AR: Enc Pk_R (s"v)

R John: (s"c)

At the end of the election period, the authority of the E-Voting sends to the registry (R) (s"v) pairs then R checks the validity of the signature for the participant (s, c) if this encrypted code is not online. The voter can complain about it to see where is his/her vote gone to. A voter can check his/her voting code to make sure this code went to his/her favourite candidate and nothing went wrong in the election times. Participants *(P)* In the voting time, V is the voter (V1, V2, V3 ... Vn) the participant would like to make sure that his/her participant in the EV proceeds to the final round and considered as a vote at the end of the election time process.

4 Results

The latest technology is one of the best ways to have secure and safe elections in a country. Such technologies also reduce the burden of administrative staff and provide them several feasibilities. Choosing the two numbers, a prime number (p) and element number (g) Z/pZ V_{ID}, Sk_v, e, f, c) $Sk_v V_{ID}$ is the ID of the voter, Skv is the secret key of the voter, e is the input of the voter request,f is the output of the request of the voting it will encrypt it, V_{ID}"r"y,

$$R \equiv Enc\, Sk_v(g) \tag{13}$$

$$Y \equiv g^x (mod\, p) \tag{14}$$

This scheme is based on Homomorphic Encryption. Let P is the plain text space, and the C is the ciphertext space, P works with them $\oplus$, and the C works also with the $\oplus$ Enc1(m1) is the encryption of the message using number r. The encryption scheme is homomorphic C1 = Enc (m1) and C2 = Enc2 (m2), C1 C2 Enc (m1 m2).

5 Comparison Analysis

A considerable research study on voting has been carried out especially online voting systems. Cryptographic mechanism of carrying out a particular task has turned into a necessary constituting part of E-voting systems. Several techniques have been proposed in order to obtain a guaranteed and secured voting systems such as secret sharing, and mix-nets. The first scheme of e-voting was put forward by Chaum [17]. which exploits anonymous unattributed channels such as mix-nets channel. More schemes were suggested later on also based on mix-nets along with different enhancement such as Aditya et al. [18], as they ameliorated the scheme adequacy which was proposed by Lee et al. [19] via an adjusted optimistic scheme. While the scheme within [20] utilize 2 types of mix-nets in order to stop the vote updating process from being discovered via coerces. Nevertheless, utilizing mix-nets implies that translucency could not be ensured. The blind signature scheme gives the ability for the authority to add their signature to an encrypted message with no need to be aware of the message area or scenery behind the context, especially when perceived as a framework for it [21]. In spite of that, it is hard to resist against mischievousness that is done by authorities, as they could acquire the results prior to being announced on the counting stage, which fail to comply with the legibility of the voting protocol. Chow et al. [22] suggested utilizing linked circular band signature, as the message signed by the self-same affiliate member could be consistent with. In spite of that, it is not marked toward the to the member. A platform merging mix-nets and blind signature are provided in [23]. Voting systems that depend on the encryption of homomorphic could keep track of the events that strongly influencing later developments by Benaloh [24]. a various number of voting platforms utilize homomorphism by relying on secret sharing [25]. Several schemes [26] exploits Shamir's threshold approach which is based on secret sharing [27], at the same time, other schemes [28, 32, 33] depend on Chinese remainder reasoning concept, while our scheme is founded on more comprehensible/intelligible scheme, the currently available voting schemes, as secret sharing usually is exploited by authorities either to accumulate their shares to acquire the vote decryption key that performs tallied votes decryption [29], or to pool all shares at one and the same time in order to retrieve the encrypted final outcome [28]. As an alternative, among our scheme, the process of secret participation is utilized within voters to partake in the secret votes and afterward to retrieve their still available votes. Furthermore, our proposed protocol does not necessitate particularized appliances also it distributed naturally because of its well-made design. The concept of randomization guarantees the privacy of the voting process, during the period of time following vote casting by the voter, he or she will be provided with a coupon which proves that the voter has participated in the voting process. In contrast to our model, where the voter shall not sight immediately that his own vote is enumerated. Also, our model can be considered as cost-efficient, as it utilizes the already existing mobile phone network to deliver the generated keys to the voters and reach their own private cell phones, and does not demand supplementary hardware equipment. Regrettably, due to rigorous and inconsistent e-voting demands [30], currently, there is no system that meets all the expectations of voting required characteristics, Security impairments exist even in quite sophisticated voting platforms [31].

6 Conclusion

The proposed scheme covers all the security required for the e-voting election system such as robustness, fairness, eligibility. To make the fair elections among the contestants to choose the best candidate for the country or any company, university, using anonymous return channel between the voter and the authority it's a one-way channel. The participant can vote with their ID after the registration. We use simple encryption to encrypt the vote and send it to secure to the server. The proposed scheme covers all the security required for the e-voting election system such as robustness, fairness, eligibility. To make the fair elections among the contestants to choose the best candidate for the country or any company, university, using anonymous return channel between the voter and the authority it's a one-way channel. The participant can vote with their ID after the registration. We use simple encryption to encrypt the vote and send it securely to the server.

References

1. Bargnani, A., Pieprzyk, J., Safavi, R.: A practical electronic voting protocol using threshold schemes. Centre for Computer Security Research, University of Wollongong, Australia (1994)
2. California Secretary of State Ad Hoc Touch screen Voting Task Force Report; from Caltech-MIT. Voting: Cal Tech-MIT Voting technology Project Report (2001). www.vote.caltech.edu/Reports

3. Vu, D., Luong, T., Ho, T., Nguyen, C.: An efficient approach for electronic voting scheme without an authenticated channel. In: 2018 10th International Conference on Knowledge and Systems Engineering (KSE), Ho Chi Minh City, pp. 376–381 (2018)
4. Halderman, J.A., Teague, V.: The New South Wales iVote system: security failures and verification flaws in a live online election. In: Haenni, R., Koenig, R., Wikström, D. (eds.) E-Voting and Identity, Vote-ID. Lecture Notes in Computer Science, vol. 9269. Springer, Cham (2015)
5. Huszti, A.: A secure electronic voting scheme. Period. Polytech. Electr. Eng. (Arch.) 51(3–4), 141–146 (2007). https://doi.org/10.3311/pp.ee.2007-3-4.08
6. Kim, K., Hong, D.: Electronic voting system using mobile terminal. World Acad. Sci. Eng. Technol. 33–37 (2007)
7. Wang, K.-H., Mondal, S.K., Chan, K., Xie, X.: A review of contemporary e-voting: requirements, technology, systems and usability. Data Sci. Pattern Recogn. 1(1), 31–47 (2017)
8. Kumar, M., Katti, C.P., Saxena, P.C.: A secure anonymous E-voting system using identity-based blind signature scheme. In: Shyamasundar, R., Singh, V., Vaidya, J. (eds.) Information Systems Security, ICISS. Lecture Notes in Computer Science, vol. 10717. Springer, Cham (2017)
9. Nassar, M., Malluhi, Q., Khan, T.: A scheme for three-way secure and verifiable E-voting. In: 2018 IEEE/ACS 15th International Conference on Computer Systems and Applications (AICCSA), Aqaba, pp. 1–6 (2018). https://doi.org/10.1109/aiccsa.2018.8612810
10. Hirt, M., Sako, K.: Efficient receipt-free voting based on homomorphic encryption. In: Preneel, B. (ed.) Proceedings of the 19th international conference on Theory and application of cryptographic techniques (EUROCRYPT 2000), pp. 539–556. Springer, Heidelberg (2000)
11. Meyer, M., Smyth, B.: Exploiting re-voting in the Helios election system. Inf. Process. Lett. 143, 14–19 (2019). https://doi.org/10.1016/j.ipl.2018.11.001. ISSN 0020-0190
12. Nair, D.G., Binu, V.P., Santhosh Kumar, G.: An improved E-voting scheme using secret sharing based secure multi-party computation. arXiv e-prints, reprint -1502.07469 (2015)
13. Cetinkaya, O., Doganaksoy, A.: A practical verifiable e-voting protocol for large scale elections over a network. In: The Second International Conference on Availability, Reliability and Security (ARES 2007), Vienna, pp. 432–442 (2007)
14. Obaidat, M.S., Maitra, T., Giri, D.: Protecting the integrity of elections using biometrics. In: Obaidat, M., Traore, I., Woungang, I. (eds.) Biometric-Based Physical and Cybersecurity Systems. Springer, Cham (2019)
15. Chaidos, P., Cortier, V., Fuchsbauer, G., Galindo, D.: BeleniosRF: a non-interactivereceipt-free electronic voting scheme. In: 23rd ACM Conference on Computer and communications security (CCS 2016), Vienna, Austria, October 2016. https://doi.org/10.1145/2976749.2978337
16. Royal-Holloway, University of London Egham, Surrey TW20-0EX – England. Technical report RHULMA (2006). http://www.rhul.ac.uk/mathematics/techreports
17. Chaum, D.: Untraceable electronic mail, return addresses, and digital pseudonyms. Commun. ACM 24, 84–90 (1981)
18. Aditya., R., Lee, B., Boyd, C., Dawson, E.: An efficient mixnet-based voting scheme providing receipt-freeness. In: Proceedings of the International Conference on Trust and Privacy in Digital Business, Zaragoza, Spain, 30 August–1 September 2004, vol. 3184, pp. 152–161 (2004)

19. Lee, B., Boyd, C., Dawson, E., Kim, K., Yang, J., Yoo, S.: Providing receipt-freeness in Mixnet-based voting protocols. In: Proceedings of the International Conference on Information Security and Cryptology, Seoul, Korea, 27–28 November 2003, vol. 2971, pp. 245–258 (2003)
20. Araújo, R., Barki, A., Brunet, S., Traoré, J.: Remote electronic voting can be efficient, verifiable and coercion-resistant. In: Financial Cryptography and Data Security: FC 2016 International Workshops, BITCOIN, VOTING, and WAHC, Revised Selected Papers, Christ Church, Barbados, 26 February 2016, pp. 224–232. Springer, Heidelberg (2016)
21. García, D.L.: A flexible e-voting scheme for debate tools. Comput. Secure. **56**, 50–62 (2016)
22. Chow, S., Liu, J., Wong, D.: Robust receipt-free election system with ballot secrecy and verifiability. In: Proceedings of the 16th Annual Network and Distributed System Security Symposium (NDSS), San Diego, CA, USA, 8–11 February 2008, pp. 81–94 (2008)
23. Ohkubo, M., Miura, F., Abe, M., Fujioka, A., Okamoto, T.: An improvement on a practical secret voting scheme. In Proceedings of the 2nd International Workshop on Information Security (ISW 1999), Kuala Lumpur, Malaysia, 6–7 November 1999, pp. 225–234 (1999)
24. Benaloh, J.: Verifiable secret ballot elections. Ph.D. thesis, Yale University, New Haven, CT, USA, 1987
25. Adewole, A.P., Sodiya, A.S., Arowolo, O.A.: A receipt-free multi-authority e-voting system. Int. J. Comput. Appl. **30**, 15–23 (2011)
26. Schoenmakers, B.: A simple publicly verifiable secret sharing scheme and its application to electronic voting. In: Proceedings of the 19th Annual International Cryptology Conference on Advances in Cryptology (CRYPTO 1999), Santa Barbara, CA, USA, 15–19 August 1999, pp. 148–164 (1999)
27. Shamir, A.: How to share a secret. Commun. ACM **22**, 612–613 (1979)
28. Iftene, S.: General secret sharing based on the Chinese remainder theorem with applications in e-voting. Electron. Notes Theor. Comput. Sci. **186**, 67–84 (2007)
29. Lee, B., Kim, K.: Receipt-free electronic voting through collaboration of voter and honest verifier. In: Proceedings of the JW-ISC, Naha, Japan, 25–26 January 2000, pp. 101–108 (2000)
30. Chevallier-Mames, B., Fouque, P.A., Pointcheval, D., Stern, J., Traoré, J.: On some incompatible properties of voting schemes. In: Chaum, D., Jakobsson, M., Rivest, R.L., Ryan, P.A., Benaloh, J. (eds.) Towards Trustworthy Elections, pp. 191–199. Springer, Heidelberg (2010)
31. Karlof, C., Sastry, N., Wagner, D.: Cryptographic voting protocols: a systems perspective. In: Proceedings of the 14th Conference on USENIX Security Symposium, Baltimore, MD, USA, 31 July–5 August 2005, vol. 5, pp. 33–50 (2005)
32. Srivastava, G., Dwivedi, A. D., Singh, R.: PHANTOM protocol as the new crypto-democracy. In: IFIP International Conference on Computer Information Systems and Industrial Management, September 2018, pp. 499–509. Springer, Cham (2018)
33. Srivastava, G., Dwivedi, A.D., Singh, R.: Crypto-democracy: a decentralized voting scheme using blockchain technology. In: ICETE (2), pp. 674–679, July 2018. By-the authors. Submitted-to-Journal-Not Specified for-possible open access publication under the terms and conditions of the Creative Commons Attribution (CC BY) license (2019). http://creativecommons.org/licenses/by/4.0/

Author Index

© Springer Nature Switzerland AG 2020
L. C. Jain et al. (Eds.): ICICCT 2019, LAIS 9, pp. 477–478, 2020.
https://doi.org/10.1007/978-3-030-38501-9